Catalysis –
Science and Technology

Volume 2

EDITORS:

PROF. DR. J. R. ANDERSON (Melbourne/Australien)
PROF. DR. M. BOUDART (Stanfort/USA)

CONTRIBUTORS:

PROF. DR. G. M. SCHWAB, PROF. DR. G. FROMENT,
PROF. DR. J. HABER, DR. L. HOSTEN,
DR. A. J. LECLOUX, PROF. DR. K. TANABE

CATALYSIS -
Science and Technology

Volume 2

Akademie-Verlag · Berlin
1983

Die
Originalausgabe erscheint
im Springer-Verlag Berlin · Heidelberg
New York

Vertrieb
ausschließlich für alle Staaten mit Ausnahme
der sozialistischen Länder:
Springer-Verlag Berlin · Heidelberg
New York

Vertrieb
für die sozialistischen Länder:
Akademie-Verlag Berlin

Erschienen im
Akademie-Verlag, DDR-1086 Berlin, Leipziger Straße 3–4

Lizenznummer: 202 · 100/524/82
Gesamtherstellung: VEB Druckerei „Thomas Müntzer“, 5820 Bad Langensalza
Umschlaggestaltung: Eckhard Steiner
Bestellnummer: 763 112 5 (6707) · LSV 1215
Printed in GDR
DDR 132,— M

General Preface to Series

In one form or another catalytic science reaches across almost the entire field of reaction chemistry, while catalytic technology is a cornerstone of much of modern chemical industry. The field of catalysis is now so wide and detailed, and its ramifications are so numerous, that the production of a thorough treatment of the entire subject is well beyond the capability of any single author. Nevertheless, the need is obvious for a comprehensive reference work on catalysis which is thoroughly up-to-date, and which covers the subject in depth at both a scientific and at a technological level. In these circumstances, a multi-author approach, despite its well-known drawbacks, seems to be the only one available.

In general terms, the scope of *Catalysis: Science and Technology* is limited to topics which are, to some extent at least, relevant to industrial processes. The whole of heterogeneous catalysis falls within its scope, but only biocatalytic processes which have significance outside of biology are included. Ancillary subjects such as surface science, materials properties, and other fields of catalysis are given adequate treatment, but not to the extent of obscuring the central theme.

Catalysis: Science and Technology thus has a rather different emphasis from normal review publications in the field of catalysis: here we concentrate more on important established material, although at the same time providing a systematic presentation of relevant data. The opportunity is also taken, where possible, to relate specific details of a particular topic in catalysis to established principles in chemistry, physics, and engineering, and to place some of the more important features into a historical perspective.

Because the field of catalysis is one where current activity is enormous and because various topics in catalysis reach a degree of maturity at different points in time, it is not expedient to impose a preconceived ordered structure upon *Catalysis: Science and Technology* with each volume devoted to a particular subject area. Instead, each topic is dealt with when it is most appropriate to do so. It will be sufficient if the entire subject has been properly covered by the time the last volume in the series appears. Nevertheless, the Editors will try to organize the subject matter so as to minimize unnecessary duplication between chapters, and to impose a reasonable uniformity of style and approach. Ultimately, these aspects of the presentation of this work must remain the responsibility of the Editors, rather than of individual authors.

The Editors would like to take this opportunity to give their sincere thanks to all the authors whose labors make this reference work possible. However, we all stand in debt to the numerous scientists and engineers whose efforts have built the discipline of catalysts into what it is today: we can do no more than dedicate these volumes to them.

Preface

Catalysis is a subject which draws upon many of the traditional scientific and technological disciplines, and its present structure has only been reached after many decades of detailed work. Nevertheless, in catalysis as in all other areas of science, experiments are carried out, new data are acquired, theories are proposed; and these things are done not in an intellectual vacuum, but in relation to previously established concepts and ideas. The history of how these guiding concepts have developed is nothing less than a skeletal history of the subject itself. In one respect at least, catalysis is similar to all other of mankind's endeavors: a failure to learn from history is a rejection of one's heritage. None of us should willingly plead guilty to this, and Professor G.-M. Schwab's opening chapter in this volume provides us with a ready means to avoid the need to re-invent what has been known for some time.

Heterogeneous catalysis is dominated by the concept of a reactant molecule entering into some type of chemical interaction with the surface of a solid. Thus, to understand catalysis it is essential to understand as fully as possible the chemical nature of the solid. In the second chapter of this volume, Professor J. Haber provides a review of the majority of solid phases which are of catalytic interest. The framework for this review is chemical crystallography, and the author goes to some pains to draw attention to features that are of particular importance in catalysis.

It is a truism that, in a relative sense, laboratory work is cheap and development work is expensive. This is certainly true in catalysis where, to run a reactor on a scale that is large enough to obtain reliable engineering design data, is a very expensive undertaking indeed. Particularly in this situation, it is essential to squeeze the last drop of useful information out of such expensive experimental data. This is essentially the problem which is addressed by Professor G. F. Froment and Dr. L. Hosten: how to model catalytic reactor kinetics with maximum efficiency.

Real heterogeneous catalysts, that is those which are used in practical applications, usually have a large surface area. A catalyst unit such as a pellet is usually porous, and parameters such as surface area and pore size distribution are often of great importance in determining the behavior of a catalyst in practice. To characterize the texture of a catalyst is a complex task bacause it requires a number of different sorts of measurements. Dr. A. J. Lecloux gives a comprehensive account of how this problem may be successfully tackled. An important question is that of selecting a combination of techniques most appropriate to a particular situation: in this sense the solution is part of the problem.

Solid acid catalysts, and to a lesser extent solid base catalysts, are of substantial industrial importance. They have a long history in catalytic science and technology, yet they remain of considerable current interest. Professor K. Tanabe gives a comprehensive account of the nature of catalysts of this sort, together with a description of the methods by which their acidic or basic properties may be characterized.

Contents

List of Contributors

Professor Dr. Georg-Maria Schwab
Institut für Physikalische Chemie
Universität München
Sophienstr. 11, D-8000 München 2, FRG

Professor Dr. Gilbert F. Froment
Laboratorium voor Petrochemische Techniek
Rijksuniversiteit Gent
Krijgslaan 271, B-9000 Gent, Belgium

Professor Dr. Jerzy Haber
Institute of Catalysis and Surface Chemistry
Polish Academy of Sciences
ul. Niezapominajek, PL-30-239 Kraków, Poland

Dr. Lucien Hosten
Laboratorium voor Petrochemische Techniek
Rijksuniversiteit Gent
Krijgslaan 271, B-9000 Gent, Belgium

Dr. André J. Lecloux
Solvay & Cie., Central Research Laboratory
Catalysis Department
Rue de Ransbeek 310, B-1120 Brussels, Belgium

Professor Dr. Kozo Tanabe
Dept. of Chemistry, Faculty of Science
Hokkaido University
Sapporo, Japan

Chapter 1

History of Concepts in Catalysis

G.-M. Schwab

Institut für Physikalische Chemie
Universität München
Sophienstr. 11, D-8000 München 2, FRG

Contents

1. Introduction

The conceptual importance of catalysis is based on its surprising nature. Whereas normally in nature the law is valid: "Causa aequat effectum" (The cause is equal to the effect), catalytic phenomena are of an entirely different nature. Some time ago the father of the energy conservation law, Julius Robert Mayer, and the father of physical chemistry, Wilhelm Ostwald, (1840 and 1900) saw clearly the following discrimination. When, e.g., in a steam engine, mechanical energy is produced from fire, this is simply causal causality. The cause — the heat energy spent — is somehow equivalent to the effect, the gained mechanical work; both are, if not equal, at least of the same order of magnitude. This is why those authors talked of "causal causality" (Verursachungskausalität). The phenomenon is entirely different when the pressure on the trigger of a rifle gives the ball an enormous velocity or when the opening of a sluice by hand makes an enormous mass of water stream down with great velocity. A similar quantitative relationship does not exist in these cases and the manual work on the trigger or the sluice wheel is negligible as compared to the mechanical effect. In these cases the expression "release causality" (Auslösungskausalität) is appropriate. Cause in the sense of causal causality is the explosive power of the rifle load or the amount and original position of the water. However, this cause will not work before a release by some incommensurably small additive cause happens.

Transferred to chemistry this means, e.g., that in a firebrand in a farm the causal cause is the combustibility of grain and hay but the release cause is the match or cigarette end, i.e. a negligible thing. Of this nature are also catalytic reactions, and here it is unimportant if the releasing substance actually enters into the reaction sequence, e.g. if the platinum forms an intermediate oxide or if it remains unchanged and acts merely by its presence in the platinum catalysed hydrogen combustion. In any case, release causality in catalysis is so surprising to the observer and so challenging for the researcher that numerous concepts have been formed to understand this type of reaction.

Already Aristotle wrote in Vol. IV, chapter 1 of his Meteorology:

"Ὅτι μὲν οὖν τὰ μὲν ποιητκὰ τὰ δὲ παθητιχά φανερόν. Διωρισμένων δὲ τούτων ληπτέον ἄν εἴη τὰς ἐργασίας αὐτῶν, αἷς ουτως ἐργάζονται τὰ ποιητικά, καὶ τῶν παθητικῶν τά εἴδη."

Αριστοτέλης, Μετεωρολογία, Δ,1

[It is clear that the one substances (we would nowadays say: the catalysts) are active, the others (the substrates) are passive. When this has been stated, one must know both the interactions by which the active substances act, and also the groups of passive substances.]

Here we see the first program to investigate active catalysts and their specifity. It is perhaps not fortuitous that, in a period when the thoughts of Aristotle were decisive for the forerunners of natural science, the catalytic idea was reborn without any respective observations as the "Philosopher's Stone" or "quinta essentia" of the medieval alchemists. From all the descriptions of this substance or the preparations thought to be such, it appears that exactly those actions were ascribed to them which are characteristic of a catalyst: To produce or to accelerate a process per se possible (spontaneous) like the transmutation of lead into gold ($^{207}Pb \rightarrow {}^{197}Au$) or the healing of a sick person. It would be interesting to investigate how far Aristotle was responsible for this widespread opinion and practice.

The clear formulation of the concept of catalysis on the basis of the manyfold observations of the 17th and 18th century is known to be the achievement of J. Berzelius (1835). Once chemical thermodynamics and independent kinetic theory were developed in the 19th century, W. Ostwald (1896) was then able to formulate the exact definition of this concept in terms of reaction rate increase. This definition, here not repeated, was in turn the catalyst for catalytic research, started at once in Ostwald's laboratory. It was stimulated by the contemporary and continuously increasing technical importance of catalysis. With the scientific research and the growing empirical material the formation of theories and theoretical concepts began at once and continues until the present. This will be the subject of this article, with heterogeneous catalysis in the foreground.

2. The Condensation Hypothesis

Great early workers like Döbereiner and Faraday thought that the density increase of gases alone, e.g. oxygen and hydrogen on the surface of platinum, would be sufficient because of the increase of impact numbers to accelerate water formation. It appears to be M. Polanyi who showed that the maximal possible density increases, even multiplied by each other, are not sufficient to produce the real velocity increases. This showed clearly that the catalysed reaction needs special reactive states of matter, and the concept of adsorption was introduced.

3. The Diffusion Hypothesis

This was the hypothesis from which the man started who was the first to make kinetic experiments in Ostwald's laboratory in order to draw conclusions on the mechanism. He was M. Bodenstein. Originally he imagined that the locus of the reaction, i.e. the surface of the catalyst, is surrounded by adsorption layers of the thickness of several molecular diameters, "atmosphere-like layers", and that the reactants have to diffuse through these layers towards the surface. This diffusion was considered as rate determining. He thought especially that by these assumptions characteristic inhibitions could be understood together with the fact that only the stoichiometrically deficient partner of binary mixtures appeared in the rate equations. Bodenstein himself gave this concept up in the twenties when I. Langmuir proved that the adsorption layers, even at high pressures, are unimolecular.

4. Adsorption Kinetics

Langmuir himself drew the first chemical consequences from his physical findings. He for the first time introduced the surface concentration of the adsorbed reactants as factors into the kinetic mass action law. For this purpose he derived his mixture adsorption isotherm. C. N. Hinshelwood applied this concept to an ample number of heterogeneous gas reactions consequently in the years after 1925. He could in many cases explain characteristic inhibitions by displacements of reactants by others or by reaction products. Schwab and others followed him.

Like in the original derivation of Langmuir's isotherm, a homogeneous surface was assumed, in which all the adsorption centers or surface atoms would have the same adsorption energy and capacity. Soon, deviations were observed from the rate equations thus formulated.

5. The Active Centers

One of these deviations is the fact that on a certain catalyst, say platinum, certain reactions, e.g. the decomposition of hydrogen iodide, are not inhibited by the hydrogen present, whereas others, e.g. the decomposition of ammonia, show a strong inhibition. Other observations not compatible with the picture of homogeneous surfaces are the dependence of the heat of adsorption on the degree of saturation, the thermal sintering of catalysts, the higher sensibility of catalysts for sintering and poisoning as compared with adsorption, the non-linearity of poisoning and the dependence of catalytic activity on starting material and pretreatment of the catalyst. All these circumstances have been made understandable with a common, simple, and evident theory, that of active centers (H. S. Taylor 1925–26). It claims that solids do not offer plane and smooth outer surfaces but a sort of mountain-and -valley landscape, in which edges, corners and peaks occur. Only these atoms in exposed positions are considered as catalytically active and not those in plain faces. Since the most active atoms have higher adsorption energies, they are covered first, and the decrease of the average adsorption energy with increasing coverage becomes understandable. The same is true for sintering below the bulk melting point by the wandering of active atoms into sites of lower energy. The non-linear poisoning is due to the successive poisoning of different groups of centers, as is the low sensitivity of the adsorption to the contribution of inactive centers. The influence of material and preparation of catalysts means the formation of various distributions of centers, and finally the deviations from Langmuir-Hinshelwood kinetics suggest the variability of adsorption coefficients and rate constants.

A variation of this theory is the adlineation theory (Schwab). It considers that the rough surface of a solid does not contain disordered atoms but microcrystallites with well indexed elements as corners and edges. Especially the crystal edges and grain boundary lines and also in mixtures phase boundaries are considered as places of "adlineation" and active places for catalysis.

Constable (1925) gave those theories a mathematical shape, assuming an exponential distribution of the active centers. He regarded the differential activation energy of a reaction as being antibatic to the potential energy of the different centers. We will discuss the consequences later.

6. Intermediate Compounds

During the development described hitherto it has always been supposed that the reactants are first bound to the catalyst by adsorption and that they react in this state. This corresponds to the much older chemical theories with the scheme

$$A + K \rightarrow AK$$
$$AK + B \rightarrow AB + K$$

$$\text{Sum:} \quad A + B \rightarrow AB$$

Schönbein had accepted this view long before the scientific period of catalysis research, whereas Döbereiner (1818) rejected it. P. Sabatier postulated his theory of intermediate chemical compounds of the catalyst to be the general basis of catalysis even when the intermediate cannot be isolated (1914–1927), a view vigourously fought by Ostwald. Sabatier based his considerations mainly on his experience with organic and inorganic homogeneous reactions, but he extended it also to the action of metals on gases where before one had only talked of adsorption. We will see later that this idea returns in all the newer concepts as a general basis and that essentially always the catalytic property of a substance is traced to its chemical affinity as a cause, the latter being expressed mostly in physical terms.

7. The Hedvall Effect

In connection with the theory of active centers we have to consider the following postulate of A. Hedvall (1934): Two allotropic modifications of a solid must differ in their catalytic properties because in the two lattices the atomic order and still more the disorder is different and therefore the activation energies should also be different. This is termed 1st kind effect. Besides, during the transition itself and at its temperature, each of the two atomic arrangements will be destroyed at least in the phase boundary, and particularly there many active centers should be present, giving the 2nd kind effect. Hitherto it has not been possible to prove the 1st kind effect, but the 2nd kind effect was found in the SiO_2-catalysis of SO_3 synthesis.

It may be mentioned here that the inverse "Valhed-effect", the acceleration of a phase transition by a simultaneous catalytic gas reaction, does not exist.

8. Magneto Catalysis

Many attempts have been made to find correlations between catalytic reactions and magnetic properties or phenomena on the assumption that the modification of electronic states, visible in the Zeeman effect, should also influence the valence electrons used in catalytic reactions. Especially a Hedvall effect' has been looked for at the Curie point of ferromagnetic catalysts, metals and oxides. The catalytic constants of the ferromagnetic and the paramagnetic phases have always been the same (with one dubious exception). However, during the magnetic transition an effect of the 2nd kind appeared regularly, tentatively ascribed to magnetostriction and hence to mechanical deformation. Mechanical-catalytic effects have been proven apart from magnetism.

A drastic exception from these facts is, however, the catalytic spin transition of hydrogen, the ortho-para hydrogen conversion. As is known, it is catalysed by the inhomogeneous magnetic fields of paramagnetic ions and radicals. This catalysis is with some lanthanide oxides, in turn, dependent on superimposed weak external magnetic fields. Normal chemical catalyses, however, are independent of even strong magnetic fields.

9. The Compensation Effect

Since 1929 it has been remarked (Schwab, Cremer) that in the representation of catalytic velocities by means of the Arrhenius equation, the two representative kinetic constants, i.e. the activation energy and the preexponential factor, in the comparison of similar reactions on one catalyst or of one reaction on similar catalysts, are found not to be independent on each other but are mostly symbatic. Since an increase of the activation energy produces a retardation of the reaction and an increase of the preexponential factor means an acceleration, the symbatic trend of these two constants effects a compensation so that "the trees of catalysis do not grow to the sky". A mathematical treatment of this fact leads to the result that at an "isokinetic" temperature θ all the members of a comparison group of reactions exhibit equal velocities. This has frequently been observed. The basis of this treatment has been the mentioned assumption of Constable of a distribution of active centers by which highly active centers of high potential energy and low activation energy are rare and those of low energy (high activation energy) are common and numerous. If additionally it is assumed that such a distribution has been established the last time during the preparation of the catalyst, it follows that θ is the preparation temperature. Indeed, up to 1959 many examples of this coincidence could be presented. Later, however, equally many contrary examples became known, corresponding to the fact that many catalysts are not in thermal equilibrium but have been formed by topochemical reactions far from equilibrium. In these cases, θ is only a parameter related to some Gaussian distribution.

As a compensation effect of similar nature is observed in various physical phenomena and even in homogeneous chemical reactions, the recent view is rather that the compensation effect is not primarily a matter of the active centers but merely a parallelism of the energy and entropy of a process. This can be understood very generally by different models. This does not change the fact that the effect is so general in catalytic reactions that for testing catalysts one does not need knowledge of both parameters, but preferentially the activation energy only which characterises the reaction independent of secondary effects, like porosity or dispersion.

10. Pore Diffusion

A catalytic reaction can be considered as a "drama in acts" (Schönbein). First the reactants must arrive at the catalyst (convection), then they must reach the active sites (diffusion) even when these are situated in the interior of narrow pores, as in porous technical catalysts, after that they must react there (reaction), then the reaction products must diffuse into the outer space and last the products must leave the catalyst by convection. In industry the scope is that the reaction is at least equally fast as the convection. However, for catalyst testing or kinetic research in the laboratory the reaction must be the slowest act of the drama, so that it may be measured. Now it may happen and happens frequently that diffusion in the narrow pores is still slower than the reaction therein. In these cases the reactants cannot penetrate into the interior of the pores and pass by, so that the active centers remain unused or nearly unused. In this case even a good catalyst produces only a small conversion. An apparatus of formulae has been derived (Wheeler and others, 1956) for calculating the true activity of the centers from geometric data of the catalyst.

11. The Chemical Viewpoint

Most of the concepts described hitherto allow one to understand certain features of the catalytic phenomenon but scarcely give the answer to the basic question why certain substances catalyse just certain reactions. The most general viewpoint in this direction seems to be the already mentioned intermediate reaction, including nonstoichiometric adsorption compounds.

It struck chemists already in early times (1926) that the transition elements play a special role in catalysis (homogeneous as well as heterogeneous), and this was ascribed to their ability for valence change and/or for complex formation.

A. Complexes and Clusters

Just this last viewpoint came distinctly into the foreground during recent years in homogeneous catalysis as well as in catalysis by surface-fixed complexes. It has been shown that in complexes of transition elements various reactions can be carried out by the exchange of ligands against reactants, rearrangement of these and back exchange against free ligands. In this respect, special importance is ascribed to polynuclear complexes i.e. clusters containing many metal atoms surrounded by appropriate ligands. Similar clusters are also found in metals spread on suitable carriers. Such clusters are also useful for theoretists for the quantum mechanical treatment of catalytic reactions because a treatment by bond theories of the chemisorption bond is easier on a cluster with localised bonds than on a crystal surface with delocalised electron states.

B. The Multiplet Hypothesis and the Geometric Factor

A. Balandin (1929) has established a special form of the theory of intermediate reactions by combining it with geometric considerations. Aromatic compounds are thought to be easily dehydrogenated on (111) faces of transition metals only when they are chemisorbed on sextets in a special configuration so that certain metal atoms diminish by attraction the distance of two neighboring ring carbons from that of a single bond to that of a double bond, whereas other metal atoms combine the respective hydrogen atoms to molecules by one-center adsorption. Similarly the hydrogenative splitting of C-X bonds is thought to presuppose chemisorption on catalysing doublets, and to depend on the difference between energies of bonds to be split and to be formed.

A certain coincidence or at least relationship between atom and ion distances of the catalyst and the reactant molecule is also postulated for other reaction types, mainly eliminations. The geometric factor is also used in cases of different catalytic activity of different crystallographic faces.

C. The Volcano Curve

In a similar way, as Balandin in his multiplet theory has estimated the series of activation energies from differences of bond energies, Sachtler (1965) has estimated the series of decomposition velocities of formic acid from bond energies of supposed chemisorption compounds of the catalysing metals. These bond energies are taken simply from the formation energies of stoichiometric compounds, e.g. for the decomposition of formic acid from those of metal formates. It is necessary to discriminate if formation or decomposition of the formates is considered rate determining. If it is the decomposition, the bond is strong and the reaction is of zero order. If the formation (chemisorption) is rate determining, the bond is weak and the reaction of first order. With increasing bond strength (formation energy of the respective formate) the zero order rate will decrease and the first order rate will increase. With the formation energy of formates as abscissa and the rate as ordinate a V-shaped curve will result, and with the temperature of a certain rate as ordinate the reverse will result, like the profile of a hill, mountain or volcano.

The same behavior is sometimes claimed also for other reactions, although its validity even with formic acid is restricted, (iron does not follow it.).

D. The Electronic Factor

We will not mention here older views on electrons emitted by the catalyst or surrounding it. An electronic factor besides the geometric and the chemical one is to be seen in a correlation of catalytic activity (expressed by activation energy and perhaps preexponential factor) and the electronic structure of the catalyst. This development (Schwab) became possible only as late as 1940 when the band theory of solids provided reliable concepts on the electronic properties of solids, especially metals and semiconductors. The previous valence theory gave such informations only for transition metals introducing the notion of *d*-character (see below).

E. The *d*-Character

This notion, applied to catalysis, says that only a portion of the *d*-electrons takes part in the interatomic *dsp* hybrides of cohesion bonds. It may be calculated from the saturation magnetisation and is related to catalysis and especially hydrogenation. In numerous papers this *d*-character is correlated to metal catalysis. It is questionable if this is informative because the numerical values of the *d*-character are all in the neighbourhood of 40% whereas the catalytic activities vary by great differences. Also, the catalysis by metals other than transition metals remains unexplained.

12. Band Theory

The method of application of the band theory to catalysis (Schwab et al. since 1940) is to vary the electronic distribution in the solid by very small additions, and to control the effect of these additions on catalysis. As test reactions distinct donor reactions are usually used in which the activation of the chemisorbed reactant involves an electron transfer from the molecule to the catalyst, e.g. dehydrogenation of formic acid or ethylene hydrogenation; acceptor reactions are occasionally used. This program succeeded for the first time with the Hume-Rothery alloys where small additions of elements of the II–IV subgroups of the Periodic Table to Cu, Ag and Au increase the occupation ratios of the first Brillouin zone (conduction band). The corresponding increases of the activation energies of donor reactions are regular and proportional to the valency of the additive minus unity, like the electric resistance of the same alloys. Hence electron holes must be considered as involved in the catalysis. In intermetallic compounds where the occupation degree of the zone depends on the lattice type, analogous rules are observed, and the gamma-phases with 90% of occupation show extreme increases in the activation energies.

The step from metals to nonmetallic catalysts has also been possible (1951). These of course must be semiconductors, since only in these can statements on the distribution of electrons and defect electrons (holes) be made, whereas in insulators only pair bonds are present, subject to other concepts. Semiconductors can be made p- or n-conductors by doping, and the conductivity of intrinsic p- and n-conductors can be modified by doping. Thus, the number of electrons in the conduction band or of holes in the valence band can be modified. Very generally it was found that increasing p-character is favorable for donor reactions and increasing n-character for acceptor reactions.

13. Synergy

In 1912 R. Willstätter observed in his studies of enzyme reactions that in cases where two different catalysts (ferments) of the same reaction were present in the system, the total conversion was higher than the sum of the separate actions. For this mutual promotion he coined the notion "synergy" or synergetic promotion. In the special cases of heterogeneous catalysis the promotion can be separated in two groups, as has been exemplified in the decomposition of nitrous oxide by oxides. One is textural promotion, obtaining when one component (the less active or even inactive one) of the mixed catalyst only produces or at least stabilises the fine dispersion of the other and prevents its sintering. This is typical for the classical industrial ammonia catalyst, Al_2O_3 stabilising the active iron. The activation energy remains unchanged in these cases. The ammonia catalyst was discovered by A. Mittasch around 1910. He himself had always declined to give explanations for promotion. The other group is synergetic promotion in which the mixture has a higher activity and also a lower activation energy than either of the components. Here, at least in the phase boundary, a chemical interaction must be assumed.

14. The Support Effect

It is often observed that a metal deposited in thin layers or small clusters on the surface of an "inert" carrier or support exhibits not only an increase and stabilisation of its own surface, but also a reduction of its characteristic activation energy for a certain reaction, a synergetic promotion. In a similar way, as the action of a simple catalyst, metal or semiconductor, has been understood by the electronic factor (see above), it is feasible to explain the synergetic promotion by an electron transfer between support and catalyst. Since metals usually show a lower work function than semiconductors, in electric contact electrons must be emitted from the metal to the support. Provided the metal is, as usual, present in relatively small amounts, its Fermi level must be lowered to that of the support. The effect should be the same as an admixture of an electronpoor metal like palladium or nickel. This effect is indeed observable in many examples, and the technical importance of binary catalysts may often be due to this effect.

15. Inverse Mixed Catalysts

Because in metals the concentration of mobile electrons is of the same order of magnitude as that of the metal cations, and in semiconductors at most that of the dope, we have in supported metal catalysts of the above type an action of only a few electrons on a great excess of electrons. It can only be understood by the fact that normally the metal is present in very small amounts. This created the idea (Schwab, 1966) to reverse the situation and to deposit a catalytically active semiconductor on a metallic carrier surface. This system was termed "inverse mixed catalyst". In this case the emission of electrons into the semiconductor has as consequence a common Fermi level and a bending of the bands of allowed states in the semiconductor downwards. The relative position of the Fermi level in respect to these bands moves upwards, and this is comparable to a doping in the n-type sense. The activation energy of a reaction taking place on the backside of the thin semiconducting layer or of small particles of it, should then be increased in the case of a donor reaction and decreased in an acceptor reaction. The first effect has been confirmed in carbon monoxide oxidation, the latter in that of sulfur dioxide. These effects are much more intense than those in normal supported metal catalysts: It may be that this concept will be helpful in the endeavour to create "tailor-made" catalysts.

16. Conclusions

In the preceding sections the most important concepts have been named and discussed which, since the beginning of exact catalysis research, have been brought forward in order to "explain" this formerly so enigmatic phenomenon, i.e. to place unrelated facts into a scientific system. Completeness has not been aimed at, but the most important points have perhaps been touched. An historian reading this "History of Concepts" will perphaps be surprised to find so few dates of actual events. No chronological series of concepts has been given, which would have indicated a logical development of the concepts from each other. Such development did not exist. While both earlier as well as more recent authors of concepts may have believed they had created a more or less generally valid "theory of catalysis", with time their concepts have gone astray more and more, and have grown rather in number than in validity. A man who has followed the history of concepts during several decades has seen that an older concept has seldom been totally replaced or refuted by a newer one, but that usually an older concept has been represented in new or modern scientific language. This is a proof of the fact that the mechanisms of catalysis are so manifold that many fields of thought must be employed — although the notion of catalysis is very unitary from the viewpoint of thermodynamics and kinetics. It is to be hoped that many of these concepts will find their confirmation or elimination by the modern methods of surface physics and their application to catalytic processes.

Chapter 2

Crystallography of Catalyst Types

J. Haber

Institute of Catalysis and Surface Chemistry
Polish Academy of Science
ul. Niezapominajek, PL-30-239 Kraków, Poland

Contents

1. Introduction

Heterogeneous catalysis is a phenomenon, in which a relatively small amount of a solid, called the catalyst, increases the rate of a chemical reaction. Accelerating effect of a catalyst may be due to:

(a) creation for the given reaction, through interaction with the substrate, of a new reaction path, usually multi-step, characterized by a lower energy barrier;

(b) lifting of the symmetry restrictions in the case of a concerted reaction.

The mechanism of catalytic reactions may be thus discussed in terms of the type of intermediate complex formed in the course of the reaction between the substrate molecule and the group of atoms at the surface of the solid, constituting the component of its lattice and called the "active center". On interaction with such an active center, the electronic structure of the reacting molecule is modified, resulting in the rearrangement of chemical bonds and transformation of the molecule.

All solids may be divided into crystalline and amorphous substances. The former are characterised by an ordered periodic arrangement of the structural elements in space, whereas in the case of the latter the structural elements do not form an ordered structure.

Irrespectively, however, of whether the structure as a whole is ordered or not, the nearest surrounding of each atom is usually ordered and its form may be explained on the basis of the type of chemical bonds linking the given atom with its nearest neighbours, which may be considered as ligands, forming a coordination polyhedron. When the bonds are covalent, the number and position of ligands are determined by hybridization of the orbitals of the central atom, when they are ionic- by the radii of ions. Very often it is convenient to consider the solid as composed of such coordination polyhedra [1, 2, 3]. In the crystal these polyhedra are linked together in an ordered mode, whereas in the amorphous substance — in a more or less disordered fashion. Thus, crystals are characterized by both short range as well as long range order, whereas in amorphous substances only short-range order is present. It must be emphasized that coordination polyhedra are not equivalent to groups of atoms which could be defined as a chemical molecule; they must be considered as arbitrary structural units which may be chosen in many different ways. The coordination polyhedra acquire a physical sense when the bonds between atoms within the coordination polyhedra are stronger than bonds between the polyhedra. This is the case when the valence of the central atom is greater than half of the sum of the valencies of ligands. In the limiting case, when the valence of the central atom is equal to the sum of the valencies of ligands, the coordination polyhedron is a chemical molecule. In such a case obviously, different chemical bonds operate within the coordination polyhedra and between them. Depending on whether the structural elements of the crystal are linked together with one type of bond or whether different types of bonds are present in the crystal, we may divide all structures into two groups: homodesmic and heterodesmic [4]. The structure is homodesmic when all atoms of which it is composed are linked together with the same type of bonds. Diamond, sodium chloride or metallic copper may serve as examples of homodesmic structures with covalent, ionic and metallic bonds respectively. The structure is heterodesmic when its structural elements are linked with at least two different types of bonds. Salts containing complex cations or anions are typical examples. In sulphates, phosphates etc., the bonds within the complex anion have predominantly covalent character, whereas the bonds between cations and complex anions are ionic. The majority of organic solids belong also to heterodesmic structures, the bonds within the organic molecules being covalent, while those operating between the molecules are van der Waals type bonds.

Let us now return to the discussion of the interaction of the substrate molecule of a catalytic reaction with the solid catalyst. It is convenient to choose the coordination polyhedra of the solid as models of the active centers. When they are located at the surface, at least one coordination site is empty and may accomodate the reacting molecule. Let us assume, as an example, that the coordination polyhedron is a chemical molecule composed of a central cation linked by ionic bonds through oxygen atoms to five octahedrally coordinated oxygenated ligands, the sixth ligand being the substrate molecule, e.g. a hydrocarbon. Obviously, this would be a typical case for homogeneous catalysis. Modification of the electronic structure of the substrate molecule, which is responsible for the change of its reactivity and thus for the catalytic effect, will depend on the type of bonding between this molecule and the central atom, properties of the central atom and properties of the ligands, the total number of electrons in the complex being fixed in the course of the whole transformation. Let us now assume that we increase the number of polyhedra by linking them together through their ligands into an ordered arrangement of a crystal of an oxide. We may still discuss the bonding of the substrate molecule to the given polyhedron as an active center on the basis of the same model, however with some important differencies [5, 6]. Firstly, it is not the number of electrons in the complex which is now fixed, but the chemical potential of electrons, determined by the position of the Fermi-level in the solid. Secondly, the bonded substrate molecule is now under the influence of all other coordination polyhedra constituting the surface of the solid, i.e. the Madelung potential at the active site must be taken into account.

When analysing the reactivity of a molecule in the course of a catalytic transformation several factors should be taken into account: readiness of the formation of the intermediate complex with the active center of the catalyst, ease of its transformation into products, and possibility of their desorption. Formation of an intermediate complex requires a vacant coordination site to be present in the coordination polyhedron of the catalyst, performing the role of the active center. The number of such vacant sites depends on the position on the surface of the crystal, where the active center is located (type of crystal plane, edges, kinks etc.) as well as on the concentration of defects in the lattice. On the other hand the defect structure of the lattice determines the position of the Fermi-level, which in turn has a decisive influence on the occupancy of the molecular orbitals of the intermediate complex and thus on its reactivity. The defect structure of the solid is also responsible for the mobility of lattice constituents, which may play an important role in enabling the transformations of the intermediate complex and facile desorption of the reaction products. This factor is of particular importance in the case of reactions, in which one of the constituents of the catalyst lattice is inserted into or abstracted from the reacting molecule and then exchanged between the lattice and the gas phase. This type comprises such important reactions as oxidation on oxide catalyst, desulphurization on sulphide catalysts, hydrogenation and dehydrogenation on metals forming hybrides etc. Analysis of the defect structure of solids is thus of importance for understanding their behaviour in catalytic reactions.

Until now we have been discussing the coordination polyhedra composed of central metal cation surrounded by simple or complex anions, characteristic for binary metal compounds or compounds which may be classified as salts. We may now visualize a series of coordination polyhedra with decreasing ionicity of bonds between the central atom and the ligands. The limiting case of this series is a coordination polyhedron composed of a central metal atom surrounded by identical metal atoms as ligands as shown in Figure 1. This is a very small metal cluster which still exhibits the properties of molecule and may be examined by molecular orbital computational procedures. On increasing the number of metal atoms in the cluster the characteristic properties of the bulk metal begin to be displayed (band structure, conductivity, cohesive energy, magnetism) which may result in drastic change of chemisorptive and catalytic properties [7].

For various practical reasons the active phase of the catalyst is usually supported on a porous, high surface area, heat resistant carrier (or support) by precipitation, impregnation, mechanical mixing or other methods [8, 9, 10].

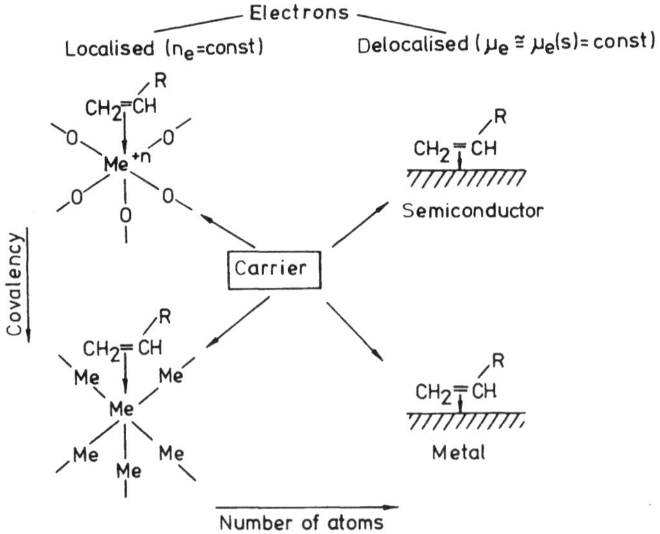

Figure 1. Properties of isolated complexes, metal clusters, solid compounds and metals

The catalyst becomes finely dispersed, large specific surface area being developed and heat resistance increased. A number of different phenomena may take place in the course of the preparation depending on the temperature of annealing as well as chemical and crystallochemical properties of both the active phase and the carrier. All these processes may be illustrated by a following scheme describing the case of a catalytically active compound A_mX_n supported on an oxide B_rO_p present in large excess and playing the role of a carrier:

$A_mX_n + B_rO_p$
— A_mX_n dispersed on the surface of B_rO_p, acting as a carrier;
— clusters of metallic A dispersed on the surface of B_rO_p acting as a carrier;
— solid solution of A_mX_n in the surface layer of B_rO_p;
— surface compounds or definite phases $aA_mX_n \cdot bB_rO_p$;

Irrespective of whether obtained by precipitation or impregnation, only when the temperature of annealing is low enough and the miscibility of the two compounds is very limited will the minority phase A_mX_n accumulate almost entirely at the surface of the carrier. Generally the active phase-on-carrier system is obtained with different degrees of dispersion depending on the conditions of preparation. In such a state, the crystallites of the active phase may show properties different from those of bulk crystals, on one hand due to the fact that collective properties of the crystal do not yet dominate over the atomic properties of the component atoms, on the other hand because of the interactions with the carrier whose effect is the more pronounced the smaller are the supported clusters, i.e. the higher is the degree of dispersion. When reducing a atmosphere is present in the course of annealing, reduction of A_mX_n may take place resulting in the formation of a metal-on-carrier catalyst. On annealing at higher temperatures, dissolution of dispersion. When reducing an atmosphere is present in the course of annealing the dissolving compound accumulates in the surface layers of the solvent crystallites. On further annealing two processes may take place simultaneously:
inward diffusion into the bulk of the crystallites, and
evaporation into the gas phase.

Depending on the relative rates of these processes the surface layers of the resulting solid solution crystallites may show different compositions. In the case when the active phase and the carrier are capable of chemically reacting with each other, surface compounds may be formed which may determine the adsorptive and catalytic properties of the system.

2. Classification of Heterogeneous Catalysts

Heterogeneous catalytic processes may be devided into two large groups: redox reactions and acid-base reactions [11, 12, 13]. The first group comprises all those reactions in which the catalyst effects the homolytic bond rupture in the reactants molecules with the appearance of unpaired electrons, and formation of homolytic bonds with the catalyst with the participation of catalyst electrons. The important step of such reactions is thus the exchange of single electrons between the catalyst and the reactants, and therefore such catalytic properties are shown by solids which have the ability to accept and donate electrons. These are transition metals and their compounds, in which the cation can easily change its valence state. As this property is also responsible for the nonstoichiometry, catalysts for redox processes are found among nonstoichiometric transition metal compounds, the most important being oxides and sulphides.

The second group includes reactions in which the reactants form with the catalyst heterolytic bonds by using the free electron pair of the catalyst or reactants, or the free electron pair formed in the course of reaction by heterolytic rupture of bonds in the reactant molecules. Examples of catalysts for such reactions are simple oxides of main group elements or their complex acids and salts showing acid-base properties of both Brönsted or Lewis type. We shall thus devide the discussion of the structural properties of catalysts into three parts:

— transition metals and their substitutional and interstitial alloys,
— nonstoichiometric transition metal oxides and sulphides,
— oxides of main group elements.

3. Metal Catalysts

A. Transition Metals

The problem of the crystallographic structure of metals is essentially that of the question: how may spheres of equal size can be packed together to fill the space as densly as possible. On a plane surface the closest packing is attained when each sphere is in contact with six others, as shown in Figure 2a. The second identical layer may now be superposed on the first so that the spheres in the upper layer are vertically above the center of the triangle formed by spheres of the lower layer (Figure 2b). On superposing the third layer we have two alternatives: the spheres of this layer may by placed in the centers of triangles formed above the spheres of the first layer or above the interstices in this layer. In the first case hexagonal close packed structure

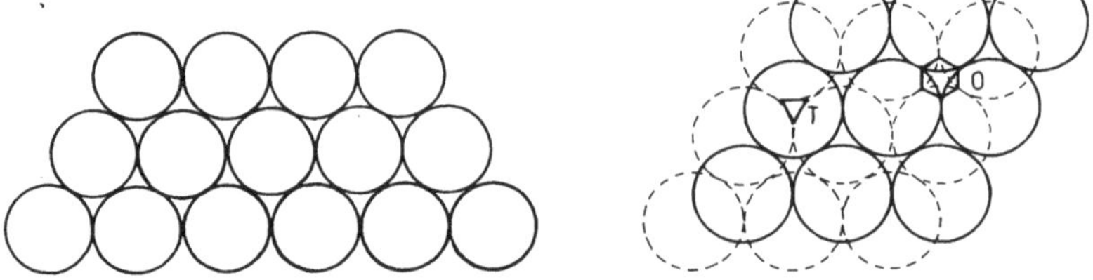

a b

Figure 2. a Layer of close-packed spheres; **b** Two superposed close-packed layers is obtained, characterized by the sequence of layers ABABA. Each sphere has 12 neighbours situated at the vertices of the coordination polyhedron shown in Figure 3a. Figure 3b shows the unit cell and Figure 3c part of the structure.

The second alternative mentioned above may be described by the sequence of layers ABC ABC. The coordination polyhedron of this structure is shown in Figure 4a. In this figure, the atoms constituting part of the unit cell shown in Figure 4b have been linked by dotted lines. Four atoms forming the face of the cube as well as the face-centering one are clearly visible. This structure, characterized also by the coordination number 12 is called cubic close-packed or face-centred cubic.

The third structure formed by the metals is the body-centred cubic. Its coordination polyhedron, unit cell and its part are shown in Figure 5. It is characterised by the coordination number 8 and is less closely packed than the two others, the space being filled only to the extent of 68 percent as compared to 74 percent in the case of hexagonal and cubic packing. The structures of transition metals are given in Table 1.

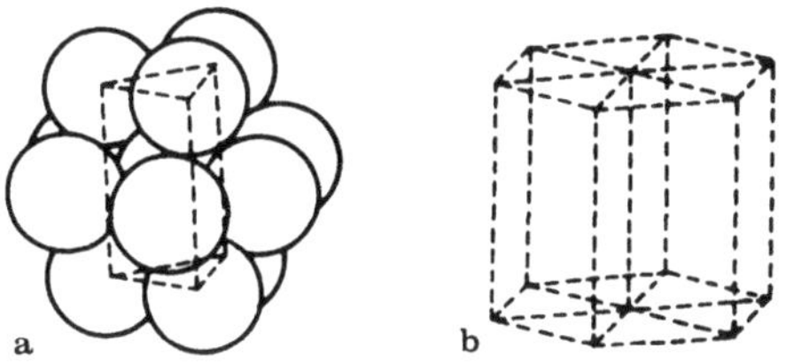

a b

c

Figure 3. Hexagonal close-packed structure. **a** coordination polyhedron; **b** unit cell; **c** part of the structure

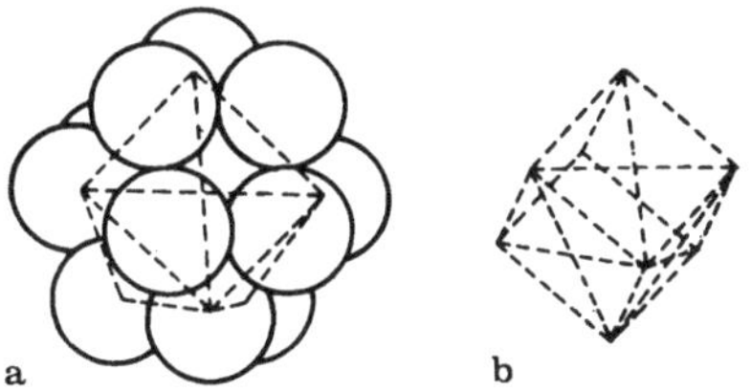

a b

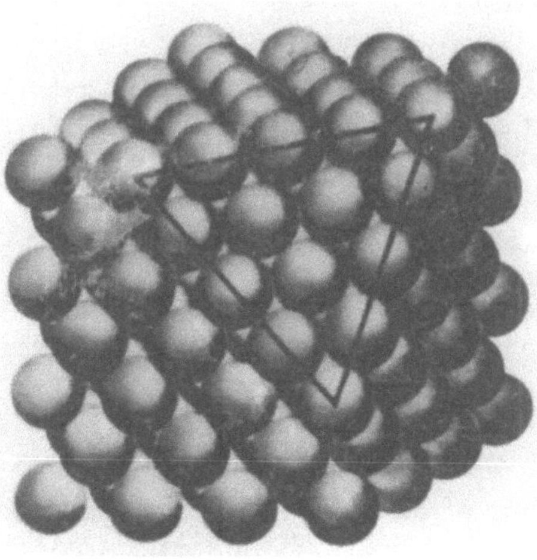

c

Figure 4. Face-centred cubic close-packed structure: **a** coordination polyhedron; **b** unit cell; **c** part of the structure

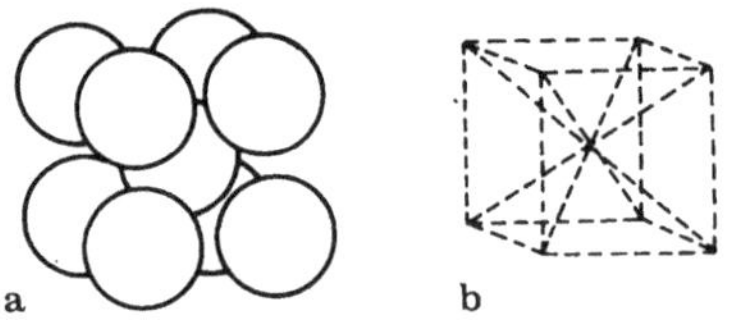

a b

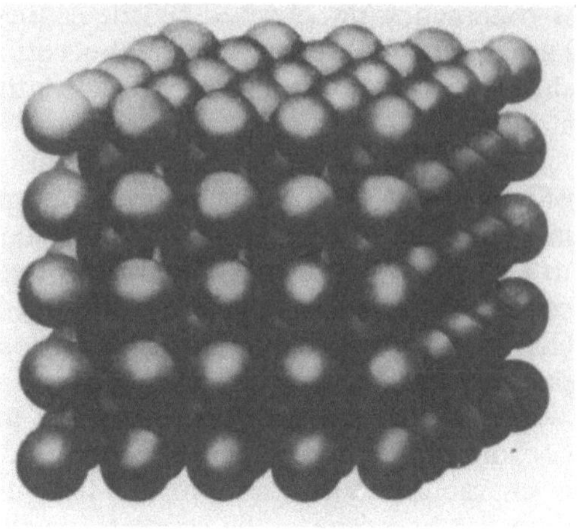

c

Figure 5. Body-centred cubic structure. **a** coordination polyhedron; **b** unit cell; **c** part of the structure

Table 1. The crystal structures of some transition metals

Metal	Structure	a/Å	c/Å	Interatomic distance/Å
α-titanium	hcp	2,950	4,683	2,89 2,95
β-titanium	bcc	3,306		2,86
vanadium	bcc	3,025		2,63
chromium	bcc	2,885		2,50
γ-manganese	ccp	3,855		2,72
δ-manganese	bcc	3,075		
γ-iron	ccp	3,647		2,58
α-iron	bcc	2,866		2,48
β-cobalt	ccp	3,554		2,52
α-cobalt	hcp	2,507	4,069	2,49 2,51
nickel	ccp	3,523		2,50
copper	ccp	3,614		2,55
niobium	bcc	3,300		2,86
molybdenum	bcc	3,147		2,72
rhenium	hcp	2,760	4,458	
ruthenium	hcp	2,705	4,281	2,64 2,71
rhodium	ccp	3,804		2,68
palladium	ccp	3,890		2,74
silver	ccp	4,085		2,88
tantal	bcc	3,302		2,86
tungsten	bcc	3,165		2,75
osmium	hcp	2,735	4,319	2,67 2,74
irydium	ccp	3,836		2,70
platinum	ccp	3,923		2,77
gold	ccp	4,078		2,88

B. Transition Metal Alloys

One of the characteristic properties of metals is the ability to form solid solutions, in which the solute atoms (minority component) and solvent atoms (majority component) are arranged at random. The conditions determining the formation of solid solutions may be summarized as follows:

— the tendency to form solid solutions is small if the metals are chemically dissimilar,
— the range of composition over which solid solutions are formed depends on the relative sizes of the two atoms. If the difference between the radii of the metals is greater than about 15 percent (of the radius of the solvent atom) there is no extensive formation of solid solutions,
— the mutual solubilities of metals are not reciprocal, the metal of lower valency usually dissolving more of the higher valence metal than vice versa.

Some alloys which are random solid solutions when quenched from the molten state undergo rearrangement when cooled sufficiently slowly. Sometimes such rearrangement takes place after suitable heat treatment has been applied. It consists in a change from the random distribution of the component atoms of the alloy into an arrangement in which there is regular alternation of atoms of different kinds throughout the structure. Such changes are called order-disorder transformations and the ordered structure formed is often described as superstructure. As an example Figure 6 shows the structure of the CuAu alloy as a random solid solution (a) and as the ordered superstructure (b). The possibility of rearrangement to form a superstructure which is a cooperative phenomenon is determined by the difference in potential energies of the ordered and disordered states and the magnitude of the energy barrier that has to be surmounted before two atoms can change places.

It should be noted that the formation of a superstructure may have an important influence on the catalytic properties. Two effects are namely considered in alloy catalysis [14, 15, 16, 17]: "electronic" and "geometric". The former describes the situation when the reactant molecule is bonded to a given surface atom and upon changing the composition of the alloy, the type of neighbouring atoms changes and this modifies the nature of the

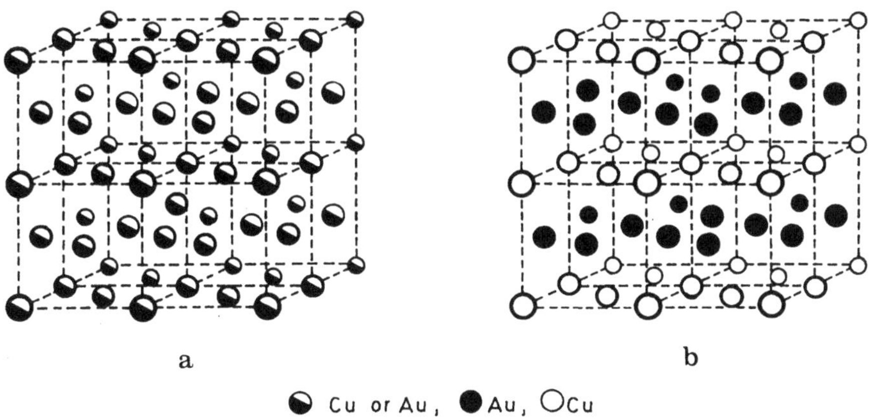

Figure 6. The crystal structure of the alloy CuAu in **a** the disordered, and **b** the ordered state

adsorbate-substrate bond. If several surface atoms of the given type are required for the formation of the reaction intermediate, alloying may change the surface concentration of such ensembles. This is described as the geometric effect. It is obvious that order-disorder phenomena may have a pronounced influence on both effects.

There is still one structure which deserved to be mentioned here which is formed by alloys of transition metals with metals of the later B subgroups and which is important for catalysis, e.g. PtSn and PtPb widely used as catalysts in petroleum reforming. By contrast with the pure metals, these crystallise in the nickel arsenide structure which, although rather resembling a simple binary compound, has, like typical alloys, a variable composition. In this structure, illustrated in Figure 7, each atom has six nearest neighbours of the other kind, but the arrangement around the two kinds of atoms is different. An arsenic atom is surrounded by six Ni atoms at the apices of a trigonal prism. A nickel atom is octahedrally coordinated by six As atoms, but there are two other Ni atoms which are sufficiently near to be considered as also bonded to the first Ni atom. Thus, the transition metal atom has eight approximately equidistant neighbours, metal-metal bonds being formed. Intermetallic phases which assume the NiAs structure are summarized in Table 2.

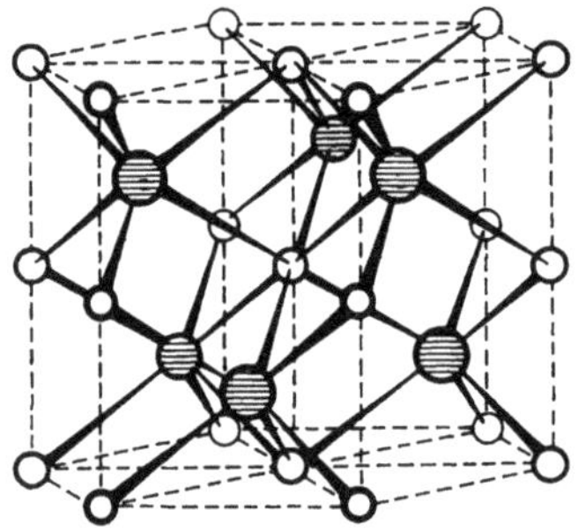

Figure 7. The structure of NiAs (As atoms shaded)

Table 2. Compounds crystallizing with the NiAs structure

	Cu	Au	Cr	Mn	Fe	Co	Ni	Pd	Pt
Sn	CuSn	AuSn			FeSn		NiSn		PtSn
Pb									PtPb
As				MnAs			NiAs		
Sb			CrSb	MnSb	FeSb	CoSb	NiSb	PdSb	PtSb
Bi							NiBi		PtBi
Se			CrSe		FeSe	CoSe	NiSe		
Te			CrTe	MnTe	FeTe	CoTe	NiTe	PdTe	PtTe

C. Interstitial Structures

Let us return to Figure 2b, showing two superposed layers of close-packed equal spheres. It may be seen that two types of interstices appear between the spheres: tetrahedral (T) and octahedral (O). Each sphere of the layer is surrounded by six interstices resulting from the superposition of the upper layer. Three of them are tetrahedral and three octahedral. Similarly, six intersticies appear due to the presence of the underlying layer. The given sphere completes one tetrahedral interstice lying below this sphere and one above it. Thus, each sphere is surrounded by $3 + 3 = 6$ octahedral interstices and $3 + 3 + 2 = 8$ tetrahedral interstices. In order to obtain the number of interstices per sphere we have to devide the number of interstices surrounding one sphere by the number of spheres forming one interstice:

$$\frac{\textit{number of octahedral interstices around the sphere}}{\textit{number of spheres forming the octahedral interstice}} = \frac{6}{6} = 1$$

$$\frac{\textit{number of tetrahedral interstices around the sphere}}{\textit{number of spheres forming the tetrahedral interstice}} = \frac{8}{4} = 2$$

Thus, in the closed-packed structures, both hexagonal and cubic, the number of octahedral holes is equal to the number of spheres, whereas the number of tetrahedral holes is twice that of the spheres. The radius of the tetrahedral hole amounts to $0.225R$ and that of octahedral hole to $0.414R$, where R represents the radius of the sphere.

The holes in the crystal of the transition metal may now be filled by small atoms of non-metals such as H, B, C, N or O, the resulting phase being called "interstitial solid solution" or "interstitial compound". Interstitial structures have many of the properties characteristic of alloys such as metallic lustre, conductivity, and from a chemical point of view, a wide range of composition. The type of holes occupied is determined by the ratio of the radii of non-metal and metal atoms.

When all octahedral holes in the cubic close-packed array are occupied the arrangement of atoms is that of the NaCl structure (cf. Figure 13). If all tetrahedral holes are occupied the CaF_2 structure results, whereas when half of these holes are filled in a symmetrical way — the zinc blende structure results. The structures based on cubic close-packed metals are summarized in Table 3, where examples of interstitial compounds are given.

The detailed knowledge of the mechanism of the formation of interstitial phases and of their properties is of great importance for understanding the mechanism of many catalytic reactions taking place at the surface of metals. Namely, in such reactions as hydrogenation or ammonia synthesis, a formation of surface or bulk hydride or nitride phases may occur, which can then completely modify the properties of catalysts. Such phenomena were demonstrated to occur e.g. in the case of hydrogenation on Pd, which in the course of the reaction may transform into PdH_2 [18]. When hydrocarbons react at the surface of the metal, deposition of carbon often takes place resulting in the formation of surface carbide phases, which show different catalytic properties from the initial metal catalyst [19, 20].

In recent years carbides, nitrides and borides have attracted attention as prospective catalysts for high temperature processes because of their well known thermal stability. Considerable catalytic activity of carbides has been found in hydrogenation and isomerization [21], but studies of catalytic properties of these compounds are yet in the very early stage.

Table 3. Interstitial structures derived from cubic close-packing [1]

Non-metal atoms in	Proportion occupied	Structure	Examples
Octahedral holes	all	NaCl	TiC, TiN, ZrC, ZrN,
	$\frac{1}{2}$		W_2N, Mo_2N
	$\frac{1}{4}$		Mn_4N, Fe_4N
Tetrahedral holes	all	CaF_2	TiH_2
	$\frac{1}{2}$	Zinc-blende	ZrH, TiH
	$\frac{1}{4}$		Pd_2H
	$\frac{1}{8}$		Zr_4H

D. Highly Dispersed Metals On Supports

The arrangement of atoms in the smallest of particles composed of several atoms to several tens of atoms and described usually as clusters, cannot be seen directly in the electron microscope and can mainly be inferred from analogy to larger crystallites. Two different types of arrangements of atoms supported on a carrier may be envisaged: two-dimensional monoatomic raft, and three dimensional particle (Figure 8). Which one of the two situations is realized depends on the relative values of the surface tensions at the respective interfaces. If the metal-metal bonds are strong as compared to the metal-support bonds so that the metal does not wet the support, the energy of the clusters is much lower in the form of the three dimensional particle than in the two dimensional raft [22]. Conversely, by very strong metalsupport bonds the formation of a two-dimensional raft can be expected.

Experiments carried out with unimetallic clusters indicate that in most cases three dimensional particles are formed. It could be expected that in

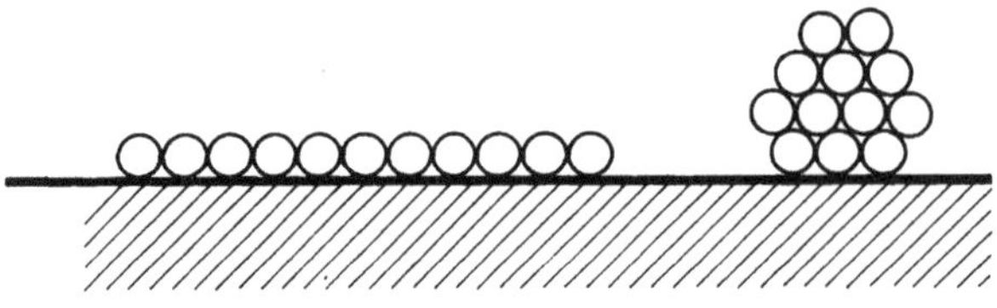

Figure 8. Schematic representation of the arrangements of atoms in the cluster supported on a carrier: **a** two dimensional monoatomic raft; **b** three dimensional particle

the case of a small number of atoms, a simple polyhedron would be formed representing a fragment of the lattice which is characteristic of the given metal. For fcc metals such as Pt or Ni these would be tetrahedron, octahedron, cube and cuboctahedron with 4, 6, 14 and 13 atoms respectively (Figure 9). It is obvious however that for clusters containing more than 5 atoms many different types of polyhedra can be constructed from the given number of atoms. Calculations of the potential energy [23] indicate that at small numbers of atoms the energy minimum does not correspond to the polyhedra representing elements of the fcc lattice, but to polyhedra which are formed by relaxation of some of the interatomic distances in the close-packed structure. It was shown, for instance, that the binding energy per atom is higher and the surface energy smaller for an icosahedron as compared to the cuboctahedron of the same number of atoms. This is shown in Figure 10 for a 13 atoms particle. The icosahedron has 42 nearest-neighbour bonds, while the corresponding cuboctahedron has only 36 bonds, the binding energy per atom being thus higher. In the case of an icosahedron only the triangular fragments of (111) crystal planes are exposed, whereas the cuboctahedron has fragments of both (111) and higher energy (100) planes, the surface energy of the icosahedron being thus smaller. This is however attained at the expence of some elastic strain introduced in the bulk of the icosahedron.

It may be thus concluded that on growing very small metal particles icosahedra are always formed [24], characterized by a five-fold symmetry un-

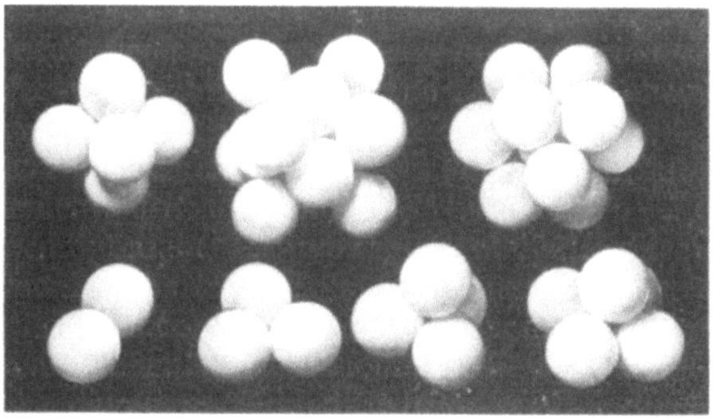

Figure 9. Clusters in form of small fragments of the fcc structure: tetrahedron, octahedron, cube and cubooctahedron

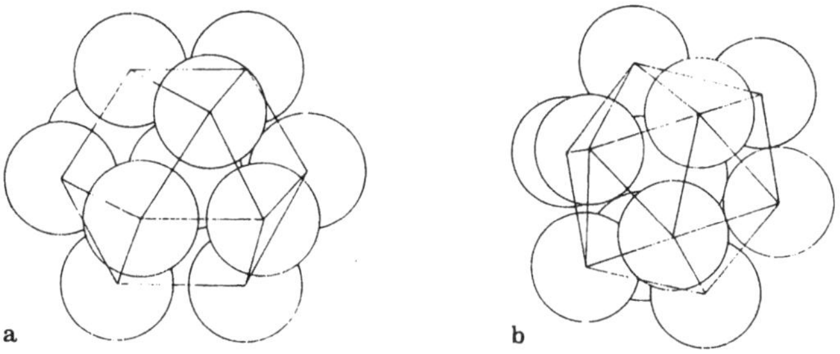

a b

Figure 10. 13-atoms particle in form of **a** cubooctahedron, **b** icosahedron known in the close-packed structures. Formation of such pentagonal particles was in fact observed (Figure 11) in the early stages of the preparation of evaporated thin films [25]. On increasing the size of the particle a transformation from icosahedral to cuboctahedral packing can be expected, when the strain energy introduced into the bulk of the icosahedron overcompensates the gain in the surface energy. Some calculations indicate [26] that this conversion might occur for particles of about 1000 nm for typical fcc metals.

As the stabilization of the icosahedral particles in respect to the cuboctahedra is mainly due to the differences in the surface energy, all factors influencing the surface energy such as adsorption, alloying or the presence of the support may change the structure of the dispersed metal. Relevance of these phenomena to catalysis is obvious in view of the fact that the behaviour of the metal in the catalytic reaction strongly depends on the arrangement of the metal atoms at the surface.

Figure 11. Electron micrograph of a pentagonal Ag particle [25]

4. Nonstoichiometric Transition Metal Oxides

A. Simple Metallic Oxides

Crystal structures of the more important oxides are summarized in Table 4. We shall discuss in more detail some of them. ReO_3, rutile, corundum and sodium chloride are built of coordination octahedra linked in different ways. In ReO_3 structure, the octahedra are sharing only vertices as shown in Figure 12. When octahedra are sharing all edges, the sodium chloride structure appears (Figure 13). In the rutile structure (Figure 14) each octahedron shares two opposite edges with the neighbouring octahedra. Chains of octahedra are formed which are linked together through the remaining free corners of the octahedra.

Because of the high positive charge on metal ions, the repulsive forces between them play an important role and cause a distortion of the octahedra, resulting in an increase of the intercationic distance. Figure 15 shows the cross-section of a regular chain of octahedra and that of a distorted chain. In the rutile lattice the surrounding of the cation has a form of a distorted octahedron, in which two oxygen ions are slightly further away than the

Table 4. The crystal structures of some metallic oxides [1]

Type of structure	Coordination numbers of M and O and formula type		Name of structure	Example
Infinite 3-dimensional complexes	6:2	MO_3	ReO_3	WO_3
	8:4		Fluorite	ThO_2, CeO_2, HfO_2, NpO_2, PuO_2, AmO_2, PoO_2, CmO_3, PrO_2, UO_2, ZrO_2
		MO_2		
	6:3		Rutile	TiO_2, GeO_2, SnO_2, MnO_2, RuO_2, OsO_2, IrO_2, CrO_2, MoO_2, WO_2, TeO_2, ReO_2, PbO_2, VO_2, (NbO_2), TeO_2
	6:4		Corundum	α-Al_2O_3, α-Fe_2O_3, Cr_2O_3, Ti_2O_3, V_2O_3, α-Ga_2O_3, Rh_2O_3
	:	M_2O_3	'A'rare-earth sesquioxide	La_2O_3, Ce_2O_3, Pr_2O_3, Nd_2O_3
	6:4		'C'rare-earth sesquioxide	α-Mn_2O_3, Sc_2O_3, Y_2O_3, In_2O_3, Tl_2O_3, Sm_2O_3, and other rare-earths oxides M_2O_3
	6:6	MO	Sodium chloride	MgO, CaO, SrO, BaO, CdO, VO, TiO, NbO, FeO, CoO, NiO, MnO
	4:4	MO	Zinc-blende	BeO
			Wurtzite	ZnO
	4:2	MO_2	Silica structures	SiO_2, GeO_2
	2:4	M_2O	Cuprite	Cu_2O, Ag_2O
	4:8	M_2O	Anti-fluorite	Li_2O, Na_2O, K_2O, Rb_2O
Layer structures				MoO_3, As_2O_3, PbO
Chain structures				Sb_2O_3, CrO_3
Molecular structural units: polymers				Sb_4O_6, As_4O_6
Molecular structural units: single molecules				All simple molecular oxides

other four. Some dioxides crystallize with a less symmetrical variant of the rutile structure in which succesive pairs of metal atoms in the string of octahedra are alternately closer together and further apart.

The corundum structure is built of face sharing pairs of octahedra interlinked through edges to form layers of six member rings (Figure 16). The coordination polyhedron of O^{2-} ions around each Al^{+3} ion is essentially regular octahedral and there is no evidence to suggest the existence of metal-metal interactions across the shared face. Thus, it is sometimes more convenient to consider the structure as a close-packed array of oxygen ions with metal ions occupying 2/3 of octahedral holes (Figure 17). The arrangement of the metal ions between each two close-packed layers of oxygen ions is that illustrated in Figure 17 (a) stacked according to the vertical projection diagram in Figure 17b.

Linking of tetrahedra takes place only through vertices and there are two

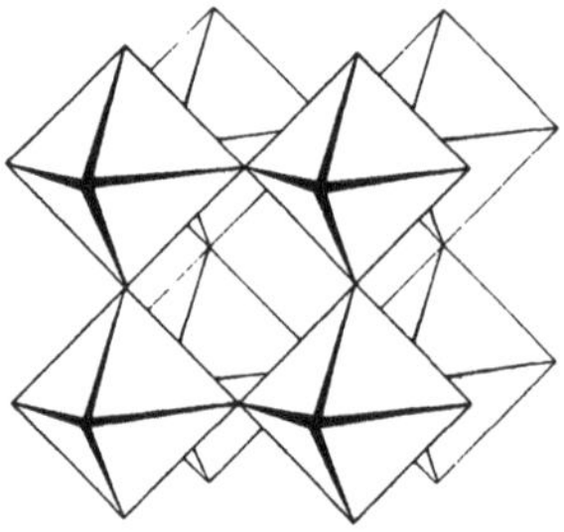

Figure 12. ReO_3 structure

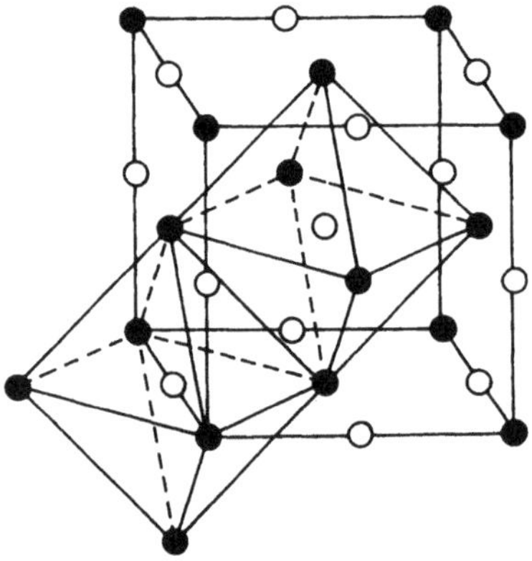

Figure 13. Sodium chloride structure

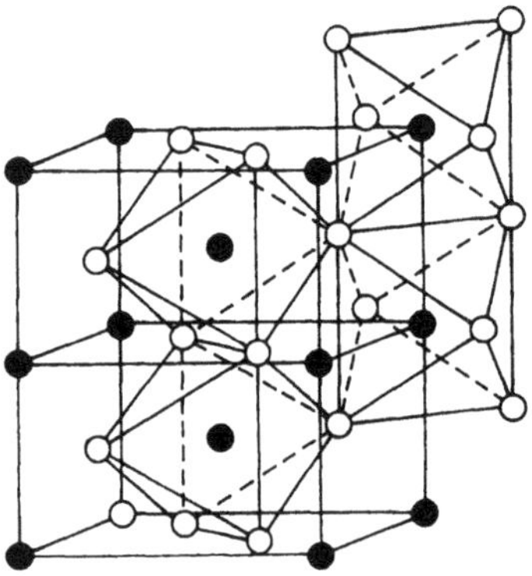

Figure 14. Rutile structure

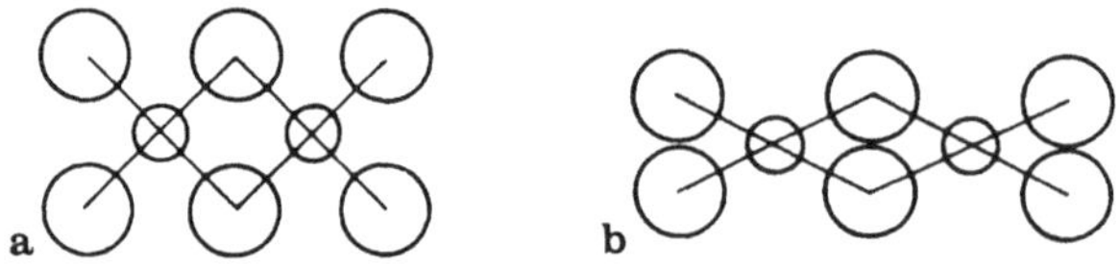

Figure 15. Chain of edge-linked octahedra in rutile structure: **a** regular, **b** distorted

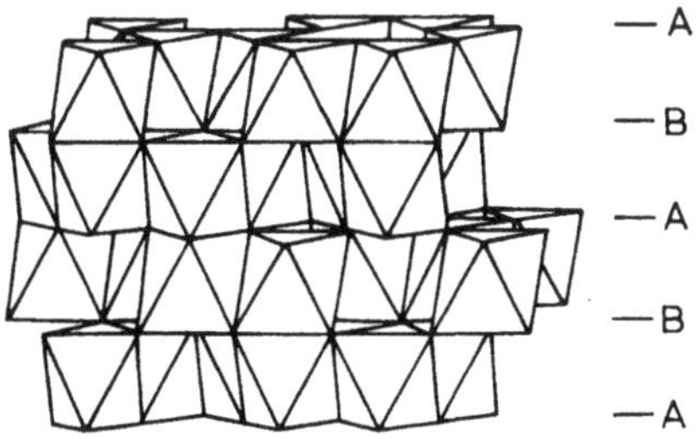

Figure 16. Packing of AlO_6 octahedra in the corundum structure

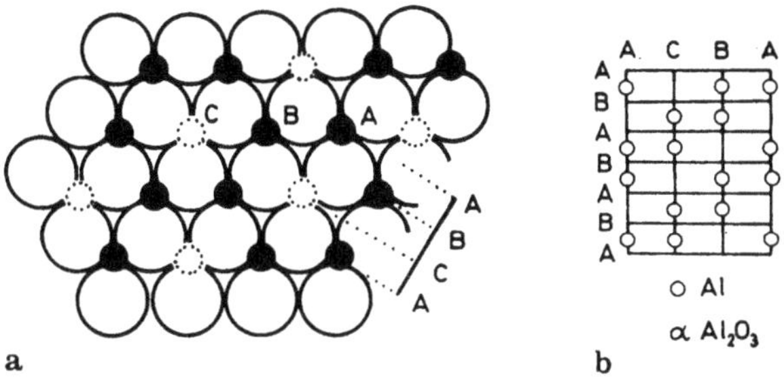

Figure 17. a pattern of the distribution of cations in the octahedral holes between pairs of close-packed oxygen layers in the corundum structure; **b** vertical projection

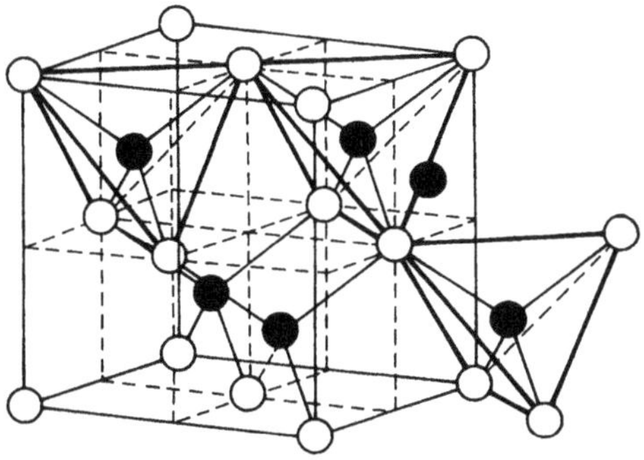

Figure 18. The structure of zinc-blende.

alternative ways of arranging them: the cubic structure of zinc-blende (Figure 18), and the hexagonal structure of wurtzite (Figure 19).

The MX_2 fluorite structure is composed of cubes sharing vertices, as shown in Figure 20, but can also be regarded as a structure in which M ions form an fcc lattice, and X ions occupy all tetrahedral holes. It must be emphasized that in the majority of compounds which adopt this structure, the smaller cations M are those which are close-packed and the larger X ions occupy the interstitial holes.

It may be noted that the X ions have tetrahedral coordination. Thus, if they were cations M, it would be customary to write the formula of the compound as M_2X and regard it as built of tetrahedra sharing all their six edges, each

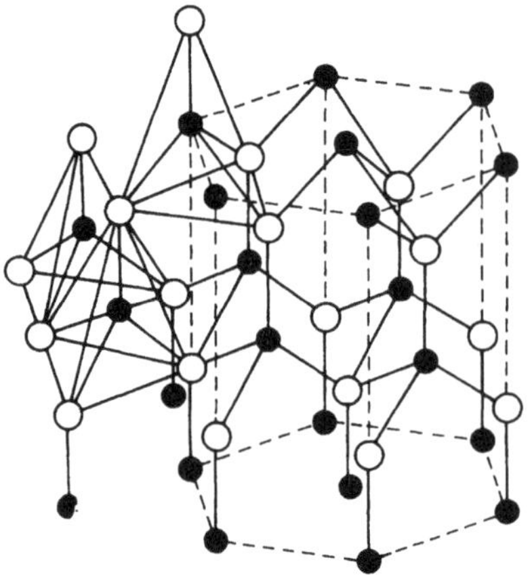

Figure 19. Wurzite structure

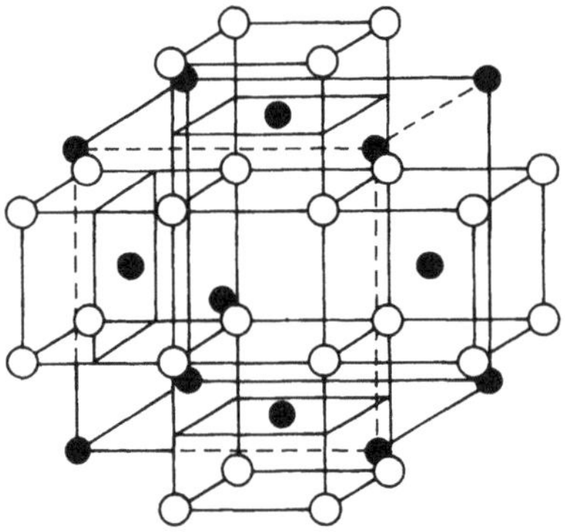

Figure 20. Fluorite structure

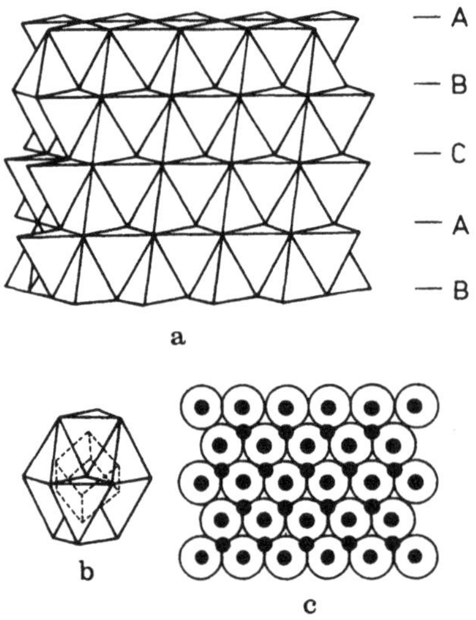

Figure 21. Antifluorite structure. **a** packing of MX_4 tetrahedra; **b** cubic coordination of the close-packed ions; **c** close-packed layer of X ions

vertex being shared by eight tetrahedra (Figure 21a). These tetrahedra are arranged so that their central atoms define a cube (Figure 21b). This structure may be thus considered as an antifluorite structure with X ions forming the close-packed array and M filling the tetrahedral holes (Figure 21c).

It can be expected that the fluorite structure in which cations are cubically coordinated by anions, will be adopted by compounds for which the ratio of cation to anion radii exceeds 0.732. In fact, fluorides of large divalent cations such as Ca, Sr, Ba, Pb, and oxides of large quadrivalent cations such as Zr, Ce, Th, U crystallize in this structure. Conversly, in the antifluorite structure the cations are coordinated tetrahedrally, and it should thus be encountered at the radii ratio 0.225–0.414. It is adopted by many intermetallic compounds.

Fluorite and antifluorite phases very often show non-stoichiometry. In the case of oxides with fluorite structure, the non-stoichiometry arises from deficiency or excess of oxygen because it is the anions which occupy the interstitial holes in the close-packed lattice of cations.

Oxides of Group IA metals assume a very peculiar structure of cuprite, consisting of two completely interpenetrating and identical frameworks without any cross-connecting bonds (Figure 22).

The stereochemistry of the metal atom in MoO_3 can be best considered as that of a distorted octahedron, although it can be easily deduced [29] from the Mo—O tetrahedron as a basic unit. Figure 23a shows an infinite string of corner sharing tetrahedra. When such strings are linked together to increase the coordination of molybdenum to 5 (Figure 23b), and other strings are brought from beneath to complete this coordination to 6 (Figure 23c), sheets are formed which are composed of octahedra sharing edges with two adjacent octahedra and corners with two other octahedra (in and out of the page of Figure 23d). A layer structure of MoO_3 appears in this way.

Closely related to the structure of MoO_3 is that of V_2O_5. The stereochemistry may be considered [30] to be either a distorted trigonal bipyramid (five vanadium-oxygen bond lengths of 1.58–2.02 Å) or a distorted octahedron (sixth vanadium-oxygen bond length of 2.79 Å). Figure 24a shows strings of double octahedra sharing edges and forming sheets by corner-sharing

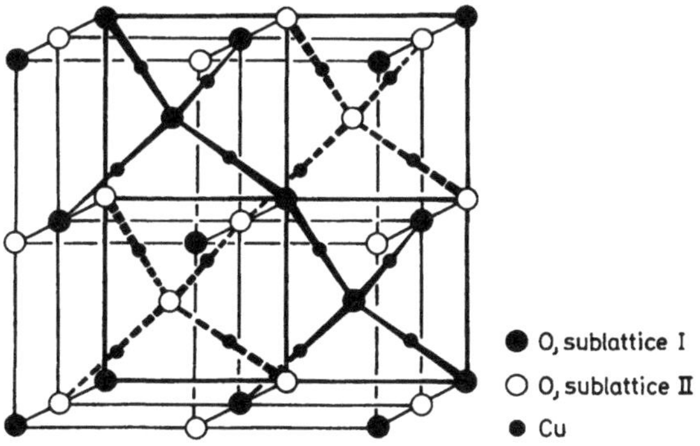

Figure 22. Cuprite structure

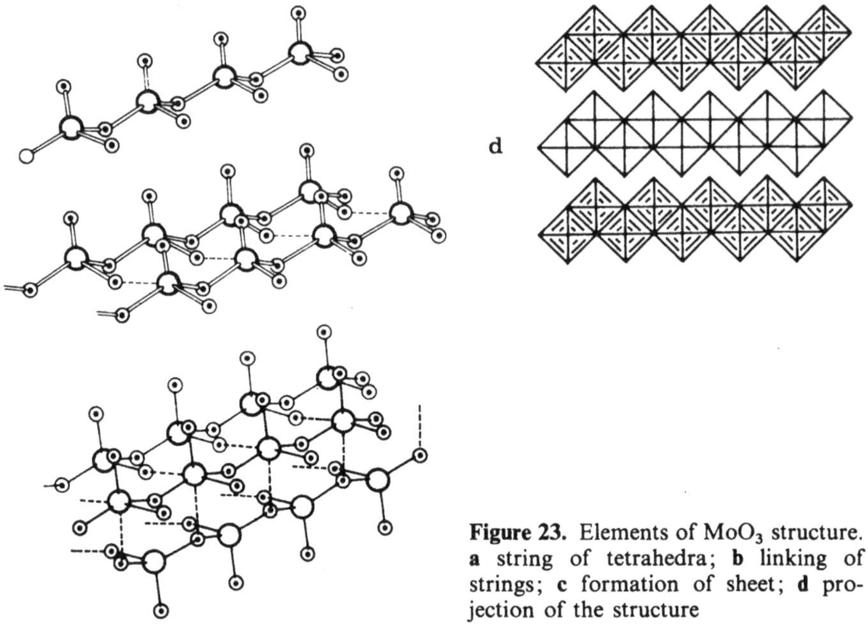

Figure 23. Elements of MoO_3 structure. **a** string of tetrahedra; **b** linking of strings; **c** formation of sheet; **d** projection of the structure

Figure 24. V_2O_5 structure. **a** sheet formed by strings of edge-sharing VO_6 octahedra linked through corners; **b** three dimensional network formed by corner-sharing of sheets; **c** the same projection as (b) represented as idealized trigonal bipyramids [30]

with the double strings on both sides. These sheets form a three-dimensional network by corner-sharing with adjacent sheets (Figure 24b). Figure 24c is the same projection as Figure 24b only drawn as idealized trigonal bipyramids. These weak vanadium-oxygen bonds which complete the octahedra give rise to perfect cleavage between the sheets. Because of the similarity between the V_2O_5 and MoO_3 structures, some of the vanadium ions can be replaced by molybdenum to form solid solutions. At a 1:1 composition the V_2MoO_8 phase is formed in which the size of the blocks of corner sharing octahedra is increased from $2 \times \infty \times \infty$ to $3 \times \infty \times \infty$ (Figure 25).

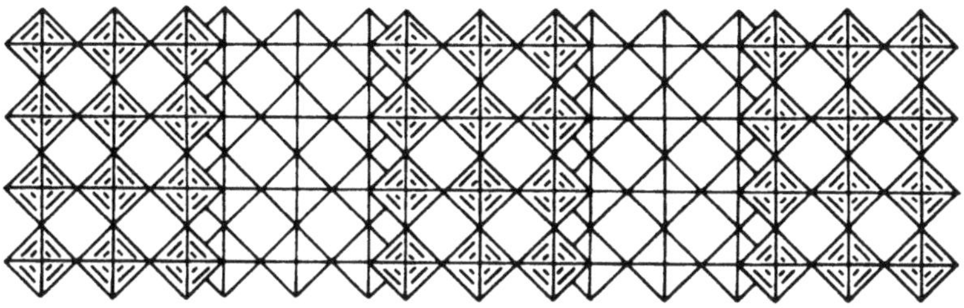

Figure 25. The structure of V_2MoO_8 [30]

B. Ligand Field Effects

When a transition metal cation with unfilled *d*-orbitals is placed as central ion in the surrounding of ligands, the electrostatic field originating from the negatively charged ligands affects to a different degree the energy of different *d* orbitals, influencing thus the geometry and stability of the structure.

As an example let us consider a transition metal ion in the octahedral surrounding of six negatively charged ligands. Such a system is shown in Figure 26 in which the shapes of *d*-orbitals of the central ion are also marked. The coordinates have been chosen in such way that they pass through the vertices of the octahedron. It may be seen that the regions of highest electron

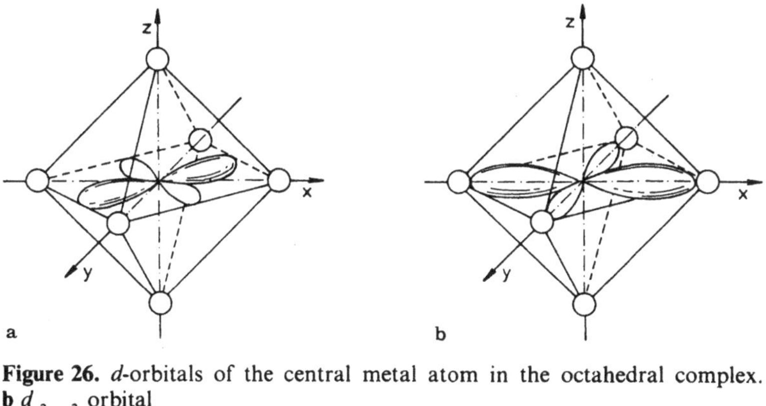

Figure 26. *d*-orbitals of the central metal atom in the octahedral complex. **a** d_{xy} orbital, **b** $d_{x^2-y^2}$ orbital

density of the $d_{x^2-y^2}$ orbital are directed towards the ligands situated along the x and y axes whereas regions of highest electron density of d_{xy} orbitals point in the directions between the ligands, thus maximally avoiding their proximity. As ligands are always negatively charged or directed towards the central ion with the negative part of the dipole, the electron on the $d_{x^2-y^2}$ orbital is subjected to a much stronger repulsion by ligands than the electron on the d_{xy} orbital. The energy of the former is thus increased in the ligand field to a higher degree than the energy of the latter. The d_{yz} and d_{xz} orbitals have in respect to the ligands the same space orientation as orbital d_{xy}, their energy is thus changed to the same degree as the energy of d_{xy} orbital. Similarly the energy of d_{z^2} orbital is changed in the same way as that of $d_{x^2-y^2}$ orbital. We may thus conclude that the system of 5 degenerate *d*-orbitals of the free ion, when placed in the octahedral ligand field are split into two groups: three orbitals d_{xy}, d_{xz}, d_{yz} of lower energy (t_{2g} orbitals) and two orbitals $d_{x^2-y^2}$, d_{z^2} of higher energy (e_g orbitals) as shown in Figure 27. The separation between the energies of e_g and t_{2g} orbitals is called the ligand field splitting and denoted as Δ or 10 Dq. The value of Δ can be calculated theoretically as well as determined experimentally by spectroscopic methods. On the basis of the data, collected for many different systems, the following generalizations are possible [31]:

— the value of Δ in complexes of ions of the given transition period in the same oxidation state and with the same ligands vary only within very narrow range;
— the value of Δ in complexes of the same ligand increases rapidly with the oxidation state of the cation;
— the value of Δ in complexes of the same ligand increases by about 30 percent on passing from the metal of the given transition period to the corresponding metal of the next one,
— the ligands can be arranged according to the increasing ligand field into such a series that the increase of the value of Δ is practically independent of the type of cation.

Obviously the value of Δ gives no information on the positions of the split energy levels relative to the energy level of the *d*-orbitals in the free ion,

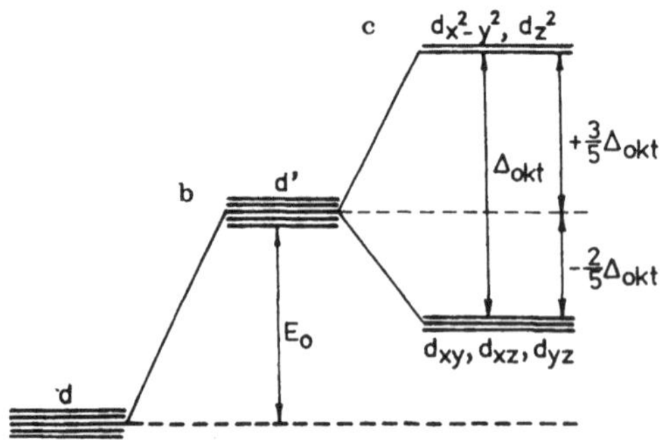

Figure 27. The influence of the ligand field on the *d*-orbitals of the central atom. **a** free atom, **b** spherical field, **c** octahedral field

therefore a reference level must be defined. Usually it is chosen as the average energy of an orbital of the *d*-shell completely filled with 10 electrons. If $E(e_g)$ and $E(t_{2g})$ are the energies of an electron on e_g and t_{2g} orbitals respectively, the average energy of an electron in the completely filled *d*-shell, which is taken as zero energy level amounts:

$$\frac{4}{10}E(e_g) + \frac{6}{10}E(t_{2g}) = 0$$

As for the octahedral complex

$$E(e_g) - E(t_{2g}) = \Delta_{oct}$$

we have $E(t_{2g}) = -\frac{2}{5}\Delta_{oct}$

$$E(e_g) = +\frac{3}{5}\Delta_{oct}$$

as shown in Figure 27.

In the case of tetrahedral coordination of ligands around the transition metal cation, the *d*-orbitals are also split into two groups; in this case however the two e_g orbitals have lower energy than the three t_{2g} orbitals. The value of Δ due to ligands arranged tetrahedrally amounts to 4/9 of the Δ caused by ligands present at the same distance in the form of an octahedron. On lowering the symmetry of the coordination polyhedron further spliting of the groups of degenerated e_g and t_{2g} orbitals takes place. Figure 28 shows schematically the way in which *d*-orbitals are split in ligand fields of various symmetries.

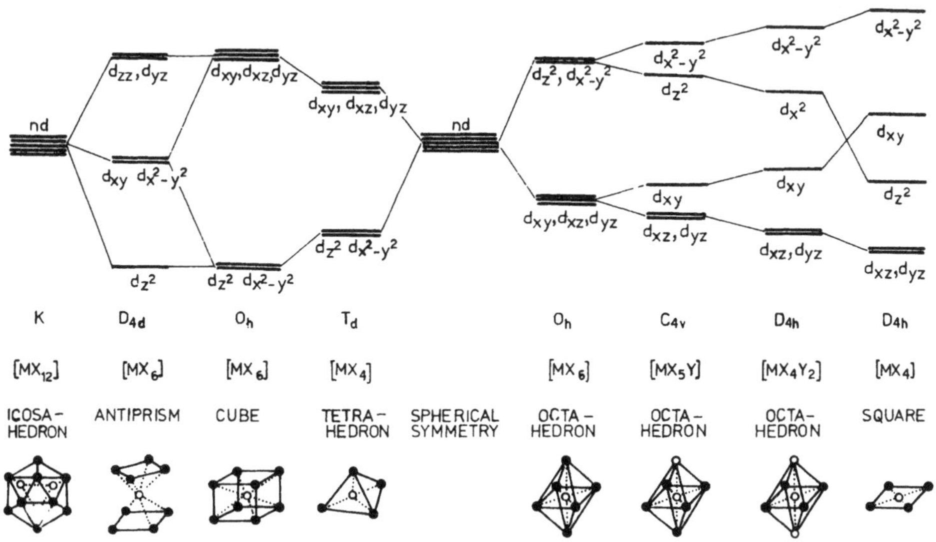

Figure 28. Splitting of *d*-orbitals of the central atom in ligand fields of various symmetries

We now have to discuss the distribution of the *d*-electrons among the available orbitals, the cation being assumed to be in its ground state. In the case of ions with 1, 2 or 3 electrons the electrons will occupy succesively the three t_{2g} orbitals because of the tendency to occupy first the orbitals of lower energy by retaining the spins parallel. The fourth electron has two alternatives: either to enter onto one of the lower lying t_{2g} orbitals with simultaneous pairing of spins, or to occupy one of the higher lying e_g orbitals retaining the spin parallel. When the ligand field is strong enough, that is, the value of Δ so large that the energy gain on occupying the lower t_{2g} level compensates for the energy loss due to the decrease of the exchange energy and the increase of the repulsion on pairing of spins, the electrons will first enter onto t_{2g} orbitals until they are fully occupied by 6 electrons, and only then they will occupy the e_g orbitals. Therefore systems with small numbers of unpaired electrons are often called systems with strong ligand field or low-spin complexes. Conversely, when the ligand field is weak, the exchange energy prevails and the electrons occupy successively the three t_{2g} orbitals and two e_g orbitals retaining their spins parallel, and only then completely fill the t_{2g} and e_g orbitals. Such systems are called systems of weak ligand field or high spin complexes. The distribution of electrons on *d*-orbitals for both types of systems is summarized in Table 5. The table contains also the values of the ligand field stabilization energy. This is the energy gain due to the

Table 5. *d*-electron arrangements in octahedral complexes

Number of *d* electrons	Arrangement in weak ligand field (high spin)		Number of unpaired spins	Ligand field stabilization energy	Arrangement in strong ligand field (low spin)		Number of unpaired spins	Ligand field stabilization energy
	t_{2g}	e_g			t_{2g}	e_g		
1	↑		1	$-\frac{2}{5}\Delta$	↑		1	$-\frac{2}{5}\Delta$
2	↑ ↑		2	$-\frac{4}{5}\Delta$	↑ ↑		2	$-\frac{4}{5}\Delta$
3	↑ ↑ ↑		3	$-\frac{6}{5}\Delta$	↑ ↑ ↑		3	$-\frac{6}{5}\Delta$
4	↑ ↑ ↑	↑	4	$-\frac{3}{5}\Delta$	↑↓ ↑ ↑		2	$-\frac{8}{5}\Delta$
5	↑ ↑ ↑	↑ ↑	5	0	↑↓ ↑↓ ↑		1	-2Δ
6	↑↓ ↑ ↑	↑ ↑	4	$-\frac{2}{5}\Delta$	↑↓ ↑↓ ↑↓		0	$-2\frac{2}{5}\Delta$
7	↑↓ ↑↓ ↑	↑ ↑	3	$-\frac{4}{5}\Delta$	↑↓ ↑↓ ↑↓	↑	1	$-1\frac{4}{5}\Delta$
8	↑↓ ↑↓ ↑↓	↑ ↑	2	$-\frac{6}{5}\Delta$	↑↓ ↑↓ ↑↓	↑ ↑	2	$-\frac{6}{5}\Delta$
9	↑↓ ↑↓ ↑↓	↑↓ ↑	1	$-\frac{3}{5}\Delta$	↑↓ ↑↓ ↑↓	↑↓ ↑	1	$-\frac{3}{5}\Delta$

splitting of d-orbitals under the influence of the ligand field. In the case of the cation with one d-electron, this electron occupies the t_{2g} orbital and the energy of the system is lowered by $-2/5\Delta_{oct}$ with reference to the energy this system would have had with completely filled electron shells. In the case of two d-electrons the stabilization energy amounts to $-4/5\Delta_{oct}$ etc. For the ground state corresponding to the configuration $(t_{2g})^m (e_g)^n$ the stabilization energy in the octahedral ligand field amounts to $\left(-\frac{2}{5}m+\frac{3}{5}n\right)$ Δ_{oct}, and in the tetrahedral field $-\left(-\frac{3}{5}n+\frac{2}{5}m\right)\Delta_{tetr}$.

It should be noticed that in the case of the octahedral field, the value of the stabilization energy attains maxima for d^3 and d^8 configurations, whereas in the case of the tetrahedral field it attains maxima for d^2 and d^7 configurations. It amounts to zero for d^5 and d^{10} configurations.

Stabilization energy is directly related to many important properties of the transition metal compounds and therefore the dependence of these properties on the number of d-electrons is expressed by characteristic curves with two extrema. We shall mention some of these properties relevant to catalysis. Figure 29 shows the dependence of the metal-oxygen interionic distance in the transition metal oxides with octahedral coordination. Along the transition period the ionic radius of metal ions should monotonically

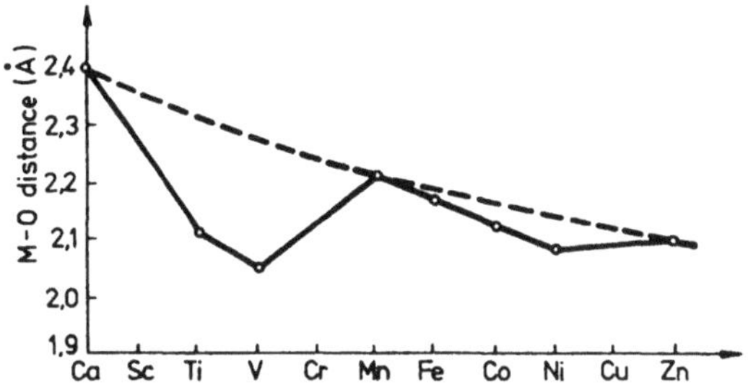

Figure 29. Metal-oxygen interionic distance in transition metal oxides with octahedral coordination

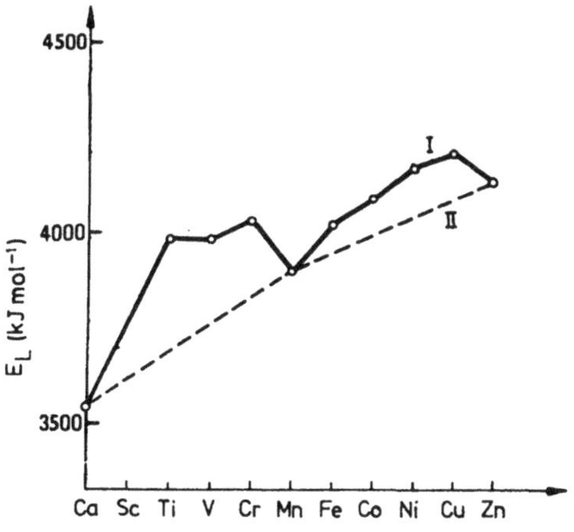

Figure 30. Lattice energy of transition metal oxides

decrease with the atomic number because the increasing nuclear charge is not fully screened by the electrons filling the d-orbitals (dotted line in Figure 29). However, screening of the core of the central cation in the direction of ligands is different for t_{2g} electrons and e_g electrons. Thus two minima are observed as expected for d^3 and d^8 ions. Figure 30 illustrates the lattice energy of the transition metal oxides of the first period. Here again the twin-peaked pattern is observed. This is of immediate importance for understanding of the behaviour of these oxides as catalysts of many reactions. In many processes of catalytic oxidation the reactant molecules are oxidized by the lattice oxygen ions leaving oxygen vacancies, which then are reoxidized very rapidly by oxygen from the gas phase. The rate of such reactions will obviously depend on the strength with which oxygen is bonded in the catalyst lattice. In fact, correlations have been described between the rate constants of such reactions as oxidation of H_2, CO, CH_4 or total oxidation of hydrocarbons and various parameters assumed as a measure of the oxygen bond strength in the lattice, for instance, heat of the formation of the oxide. As an example, Figure 31 shows the rate constants of the total oxidation of H_2 and CH_4 on transition metal oxides of the first period as catalysts [34]. The twin peaked pattern is easily recognized.

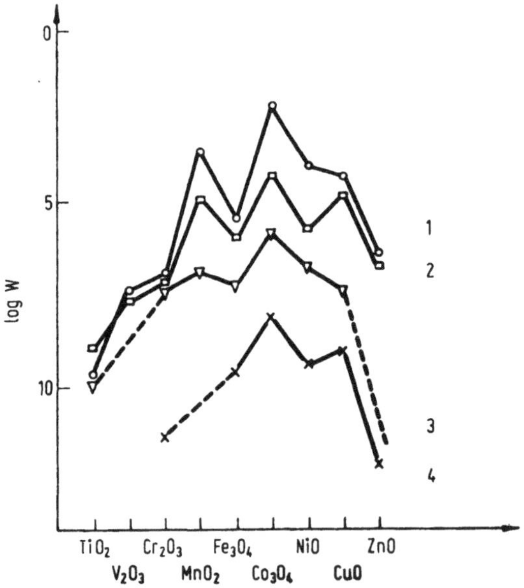

Figure 31. The catalytic activity of oxides of the elements of fourth period in the homomolecular exchange of oxygen (1), oxidation of hydrogen (2), oxidation of methane (3) and nitrogen oxide decomposition (4) at 573 K

C. Defect Structure of Transition Metal Compounds

1. Point Defects

Very rapid development of studies on the properties of crystals, due to their increasing applications in many new fields of modern technology showed clearly that the model based only on the concept of the crystal lattice is not sufficient to explain many important properties of solids such as electrical, magnetic, optical, mechanical as well as catalytic properties. In fact crystals with perfect periodical arrangement of all structural elements cannot exist, and real crystals show the presence of various imperfections, described as defects of the crystal structure [35, 36, 37]. They are due to the displacement of atoms from the lattice sites normally occupied by these atoms into other sites or interstitial positions, the presence of some vacant sites, the displacement of part of the crystal in respect to the other part along a crystal plane, etc. The defects can be best systematized according to their dimensions as shown in Table 6. As until now only the presence of point defects and crystallographic shear has been shown to be important for catalysis, these two types of defects will be discussed here in more detail.

From the thermodynamical point of view, ideal crystals can exist only at the temperature of absolute zero. At every other temperature thermal oscillations of atoms result in the formation of different kinds of point defects, because the thermodynamic potential of the defected crystal becomes smaller than that of the ideal crystal. In order to illustrate this situation let us consider a simple crystal of an element, composed of *N* atoms, in which vacancies have been formed by transferring a certain number *n* of atoms from their lattice sites to the surface of the crystal, and the formation of a new crystal plane.

The change of the Gibbs free energy in the course of this process is given by:

$$\Delta G = G - G_{id} = n\,\Delta H - T(\Delta S_c + n\,\Delta S_{osc}) \quad (1)$$

where G and G_{id} are the thermodynamic potentials of the defected and ideal crystal respectively, ΔH the enthalpy of the formation of one vacant site, ΔS_{osc} the change of the oscillation entropy of atoms surrounding the vacancy and ΔS_c the change of the configuration entropy of the system on the for-

Table 6. Classification of defects in solids

1. Zero-dimensional (point) defects
   a) vacancies
   b) interstitial atoms (ions)
   c) foreign atoms (ions)
2. One-dimensional (linear) defects
   a) edge dislocation
   b) screw dislocation
3. Two-dimensional (flat) defects
   a) antiphase boundary
   b) shear plane
   c) low-angle tilt boundary
   d) low angle twist boundary
   e) grain boundary
   f) surface
4. Three-dimensional (spatial) defects
   a) pores
   b) foreign inclusions

mation of n vacancies. The latter is equal to the difference between the entropy of a crystal containing n vacancies and that of the ideal crystal:

$$\Delta S_c = S_{c(def)} - S_{c(id)} = k \ln W_{def} - k \ln W_{id} = k \ln \frac{W_{def}}{W_{id}}$$

where W_{def} and W_{id} are the thermodynamic probabilities of the defected and ideal crystal respectively. $W_{id} = 1$, because there is only one way to distribute N atoms on N sites. When n atoms are transfered to the surface, the number of sites increases to $N + n$. W_{def} is equal to the number of ways in which n vacancies and N atoms can be distributed on $(N + n)$ sites. We obtain

$$\Delta S_c = -k \left[N \ln \frac{N}{N+n} + n \ln \frac{n}{N+n} \right] \tag{2}$$

and finally

$$\Delta G = G - G_{id} = n \, \Delta H - nT \, \Delta S_{osc} + kT \left[N \ln \frac{N}{N+n} + \ln \frac{n}{N+n} \right] \tag{3}$$

Formation of a vacancy requires always certain energy, ΔH is thus always positive. The change of the oscilation entropy ΔS_{osc} is always negative, so the first two terms ($n \, \Delta H - nT \, \Delta S_{osc}$) are always positive and increase propertionally to n. This is shown by curve I in Figure 32. Thus, if not for the change of configurational entropy, the formation of defects would be accompanied by an increase of the Gibbs free energy and would never occur spontaneously. However, as shown by equation 2 the formation of vacancies entails always an increase of the configurational entropy, because the numerator in the fraction under the logarithm is always smaller than the denominator, and the value of the logarithm is negative. The last term in equation 3 is thus the more negative the greater is n, (Curve II in Figure 32). Consequently the function describing the dependence of the Gibbs free energy on the number of defects (curve III in Figure 32), passes through a minimum at certain

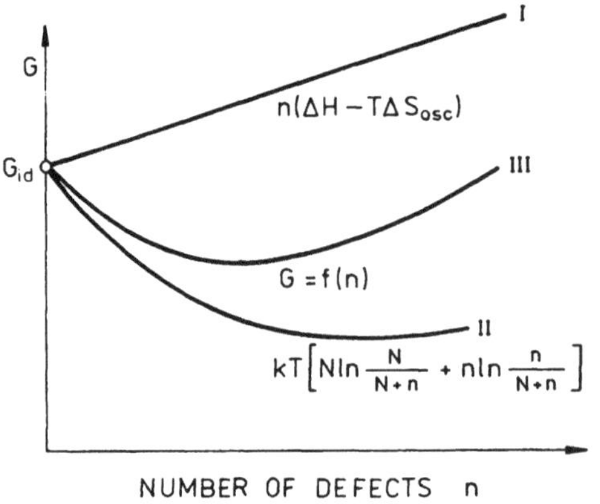

Figure 32. Gibbs free energy of a crystal as function of the defect concentration

value of n, which gives the equilibrium concentration of defects. It can be calculated from the equilibrium condition

$$\frac{\partial G}{\partial n} = 0$$

and is given by the formula:

$$\frac{n}{N - n} = N_v = \exp\left(\frac{\Delta S_{osc}}{k}\right) \exp\left(-\frac{\Delta H}{kT}\right) \tag{4}$$

where N_v is the concentration of vacancies expressed in molar fraction.

Let us now consider a simple stoichiometric binary compound of composition MX. At each temperature a certain number of atoms M and/or X leave the positions they normally occupy in the lattice and pass either into interstitial positions or enter into positions normally occupied by other type of atoms or migrate to the surface where they form new lattice sites, leaving behind vacancies (Figure 33). Because in the case of a stoichiometric compound the ratio of the number of atoms M to X must remain constant, the crystal must always contain at least two types of point defects. The following combinations are possible:

(a) a pair of vacancies in sublattice M and sublattice X, denoted by symbols V_M and V_X respectively and called Schottky defects:

$$0 \rightleftharpoons V_M + V_X$$

(b) an interstitial atom M_i and a vacancy V_M in the sublattice M, or an interstitial atom X_i and a vacancy V_X in the sublattice X. They are called Frenkel defects:

$$M_M \rightleftharpoons M_i + V_M$$

$$X_X \rightleftharpoons X_i + V_X$$

(c) a pair of interstitial atoms M_i and X_i:

$$M_M + X_X \rightleftharpoons M_i + X_i$$

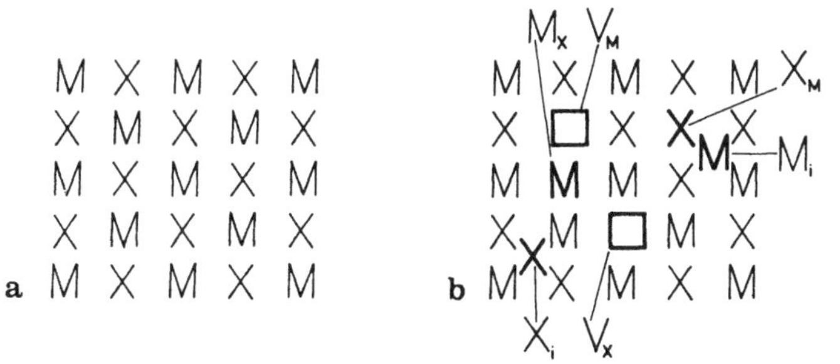

Figure 33. Schematic representation of the MX lattice **a** ideal crystal, **b** real crystal containing different types of point defects

(d) atom M on the site normally occupied by atom X (symbol M_X) and atom X on the site normally occupied by atom M (symbol X_M), called antistructural defects:

$$M_M + X_X \rightleftharpoons M_X + X_M$$

The presence of point defects entails the appearance of new energy levels in the energy spectrum of the crystal, which are usually situated in the forbiden gap. Abstraction of the valence electron from the interstitial atom M is usually easier than from the completely filled electronic shell of the ion M^{+m} occupying the lattice site, that is, the transfer of an electron from the valence band to the conductivity band. The presence of such an interstitial atom thus entails the appearance of an isolated energy level which is situated above the valence band and can easily supply an electron (donor level):

$$M_i \rightleftharpoons M_i^{\bullet} + e' - E_1$$

where E_i is the energy change. If the interstitial metal atom is divalent, a second ionization process may take place

$$M_i^{\bullet} \rightleftharpoons M_i^{\bullet\bullet} + e' - E_2$$

These levels are illustrated in the left part of Figure 34.

Conversely, vacancies in the cationic sublattice show rather the tendency to capture electrons, that is, to from acceptor levels. The ionization of these levels may be expressed by equations:

$$V_M \rightleftharpoons V_M' + h^{\bullet} - E_3$$

$$V_M' \rightleftharpoons V_M'' + h^{\bullet} - E_4$$

where $h^{\bullet}$ is a hole in the valence band. They are shown in the right part of Figure 34.

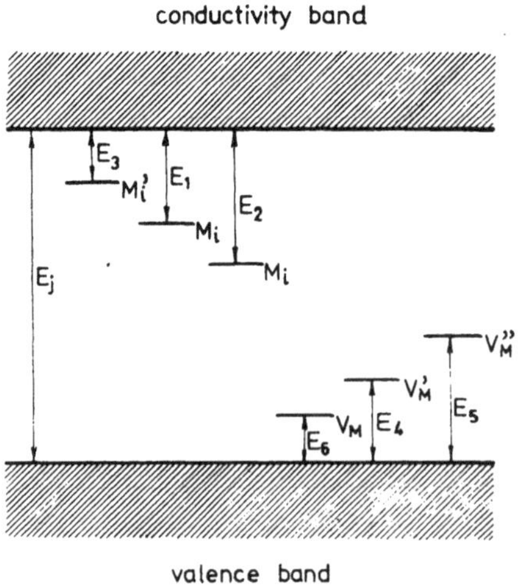

Figure 34. Local levels of defects in the energy spectrum of the crystal

It should be remembered that the intrinsic ionization process, that is, the excitation of electrons from the valence band to the conductivity band must also be always taken into account:

$$0 \rightleftharpoons e' + h^{\cdot} - E_i$$

In the case of transition metal compounds the situation is different. Transition elements are characterized by low values of the first few ionization potentials and form therefore with Group V–VIIA elements usually a number of compounds in which they are present in different oxidation states. Among them oxides and sulphides are of particular interest for catalysis. The appearance of different oxides of the given element depends on the oxygen pressure. In certain pressure ranges reduction or oxidation of the oxide of the given valence may already begin, the oxides of lower (or higher) valence forming a solid solution in the starting oxide. As the electroneutrality condition of the lattice must be satisfied, the change of the valence state of some cations entails a change of the ratio of the number of oxygen ions to the number of cations. Thus, interaction of the crystal with the gas phase results in the appearance of nonstoichiometry. This nonstoichiometry may be realized in the crystal either by formation of point defects, or by changing the mode of linkage between the coordination polyhedra and formation of extended defects, e.g. shear planes.

From the thermodynamic point of view, a crystal containing point defects can be considered as a solid solution of these defects in normal structural elements of the crystal. As the concentration of defects is usually very small, the solution can be treated as very diluted and conforming to the law of mass action. In expressing this law concentrations of various defects can be thus used, whereas the concentrations of normal structural elements can be assumed as constant and included into the value of the equilibrium constant K. The constant K is then a function of the standard chemical potentials of the components of the crystal.

When a crystal contains a considerable number of defects, their concentrations no more can be used in thermodynamic calculations and must be replaced by activities. Interactions between defects appear, leading to the formation of clusters of various defects or complex defects. As the result of such interactions the random distribution of defects in the lattice may transform into an ordered arrangement in which the defects are distributed periodically over certain defined lattice positions forming a superstructure. Such ordering of defects may be accompanied also by structural transformations, resulting in the annihilation of certain types of point defects. The nonstoichiometry associated with the presence of these defects is in such cases retained by changing the mode in which the coordination polyhedra are linked together. This is the case in the formation of extended defects such as shear planes.

2. Defect Structure of Nonstoichiometric Transition Metal Oxides

Let us consider a crystal of transition metal compound MX in equilibrium with the gas phase. In the gas phase the compound may be present in form of molecules MX, which may undergo further transformations such as association into complexes $(MX)_n$, or dissociation into atoms of the component elements which in turn may form molecules of different sizes. All these processes may be summarized in form of a following scheme:

$$
\begin{array}{ccccccc}
 & & & & M_m & & X_n \\
 & & & & \upharpoonleft\downharpoonright & & \upharpoonleft\downharpoonright \\
\frac{1}{a}(MX)_a & \rightleftharpoons & MX_{(g)} & \rightleftharpoons & \frac{1}{m}\,mM & + & \frac{1}{n}\,nX \\
 & & \downharpoonleft\upharpoonright & & & & \\
 & & MX_{(solid)} & & & &
\end{array}
$$

When the gas phase is composed of molecules MX, it has no influence on the composition of the solid. In the case of transition metal compounds such as oxides or sulfides dissociation of the molecules in the gas phase into atoms takes place. The vapour of the metallic component is composed of atoms, whereas the vapour of the nonmetallic component is composed of molecules, diatomic in case of oxygen and multiatomic in the case of sulphur. Both components independently can be exchanged between the gas phase and the solid until the respective equilibria are established, which may correspond to deviations from the stoichiometric composition. As an example let us describe the interaction of an oxide with oxygen in the gas phase, illustrated in Figure 35, resulting in the incorporation of oxygen into the lattice. The crystal becomes nonstoichiometric because of the deficiency of metallic component due to the formation of vacancies in the cationic sublattice. Simultaneously the increase of the oxidation state of the corresponding number of cations in the crystal must take place. Such a process will take place when the oxygen pressure is high enough. When it is low, the reverse process may occur in which oxygen ions from the lattice pass into

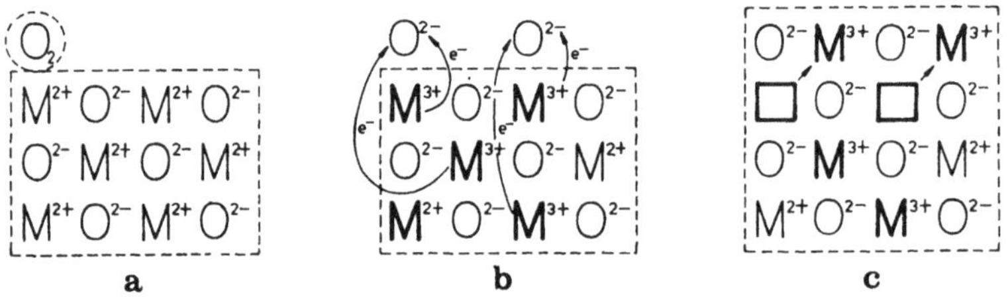

Figure 35. Incorporation of oxygen from the gas phase into the crystal lattice resulting in the formation of cationic vacancies

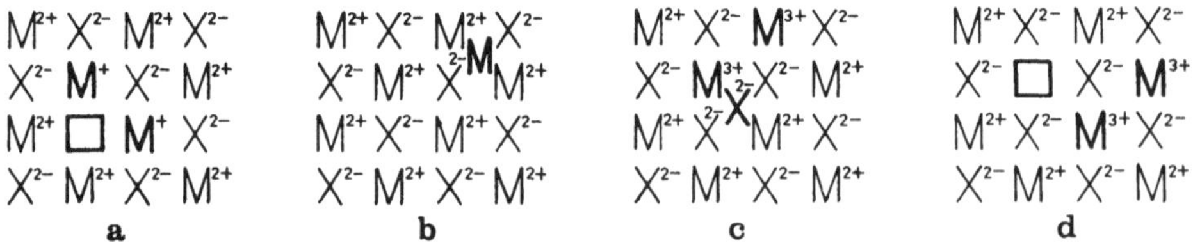

Figure 36. Four basic types of nonstoichiometric MX compounds

the gas phase, leaving behind a crystal with a deficiency of the non-metallic component. Depending on the kind of defect structure formed, the non-stoichiometric crystals of a compound MX may be devided into four basic types (Figure 36):

(1.) crystals with an excess of metal, resulting from the presence of anionic vacancies. Such a structure may be formed by transfering a certain number of anions from the crystal to the gas phase. Electroneutrality of the crystal is secured by transfering the electrons which were linked with the anions leaving the structure, to the adjacent cations. The process may be thus considered formally as reduction of the corresponding number of cations (Figure 36a).

(2.) crystals with an excess of metal resulting from the presence of interstitial cations. Such a structure may also be formed by transferring a certain number of anions from the crystal to the gas phase, the excess cations being simultaneously displaced into interstitial positions. The electrons from the leaving anions reduce the oxidation state of a corresponding number of cations, usually the interstitial ones (Figure 36b).

(3.) crystals with a deficiency of metal due to the presence of excess anions in interstitial positions. The corresponding number of electrons pass from the cations to the incorporated interstitial atoms of the non-metal, which become ionized. Formally this process is equivalent to the increase of the oxidation state of the corresponding number of cations (Figure 36c).

(4.) crystals with the deficiency of metal due to the presence of cationic vacancies, which are formed as the result of the incorporation of a certain number of non-metal atoms into lattice sites. Transfer of electrons to transform these atoms into anions results in the increase of the oxidation state of the corresponding number of cations (Figure 36d).

As already mentioned the equilibrium state between the crystal MX of a transition metal oxide and the gas phase may be described by the following equation:

$$MX_{(s)} \rightleftharpoons MX_{(g)} \rightleftharpoons M_{(g)} + \frac{1}{2}X_2 \qquad (5)$$

for which the equilibrium constant is given by

$$\frac{p_M p_{X_2}^{1/2}}{p_{MX}} = K_d \qquad (6)$$

As the molecules MX in the gas phase are in equilibrium with the solid, their vapour pressure p_{MX} is at a given temperature constant, thus

$$p_M p_{X_2}^{1/2} = K_{MX} \qquad (7)$$

According to the phase rule the system has two degrees of freedom because it is composed of two phases and two independent components (we are considering nonstoichiometric crystals). At a constant temperature we may thus freely change p_M or p_{X_2} and in this way change the defect state (nonstoichiometry) of the crystal through interaction of the given sublattice with the respective component in the gas phase:

$$M_M \rightleftharpoons M_{(g)} + V_M \tag{8}$$

$$X_X \rightleftharpoons \frac{1}{2} X_{(g)} + V_X \tag{9}$$

When the concentration of defects is small we may assume that the concentration of the cations M_M and anions X_X in the lattice sites is practically constant. We obtain thus

$$p_M[V_M] = K_o \quad \text{and} \quad [V_M] = K_o p_M^{-1} \tag{10}$$

$$p_{X_2}^{1/2}[V_X] = K_r \quad \text{and} \quad [V_X] = K_r p_{X_2}^{-1/2} \tag{11}$$

the square brackets representing the concentration of the given defect. As the vapour pressures of the components of the gas phase are related by equation 7, the concentrations of all defects may be always expressed either as a function of the pressure of nonmetal or that of the metal. Substituting, for instance, $p_M = K_{MX} p_{X_2}^{-1/2}$ into equation 10 we obtain

$$[V_M] = \frac{K_o}{K_{MX}} \cdot p_{X_2}^{1/2} \tag{12}$$

Multiplying equations 12 and 11, and dividing by equation 7 leads to

$$[V_M][V_X] = \frac{K_o K_r}{K_{MX}} = K_s \tag{13}$$

which represents the equilibrium constant of the Schottky defects.

Let us now formulate equations describing the formation of the four types of nonstoichiometric crystals described above. Obviously formation of type I is given by equation 9. The equation for type II may be written in form

$$M_M + X_X \rightleftharpoons M_i + \frac{1}{2} X_{2(g)} \tag{14}$$

for which the equilibrium condition is

$$[M_i] = K_R p_{X_2}^{-1/2} \tag{15}$$

Comparing equations 12 and 15 we obtain

$$[M_i][V_M] = \frac{K_R K_O}{K_{MX}} = K_F \tag{16}$$

which represents the equilibrium constant of the Frenkel defects.

Formation of type III nonstoichiometric crystals may be described by

$$\frac{1}{2}X_2 \rightleftharpoons X_i \tag{17a}$$

$$[X_i] = K_O p_{X_2}^{1/2} \tag{17b}$$

whereas that of type IV by

$$\frac{1}{2}X_2 \rightleftharpoons X_X + V_M \tag{18}$$

for which the equilibrium condition is given by equation 12.

As already mentioned, from the chemical standpoint reactions 9 and 14, leading to the appearance of excess of metal may be considered as reduction, whereas reactions 8 and 17a, resulting in the appearance of the deficiency of metal may be considered as oxidation. Therefore the equilibrium constants have been marked by indexes "r" or "R" (reduction) and "o" or "O" (oxidation) respectively. When the crystal forms Schottky defects on interaction with the gas phase, either anionic vacancies (type I) or cationic vacancies (type IV) are formed, whereas in the crystals showing Frenkel defects the interaction with the gas phase results in the formation of interstitial cations (type II) or interstitial anions (type III).

As an example let us consider in more detail nickel oxide which has been used in many catalytic studies as a model catalyst. Measurements of the electrical conductivity, Hall effect, Seebeck effect, diffusion and other properties [38, 39] have shown that it contains Schottky defects. At each temperature metal and oxygen vacancies exist in concentrations given by the equilibrium condition of the Schottky defects:

$$0 = V_{Ni} + V_O, \qquad [V_{Ni}][V_O] = K_S \tag{19}$$

Already at moderate temperatures these vacancies undergo double ionization to the degree, described by formulae:

$$V_{Ni} \rightleftharpoons V_{Ni}'' + 2h^{\cdot}, \qquad \frac{[V_{Ni}'']p_h^2}{[V_{Ni}]} = K_a \tag{20}$$

$$V_O \rightleftharpoons V_O^{\cdot\cdot} + 2e', \qquad \frac{[V_O^{\cdot\cdot}]n_e^2}{[V_O]} = K_d \tag{21}$$

where n_e and p_h are, respectively, the electron and positive hole concentrations. At each temperature a corresponding equilibrium of the intrinsic ionization is established, that is, a certain number of electrons pass from the valence to the conductivity band:

$$0 \rightleftharpoons e' + h^{\cdot}, \qquad n_e p_h = K \tag{22}$$

It follows from equations 19 to 21 that also the following equilibrium conditions must be fulfilled:

$$0 \rightleftharpoons V_{Ni}'' + V_O^{\cdot\cdot}, \qquad [V_{Ni}''][V_O^{\cdot\cdot}] = K_S'' \tag{19'}$$

which expresses an equilibrium of Schottky defects in the case of complete ionization of the vacancies. As the ionization energy of vacancies in nickel oxide is low, it may be assumed that already at moderate temperatures they are completely ionized and equation 23 may be used in further considerations.

Apart from the above equations describing the intrinsic defect structure of the crystal, we must also take into account the interaction of the crystal with the gas phase, which can be expressed by equation 18:

$$\frac{1}{2}O_2 \rightleftharpoons V_{Ni} + O_O; \qquad [V_{Ni}] = \frac{K_o}{K_{NiO}} p_{O_2}^{1/2} \tag{23}$$

At last it must be remembered that the defects must conform always to the equation expressing the electroneutrality condition of the crystal:

$$n_e + 2[V''_{Ni}] = p_h + 2[V^{\cdot\cdot}_O] \tag{24}$$

In order to fully describe the NiO crystal we have thus to determine seven variables, i.e.: $[V_{Ni}]$, $[V_O]$, $[V''_{Ni}]$, $[V^{\cdot\cdot}_O]$, n_e, p_h and p_{O_2}. They must fullfil six equations, 19–24, one of the parameters is thus an independent variable. As in the majority of experiments it is the oxygen pressure which can be easily changed and measured, this parameter will be chosen as the independent variable and the concentrations of all defects expressed as its function.

Exact solution of the above system of equations is very difficult, it may be however easily solved using an approximate method [40]. Namely, all equations expressing the equilibrium constants transform into linear relationships when taken in the logarithmic form:

$$\log [V''_{Ni}] + \log [V^{\cdot\cdot}_O] = \log K''_S \tag{19a}$$

$$\log [V''_{Ni}] + 2 \log p_h - \log [V_{Ni}] = \log K_a \tag{20a}$$

$$\log [V^{\cdot\cdot}_O] + 2 \log n_e - \log [V_O] = \log K_d \tag{21a}$$

$$\log n_e + \log p_h = \log K_i \tag{22a}$$

$$\log [V_{Ni}] = \log \left[\frac{K_o}{K_{NiO}} \cdot p_{O_2}^{1/2}\right] \tag{23a}$$

The main difficulty lies with equation 24 describing the neutrality condition, because it cannot be expressed in linear logarithmic form. However, as suggested by Brouwer [40] it can be approximated if one takes into account that in real crystals usually one type of defects prevails, this majority type depending on the conditions of experiment. Thus, in the electroneutrality condition the concentrations of all defects except the majority ones may be neglected. In the discussed case of NiO, at low oxygen pressure mainly vacancies $[V^{\cdot\cdot}_O]$ in the anionic sublattice will be present and the approximate neutrality condition can be written as

$$n_e = 2[V^{\cdot\cdot}_O] \tag{25}$$

$$\log n_e = \log [V^{\cdot\cdot}_O] + \log 2 \tag{25a}$$

The system of equations 19a–23a and 25a can now be solved and concentrations of all defects expressed as function of p_{O_2} or — as it is usually done — as function of the product

$$R = \frac{K_o}{K_{MX}} \cdot p_{O_2}^{1/2}$$

Following relationships are obtained:

$$\log n_e = \frac{1}{3} \log \left[\frac{2K_i^2 K_s''}{K_a}\right] - \frac{1}{3} \log R \tag{26}$$

$$\log p_h = \frac{1}{3} \log \left[\frac{K_i K_a}{2K_s''}\right] + \frac{1}{3} \log R \tag{27}$$

$$\log [V_{Ni}''] = \frac{1}{3} \log \left[\frac{4K_a (K_s'')^2}{K_i^2}\right] + \frac{1}{3} \log R \tag{28}$$

$$\log [V_O^{\cdot\cdot}] = \frac{1}{3} \log \left[\frac{K_i^2 K_s''}{4K_a}\right] - \frac{1}{3} \log R \tag{29}$$

$$\log [V_{Ni}] = \log R \tag{30}$$

$$\log [V_O] = \log \left[\frac{K_i K_s''}{K_a K_d}\right] - \log R \tag{31}$$

They can be represented in form of a diagram shown in Figure 37b (range I) in which their representations are straight lines with slopes 0, $\pm \frac{1}{3}$ and $\pm$ 1.

As indicated by equations (26)—(31), the concentrations of all defects are functions of the oxygen pressure p_{O_2}, the concentrations of V_O, $V_O^{\cdot\cdot}$ and e′ decreasing and those of V_{Ni}, V_{Ni}'' and $h^{\cdot}$ increasing with rising oxygen pressure. As a result, $[V_{Ni}'']$ and p_h become at a certain value of R so important that they begin to dominate in the electroneutrality condition, equation 24. Two cases may arise. Either p_h increases more rapidly than $[V_{Ni}'']$ and exceeds the value of $[V_O^{\cdot\cdot}]$ more rapidly than does $[V_{Ni}'']$ the value of n_e. Electroneutrality condition assumes then the form

$$n_e = p_h \tag{32a}$$

In the reverse case, when $[V_{Ni}'']$ increases more rapidly than p_h the electroneutrality condition has the form

$$[V_{Ni}''] = [V_O^{\cdot\cdot}] \tag{32b}$$

On further increase of oxygen pressure, both $[V_{Ni}'']$ becomes greater than n_e and p_h greater then $[V_O^{\cdot\cdot}]$. The electroneutrality condition can then be written as

$$2[V_{Ni}''] = p_h \tag{33}$$

Solving the system of equations 19a–23a by taking into account equations 23, 33 or 34 we can find the dependences of the concentrations of different defects on oxygen pressure for these approximate electroneutrality conditions. They are shown in Figure 37b (range II and III).

We may thus conclude that there exist three ranges of oxygen pressure, in which different defects dominate in the crystal. At low oxygen pressures the crystal contains mainly oxygen vacancies and electrons; at moderate pressures, either electrons and holes or ionized vacancies in both sublattices; and at high pressures, cation vacancies and holes.

Figure 37 may also serve to calculate the deviation from stoichiometry as a function of oxygen pressure. Let us denote this deviation by ΔC and consider the excess of metal as positive and its deficiency as negative. We may write

$$\Delta C = [V_O] + [V_O^{\cdot\cdot}] - [V_{Ni}] - [V_{Ni}''] \tag{34}$$

At high values of oxygen pressure, when nickel vacancies dominate, the crystal contains a deficiency of metal practically equal to $[V_{Ni}'']$, whereas at low values of oxygen pressure the oxygen vacancies dominate and the crystal contains the excess of metal practically equal to $[V_O^{\cdot\cdot}]$. The change from one region to the other takes place abruptly at the value of R equal

$$R_{\Delta c=0} = \left(\frac{K_i K_s''}{K_a}\right)^{1/2} \tag{35}$$

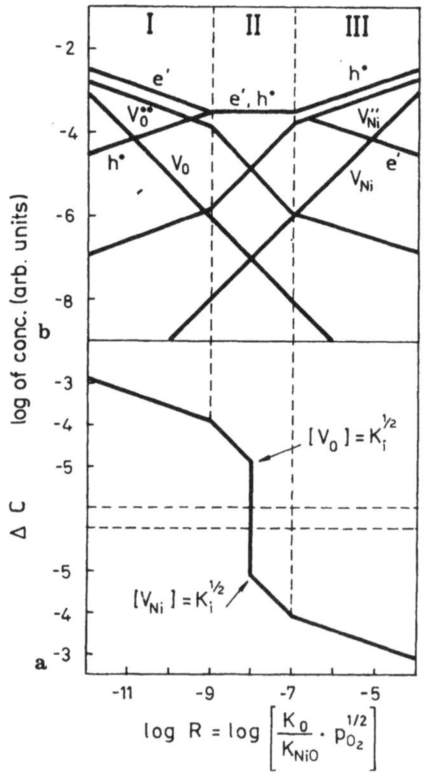

Figure 37. Concentrations of various defects in NiO (**a**) and the deviation from stoichiometry (**b**) as a function of the oxygen pressure

At this pressure the crystal abruptly changes its composition from an excess of metal to a deficiency of metal, the magnitudes of both excess and deficiency being given by $K_i^{1/2}$. This is illustrated in Figure 37a. Such dependence of the nonstoichiometry on pressure has important practical consequences because it shows that it is impossible to prepare a crystal with nonstoichiometry smaller than the limiting one, equal to $K_i^{1/2}$. Below this defect concentration, inhomogeneous crystals are obtained composed of domains with excess of metal and domains with excess of oxygen.

3. Solid Solution of Altervalent Foreign Ions in Transition Metal Compounds

Introduction of foreign atoms into the crystal lattice affects considerably many important physical and chemical properties of solids. Their nature and concentration may completely determine the electric conductivity of the solid, its magnetic properties, the rate of diffusion through the solid, the rate of its sintering, mechanical strength etc. It is also a general practice in catalyst production to introduce small amounts of various additives, described as promoters, in order to improve the catalytic properties of a solid catalyst or to modify those physicochemical properties which are of importance from the point of view of its technological exploation. The mechanism of the incorporation of foreign additives into a crystal lattice and their role in determining various properties of the solid are thus of great interest for both theory and practice of catalysis.

When foreign atoms are introduced into the crystal lattice of a certain compound and these foreign atoms are characterized by a valence different from that of the ions of host lattice (they will be called altervalent ions), new local electronic levels appear in the energy spectrum of the crystal. The alteration of the electronic structure of the crystal results in turn in a change of the concentrations of various defects. All those properties of the crystal which depend on the defect structure are thus also affected, and can be purposefully modified by the introduction of additives. On the other hand, the type of the intrinsic defect structure very often has a decisive influence on the mode in which the foreign atoms become incorporated, and on their solubility in the host lattice.

By considering the formation of such solid solutions three cases can be distinguished:

(1.) Both the host compound, as well as the compound formed by the doping ion with the second component of the lattice, have different but stoichiometric compositions. This is case, for instance, by the introduction of Al ions into an MgO lattice, which can be considered as a dissolution of Al_2O_3 in MgO. Both compounds show practically no deviations from stoichiometry. In such a case, there are two possibilities of incorporation of Al ions into the MgO lattice: incorporated Al ions are accompanied by the formation of cationic vacancies V_{Mg} or interstitial oxygen ions O_i. In both cases, the concentration of atomic defects in the crystal is determined by the concentration of the introduced dopant. This offers a possibility of purposeful modification of such properties of the crystal as ionic conductivity, diffusion coefficient etc. Therefore this mechanism of incorporation is called "controlled atomic imperfections".

(2.) Cations of the host compound can assume different oxidation states. Incorporation of lithium into the NiO lattice may serve as an example. If as the result of the incorporation of a given number of Li ions into the lattice, an equal number of Ni^{+2} ions undergoes oxidation to the valence +3, the formation of the solid solution takes place without the formation of new atomic defects. As already mentioned, the appearance of Ni^{+3} ions in the NiO lattice is equivalent to the generation of the appropriate number of holes, the concentration of which determines such properties as electric conductivity and the position of the Fermi level. Thus adsorptive and catalytic properties can, in this way, be purposefully modified. This type of incorporation is called "controlled electronic imperfections".

(3.) Foreign atoms, which form compounds of composition different from that of the host compound may, after incorporation, change their oxidation state into a new one which is identical with that of substituted host cations. For instance, when Mn ions are introduced into Al_2O_3 they assume the valence state of +3, whereas on incorporation into TiO_2 their valence state becomes +4. This phenomenon is called an "induced valence state".

The type of mechanism operating by incorporation of the additive depends not only on the properties of this additive and those of the host compound, but also on the conditions of the experiment, in particular on the pressure of the nonmetallic component. We shall illustrate the correlations between different parameters of such systems using again the example of NiO into which Li ions are incorporated.

In order fully to describe the defect structure of NiO crystal with the addition of lithium we must, besides the equations describing the equilibria of defects in the pure compound as discussed above, take into account still one more equation, expressing the process of incorporation of lithium from the gas phase into the NiO lattice:

$$Li_{(g)} \rightleftharpoons Li_{Ni} + V_O \quad (36)$$

As the lithium atom has only one valence electron, incorporated into the NiO lattice it behaves as an acceptor center and becomes ionized according to the equation

$$Li_{Ni} \rightleftharpoons Li'_{Ni} + h^{\cdot} \quad (37)$$

All these defects must conform to the electroneutrality condition:

$$n_e + 2[V''_{Ni}] + [Li'_{Ni}] = 2[V_O^{\cdot\cdot}] + p_h \quad (38)$$

We have thus 10 variables: $[V_{Ni}]$, $[V_O]$, $[V''_{Ni}]$, $[V_O^{\cdot\cdot}]$, n_e, p_h, $[Li_{Ni}]$, $[Li'_{Ni}]$ and p_{Ni} or p_{O_2} and p_{Li}. They must fulfill 8 equations: 19–23 and 36–38, two variables remain thus independent. As in the conditions of the experiment there are usually the vapour pressures of the components which may be changed, these parameters are usually chosen as independent variables. When the dopant is introduced in form of a solid, e.g. Li_2O, and the amount added is completely dissolved in the host compound, instead of p_{Li} the total amount of the incorporated foreign atoms $[Li_{tot}]$ may be taken as one of the independent variables. The total concentration of the incorporated lithium must be of course equal to the sum of the unionized and ionized centers. Thus, instead of equation 36 we may write:

$$[Li_{tot}] = [Li_{Ni}] + [Li'_{Ni}] \tag{39}$$

As the ionization energy of Li-centres is small, they are fully ionized at a comparatively low temperature, and equation 39 simplifies to

$$[Li_{tot}] \approx [Li'_{Ni}] \tag{40}$$

The system of equations may be solved by the approximate method already described, dividing the range of variation of p_{O_2} and $[Li_{tot}]$ into a set of subranges in which approximate neutrality conditions are valid. When the concentration of the introduced lithium is very small, the system behaves as pure NiO, described in the preceding paragraph. The type of approximate electroneutrality condition then depends on oxygen pressure. Let us assume that this pressure is high enough for the electroneutrality condition to be given by equation 33:

$$2[V''_{Ni}] = p_h \tag{41}$$

When the lithium concentration attains the value $[Li'_{Ni}] = 2[V''_{Ni}]$ the next range begins in which the electroneutrality condition assumes the form:

$$[Li'_{Ni}] = p_h \tag{42}$$

As the concentration of holes representing the majority current carriers is determined by the concentration of the impurity introduced into the lattice, this range represents the range of controlled electronic imperfections. The same electroneutrality condition holds also in the next range in which, however, the major part of the introduced lithium remains unionized, and it can be assumed that $[Li_{tot}] \approx [Li_{Ni}]$. On further increase of lithium concentration the range is given by

$$[Li'_{Ni}] = 2[V_O^{\bullet\bullet}] \tag{43}$$

that is the range of controlled atomic imperfections. The dependence of the concentrations of various defects on the concentration of lithium in the four ranges of electroneutrality condition are shown in Figure 38 for oxygen pressure within the range corresponding to the electroneutrality condition of pure NiO given by equation 33. Analogous diagrams may be constructed for other ranges of oxygen pressure. A full description of the behaviour of a nonstoichiometric transition metal oxide doped with altervalent ions thus requires the construction of a three-dimensional diagram with p_{O_2} and p_{dopant} as two independent variables.

It is however more convenient to use a projection of such a diagram on the $p_{O_2} - p_{dopant}$ plane as shown in Figure 39.

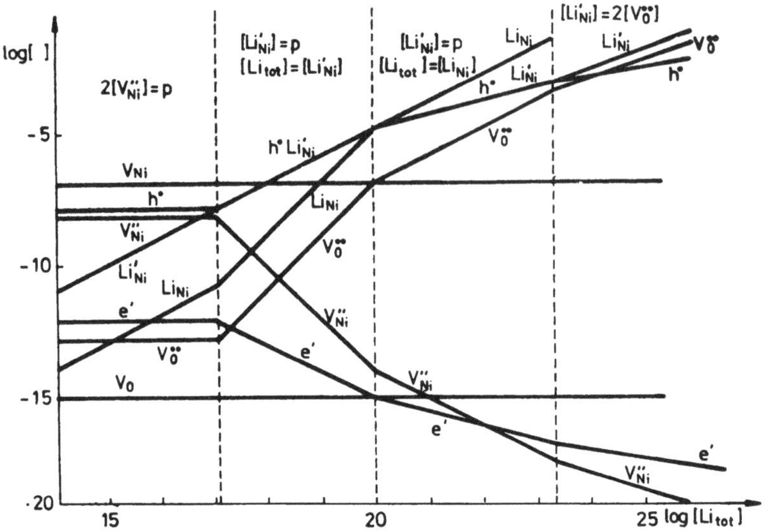

Figure 38. Concentrations of various defects in NiO doped with lithium as function of the concentration of the dopant at constant oxygen pressure

As an illustration of the relevance of the analysis of defect structure to catalysis we shall discuss the studies of the oxidation of CO on NiO doped with lithium [41]. Figure 40 shows the dependence of the values of the work function φ and the Fermi energy μ (both measured in vacuum) on the concentration of lithium. It may be seen that the μ value, representing the position of the Fermi energy with respect to the d^8 level in the bulk, decreases monotonically in agreement with the theoretical calculations given in Figure 38 (c.f. the curve describing the concentration of electrons). The curve of the work function passes through a minimum indicating that above certain concentration of Li the behaviour of Li at the surface begins to differ from that in the bulk (c.f. section 6), the value of the work function reflecting the position of the surface Fermi energy. A correlation exists between this position and the activation energy of the oxidation of CO as illustrated by Figure 41. The existence of such a correlation was predicted by the electronic theory of catalysis [95, 96].

4. Solid Solution of Transition Metal Ions in Diamagnetic Matrixes

When reacting molecules interact with transition metal ions at the surface of an oxide catalyst, the transfer of electrons takes place between the adsorbed layer and the solid. A fundamental question arises as to what extent the type of bonding between the molecule and the transition metal ion depends on the local properties of this ion such as the d-electron structure, the character and arrangement of surrounding oxygen ions, and how relevant in this respect are the collective properties of the solid, in particular the chemical potential of electrons as given by the position of the Fermi level, and the

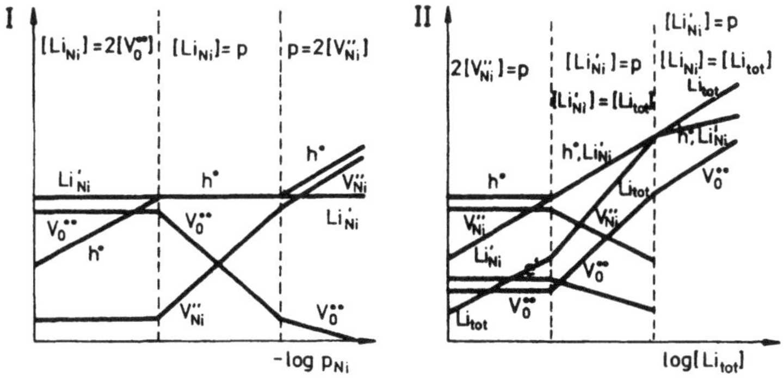

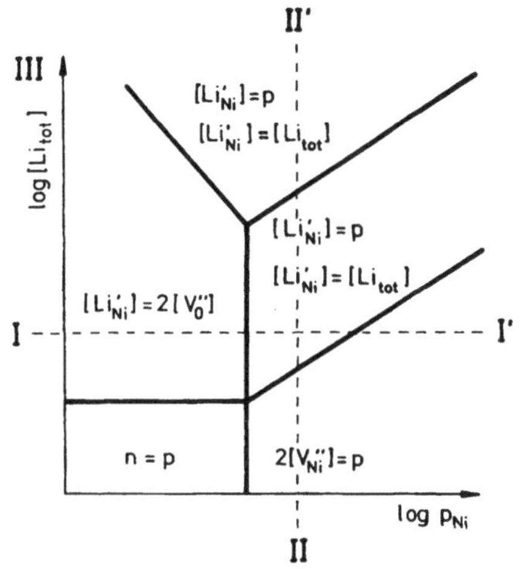

Figure 39. Schematic representation of the defect equilibria in NiO doped with Li. Diagrams I and II are cross-sections of diagram III along I–I' and II–II' lines

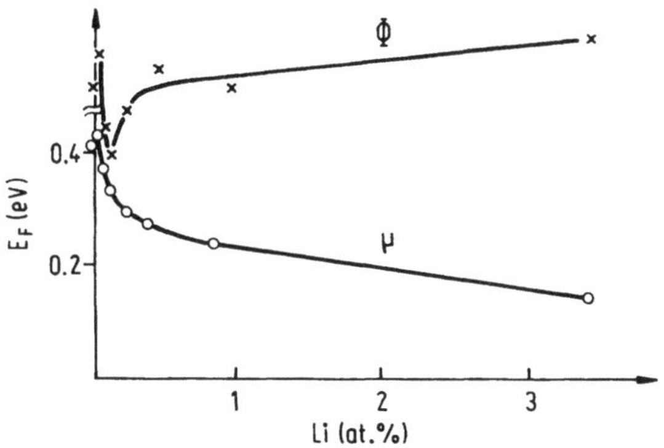

Figure 40. Work function Φ and Fermi energy μ of NiO doped with Li as function of Li concentration [41]

ability of the solid to transport the electrons from one active center at the surface to another, expressed by its electric conductivity.

One of the methods of discrimination between the influence of localised and collective properties of the oxide for its behaviour in catalytic reactions, is the use as model catalysts of solid solutions of transition metal ions in an

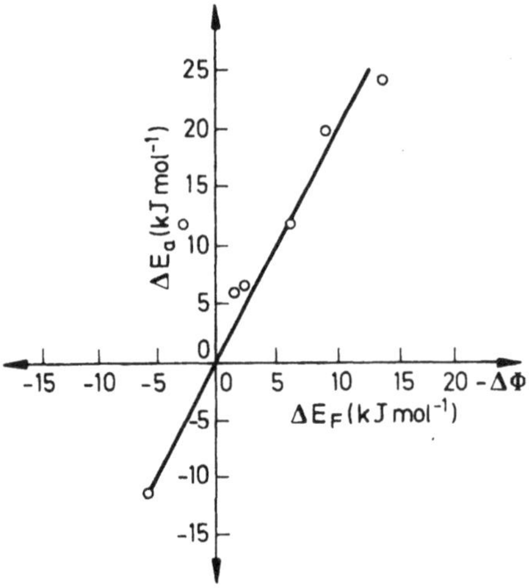

Figure 41. Apparent activation energy of CO oxidation as function of the $\Delta\Phi = \Phi_{doped} - \Phi_{pure}$ of NiO doped with Li [41]

insulating, diamagnetic oxide as solvent [42—44]. At very small concentrations it can be assumed that practically no interactions operate between the transition metal ions, and the behaviour of isolated ions in adsorption and catalysis can be thus studied. On increasing the concentration one may study the influence of the strength of interaction between the transition metal ions on their properties in chemisorption and catalysis. The significance of the following parameters may be in this way studied:

(1) the effect of *d*-electron population on the properties of the isolated ion, by comparing the activity of different transition metal ions in the same inert matrix at the same concentration;

(2) the influence of the type of coordination of the transition metal ion, by comparing the activity of the same transition metal ion in matrices of differing cation site symmetry;

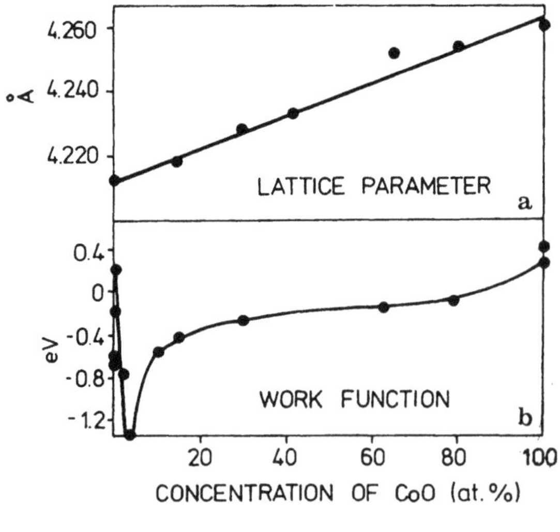

Figure 42. a Lattice parameter and **b** work function of CoO—MgO solid solutions as function of the concentration of CoO

(3) the role of the interactions between the transition metal ions, that is, of collective properties, by comparing the activity of the given transition metal ion in the given matrix at different concentrations.

MgO as a matrix offering sites with octahedral symmetry, and ZnO with tetrahedral symmetry, are usually used for such studies. Solid solutions in α-Al_2O_3 with octahedral sites and in $MgAl_2O_4$ with both octahedral and tetrahedral sites have also been investigated. The development of interactions between cations has been found in the case of solid solutions in MgO to start at the concentration of 1 %. It is usually assumed that conformity to Vegard's law may be taken as an indication of the formation of a random solid solution. In recent years however it was shown that this is not necessarily the case, as is illustrated by Figure 42, in which lattice parameter and work function of the CoO—MgO solid solutions are plotted as a function of the concentration of CoO [45]. It may be seen that although the changes of the lattice parameter follow Vegard's law, the values of the work function pass through a minimum at very low concentration, indicating that a new surface structure is formed at first. This problem will be discussed in more detail in section 6.

5. *Shear Structures*

Until now we have described the deviations from stoichiometry caused by the presence of point defects distributed randomly in the crystal lattice. Development of modern physical techniques, in particular x-ray diffraction and electron microscopy, which allow a study of structural transformations of solids on an atomic scale, has led in recent years to the conclusion that even very considerable deviations from stoichiometry may appear in the crystal without the necessity of the simultaneous presence of point defects. Namely they may be due to the phenomenon called crystallographic shear (CS), resulting in the formation of shear planes which may be arranged in the crystal into an ordered array of equidistant planes forming an infinite homologous series of related phases called shear structures. These structures are formed in oxide systems of Group IV, V, VI and VII transition elements and make the structural chemistry of these compounds very complex and versatile [46, 47, 48].

We shall demonstrate the formation of a shear plane with the example of the ReO_3 structure, which is adopted by such oxides as WO_3 (distorted version) or (Mo, W) O_3. This structure is built of octahedra linked through vertices and forming a face-centred cubic lattice (Figure 43a). On reduction, oxygen vacancies are formed at the surface. When the concentration of these vacancies surpasses a certain critical value, they aggregate into a vacancy disc (Figure 43b). Part of the crystal may now shear so that along the shear plane the linkage between the octahedra is changed from corner-sharing into edge-sharing (Figure 43c), leading to the annihilation of vacancies. The resulting element of the shear plane is bounded by a dislocation loop which, being a sink for vacancies, climbs through the crystal and extends the crystallographic shear very rapidly to the whole crystal plane, which becomes a crystallographic shear plane. It constitutes an extended two

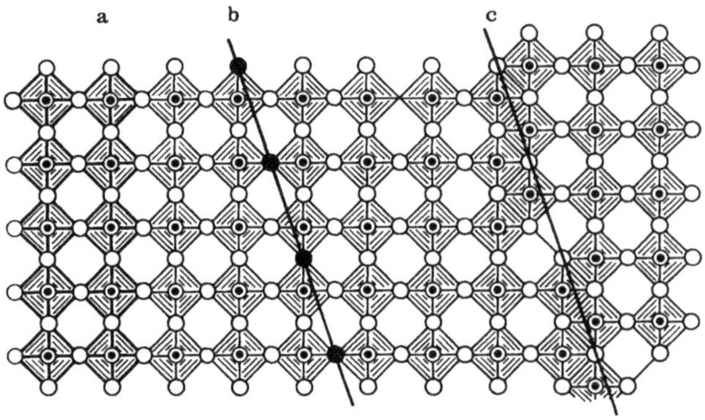

Figure 43. Formation of a shear plane in WO_3 crystal. **a** projection of the idealized WO_3 lattice, **b** aggregation of vacancies, **c** shearing of the part of crystal and formation of a shear plane

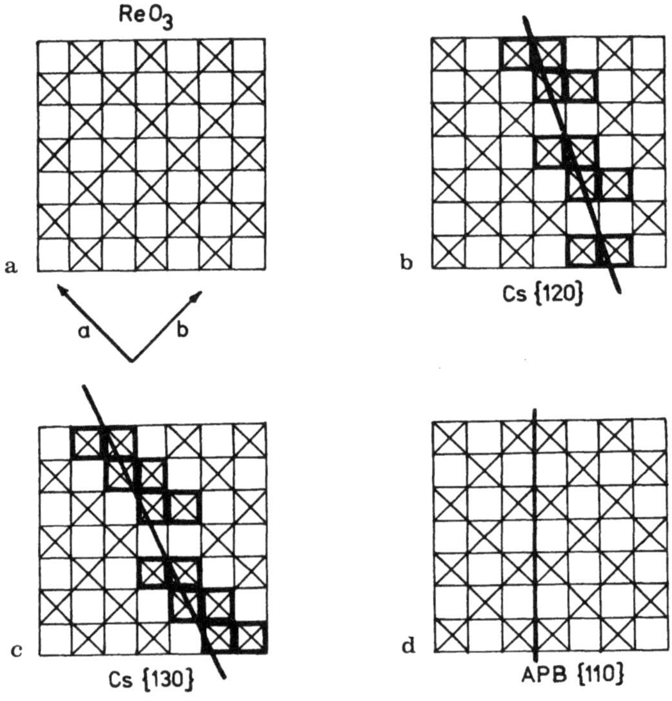

Figure 44. Crystallographic shear on {120} and {130} in ReO_3 structure; shear displacement with the same displacement vector on {110} produces an antiphase boundary

dimentional defect. The most commonly observed are the {102} and {103} CS planes (Figure 44b and c) along which blocks of four or six edge sharing octahedra respectively are formed. The crystal can be considered to be composed of slabs of corner-sharing octahedra of more or less undistorted parent structure, separated by CS planes. If the spacing of CS planes is completely regular, the result is to generate from the parent structure of MO_3 a homologous series of stoichiometrically defined, intermediate, mixed valence compounds with the general formula M_nO_{3n-p}, where n is the width of the slabs of parent structure as given by the number of coordination octahedra in unbroken rows between CS planes, and p measures the number of anionic sites eliminated by the crystallographic shear describing the resulting change of the O:M ratio. It amounts to 1 for {102} planes and 2 for {103}.

It was found in many cases that the spacing between the CS planes is not regular. Such crystals could show continuous changes in composition due to the alterations in the distribution of CS spacings. They could be considered as nonstoichiometric compounds. On the other hand such crystals could be regarded as made up of lamellae of different compositions in coherent intergrowth.

Table 7 summarizes as an example a series of ReO_3 based structural homologues found in the system of mixed oxides of tungsten and molybdenum. They contain {102} CS planes and their composition may be described by the formula $(Mo, W)_nO_{3n-1}$. The CS planes are composed of blocks of four edge-sharing octahedra, and the spacing between the CS planes $n = 8, 9, 10, 11, 12$ and 14. The structure of one of the homologues, Mo_9O_{29}, with spacing $n = 9$ is shown in Figure 45.

Shear structures based on edge sharing of MoO_3 elements are more complex. As shown in Figure 23, MoO_3 is built of double ribbons of edge-sharing

Table 7. Structural homologues $(Mo, W)_nO_{3n-1}$

Compound	MO_x	Approximate Mo/W ratio
WO_3	3.00	0
$(MO, W)_{14}O_{41}$	2.929	1/1
$(Mo, W)_{12}O_{35}$	2.917	1/2
$(Mo, W)_{11}O_{32}$	2.909	
$(Mo, W)_{10}O_{29}$	2.900	1/4
Mo_9O_{26}	2.889	
Mo_8O_{23}	2.875	

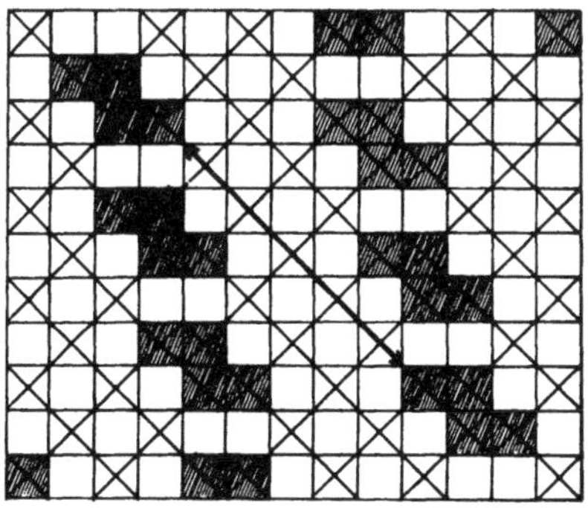

Figure 45. Schematic representation of the structure of Mo_9O_{26}, the homologue of M_nO_{3n-1} series with $n = 9$

octahedra, which form layers by corned sharing. On reduction, crystallographic shear takes place whereby groups of octahedra of adjacent ribbons are made to share edges instead of corners. The main product of reduction is $Mo_{18}O_{52}$ in which slabs of the parent structure of MoO_3 along the ribbons (c direction) are 18 octahedra long, the last three of which share edges with similar octahedra across the shear plane. Across the a-direction the slabs are six octahedra wide (Figure 46). It may be considered as a member of the Mo_nO_{3n-2} series with {311} CS planes. The existence of members of other series: $Mo_{13}O_{38}$ (Mo_nO_{3n-1}) with {211} CS planes, and $Mo_{26}O_{75}$ (Mo_nO_{3n-3}) with {411} CS planes has been reported [46].

The third large group of shear structures is based on rutile as the parent structure and can be described by the general formula Ti_nO_{2n-1}. The type of shear plane depends upon the stoichiometry of the oxide attained on reduction.

At high values of n between 36 and 16, that is, in the composition range from $TiO_{1.97}$ to $TiO_{1.93}$ shear planes are formed with {132} orientation. They are composed of octahedra sharing additional edges instead of corners as in rutile. As an example, Figure 47 shows the structure of $Ti_{16}O_{31}$. Only even values of n have been found for these {132} shear planes. They are also usually twinned, alternating {$\bar{1}$32} and {132} planes being observed.

For values of n between 9 and 4, that is, in the composition range $TiO_{1.89}$ to $TiO_{1.75}$ the shear planes are parallel to {121} orientation.

The composition range between $TiO_{2.00}$ and $TiO_{1.75}$ may be thus divided into six structural ranges [46].

1) TiO_2 to TiO_{2-x}
   rutile with point defects
2) TiO_{2-x} to $TiO_{1.999}$
   separate CS planes at spacing too great to result in realinment and aggregation.
3) $TiO_{1.999}$ to $TiO_{1.9730}$
   {132} CS planes grouped together in lamelle,
4) $TiO_{1.9775}$ to $TiO_{1.9375}$
   {132} CS phases Ti_nO_{2n-1}, $16 < n < 37$
5) $TiO_{1.9375}$ to $TiO_{1.90}$
   coexistence of {121} and {132} CS planes

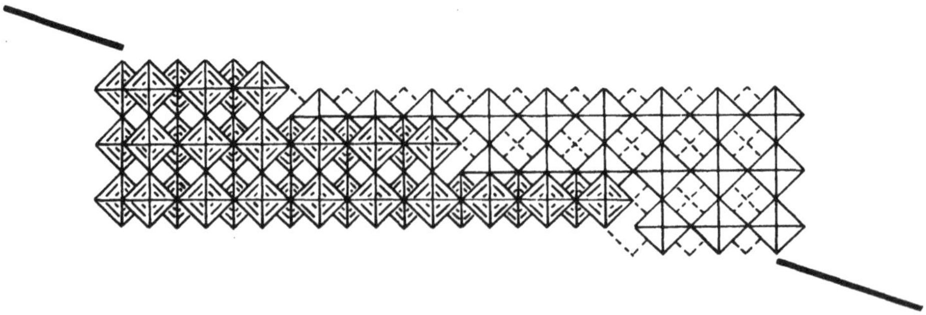

Figure 46. The structure of the shear plane in $Mo_{18}O_{52}$

6) $TiO_{1.90}$ to $TiO_{1.75}$
{121} CS phases Ti_nO_{2n-1}, $4 < n < 10$

We shall mention here still one example of shear structure formation because of its great importance for catalysis. When vanadium pentoxide is reduced, it very readily loses oxygen from one third of the {010} planes transforming into V_6O_{13}. The latter may be considered as one of the shear phases of the family $V_{2n}O_{5n-2}$ with $n = 3$. The structure of V_6O_{13} shows shear planes forming $\infty \times 3 \times \infty$ slabs of undistorted V_2O_5 (Figure 48).

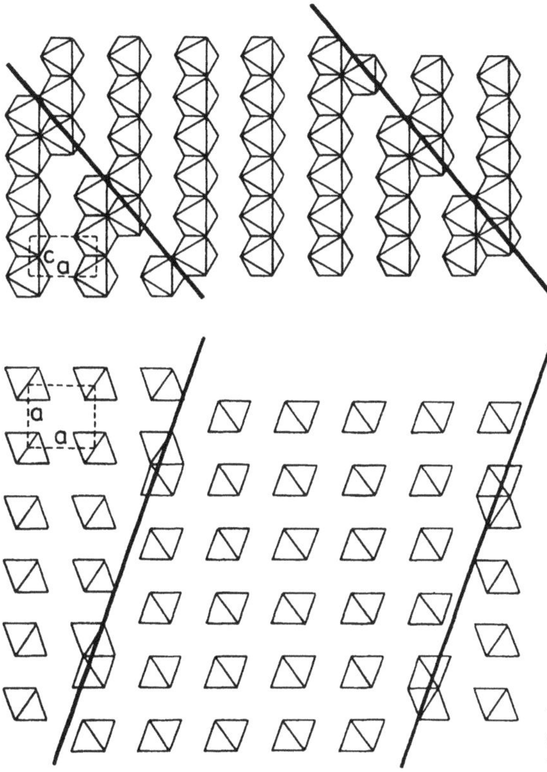

Figure 47. The structure of Ti_6O_{31} viewed down the a-axis (top) and c-axis (bottom) of the rutile structure with unit cell shown by broken lines

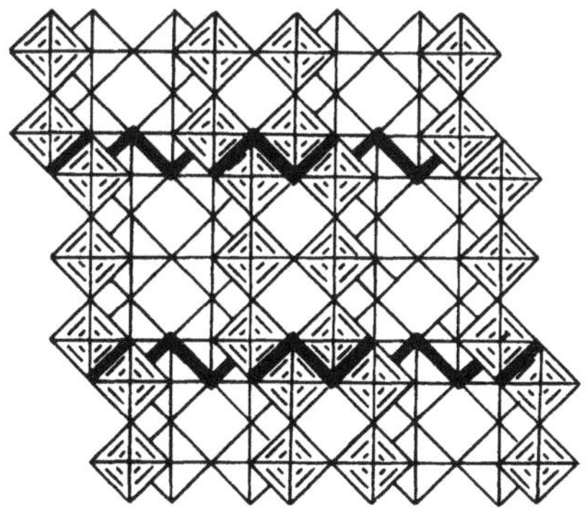

Figure 48. The structure of V_6O_{13}

It is interesting that V_6O_{13} is the only shear phase found in the composition range V_2O_5—VO_2 [49]. The readiness of exchange of oxygen between the gas phase and the V_2O_5—V_6O_{13} system is one of the important features which determine its catalytic properties in oxidation reactions [50]. Stronger reducing conditions lead to the formation of VO_2, which further loses oxygen to from a number of shear phases of composition V_nO_{2n-1} based on rutile parent structure and analogous to the Ti_nO_{2n-1} phases described above.

The existence of one or other of the discrete structures depends on the value of the thermodynamic potential of its formation, i.e. oxygen pressure. Figure 49 shows the dependence of the thermodynamic potential and equilibrium pressure on the oxygen-to-metal ratio in the oxide for the Ti—O and V—O systems [51].

The concept of crystallographic shear has a much wider significance and may also be used to demonstrate topological relationships between the compounds which form CS phases and those that do not [47]. As an example we shall describe the relationship between WO_3, which forms a distorted ReO_3 structure and complex tunnel structures of the tungsten bronze type,

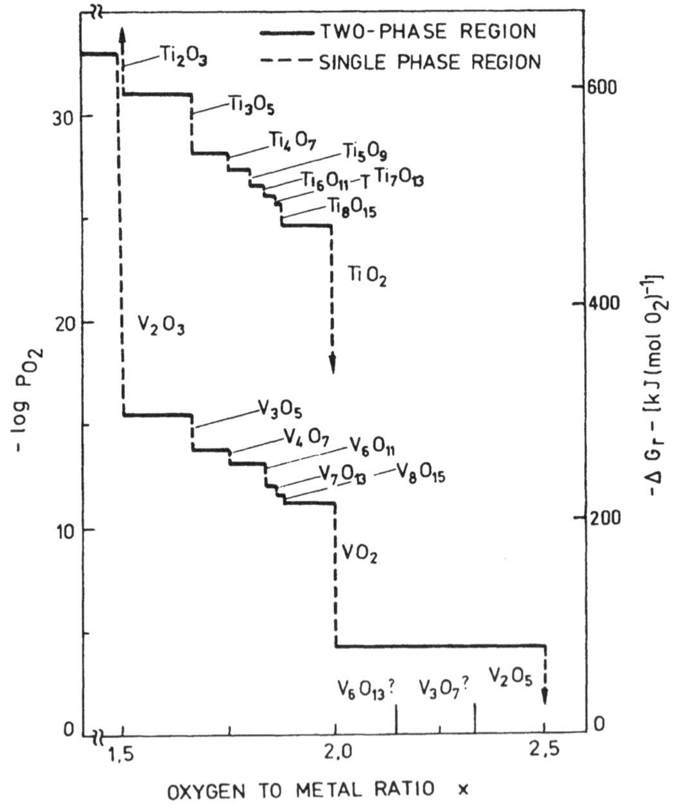

Figure 49. Equilibrium oxygen pressures in the Ti—O and V—O systems at 1000 K as function of the composition [51]

formed on doping with alkali metal. It is illustrated in Figure 50. The basic operation required is the rotation of a column of ReO_3 octahedra by 45°. This operation produces pentagonal tunnels preserving however the WO_3 stoichiometry of the parent. The real structure differs from this idealized framework in that the octahedra are distorted and the tunnels contain additional cations which may be important in stabilizing the more open skeletal structure.

Similarly, a rotary crystallographic shear operation on rutile structure produces a series of compounds, analogous to tungsten bronzes and called hollandites (Figure 51), of which the typical representative is $A_xMn_8O_{16}$.

The ability of Group V, VI and VII metal oxides to form shear structures may play an important role in determining their catalytic properties. Easy evolution of one oxygen ion on the transformation from corner-linked to edge-linked arrangement of metal-oxygen octahedra may be one of the factors creating the ability of these structures to insert oxygen into an organic molecule in processes of selective oxidation of hydrocarbons [52, 53]. By contrast with other oxides in the case of which the desorption of an oxygenated

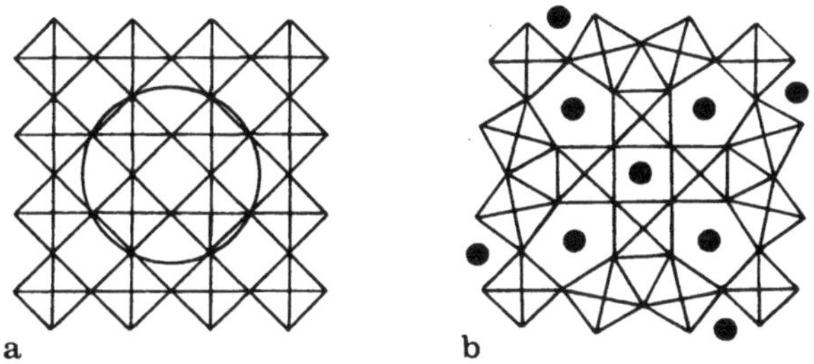

a b

Figure 50. The formation of the framework of the tetragonal tungsten bronze structure (**b**) from ReO_3 structure (**a**) by rotation of the circled unit by 45°. ●-alkali metals [47]

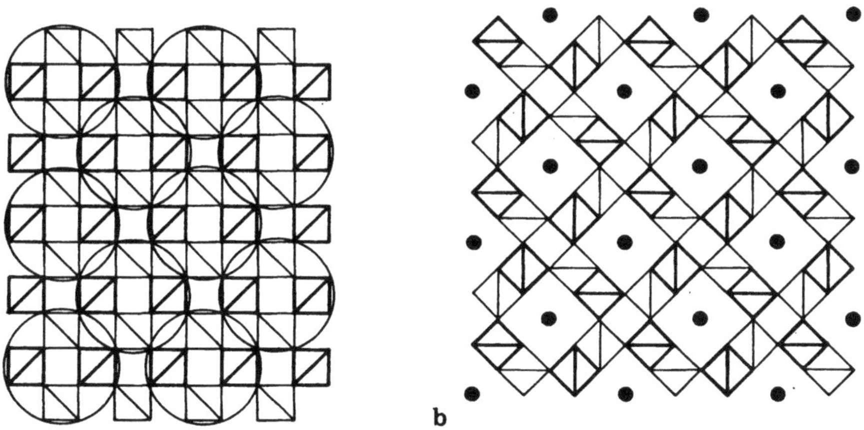

a b

Figure 51. The formation of the hollandite structure (**b**) from the rutile structure (**a**) by rotation of the circled units by 45° [47]

product must result in the generation of an oxygen vacancy and requires considerable energy, in the case of the CS systems the desorption is accompanied by the simultaneous rearrangement of octahedra, which is known to proceed readily.

D. Complex Oxides and Oxysalts of Transition Metals

Compounds containing two or more elements besides oxygen may be roughly grouped into homodesmic and heterodesmic structures depending on whether there is a smaller or larger difference between the electronegativities of the elements. Thus, in case of compounds $A_xX_yO_z$, since those containing two very electronegative elements are not numerous, we have two main groups:
— A and X both comparatively electropositive (homodesmic structure),
— A electropositive and X electronegative (heterodesmic structure).

Compounds of the first group, which may be called complex oxides are best regarded as close-packed array of oxygen ions, the A and X ions being distributed in tetrahedral and octahedral holes. The type of coordination chosen by the given ion depends on its size and ligand field site-preference energy. If the A ions have approximately the same size as oxygen ions, the structure may be regarded as a close-packed array of O and A ions with X ions in tetrahedral or octahedral holes. Compounds of the second group may be considered as composed of A ion and complex XO_m ions. The complex ion is a charged group of atoms (which may be finite or infinite in extent), within which the bonds between the central atom X and the surrounding oxygen atoms are essentially covalent. Binding between these complex ions and the A ions is essentially ionic. Such compounds are thus called oxy-salts.

1. Complex Oxides

Two main types may be distinguished: structures with close-packed oxygen ions and structures, in which oxygen and A ions are in close-packed arrange-

Table 8. Close-packed complex oxides structures [1]

Close-packed atoms	Formula	Type of structure	Holes occupied	
			tetrahedral.	octahedral
1. O ions	A_2XO_4	phenacite Be_2SiO_2	all	
	AXO_2	$LiNiO_2$		all
	AXO_3	$FeTiO_3$ (ilmenite)		2/3
		$LiSbO_3$		
	AXO_4	$MgWO_4$		1/2
	AX_2O_6	niobite		
	A_2XO_4	olivine	1/8	1/2
	AX_2O_4	spinel	1/8	1/2
	$A_2X_3O_8$	$M_2^{II}Mo_3^{IV}O_8$	1/16	1/2
2. A+ 3O ions	AXO_3	perovskite		all

ment (Table 8). We shall describe some of these oxides, which have found wide applications in catalysis.

One of the important types of complex oxides most commonly used as a catalyst is spinel. Many oxides crystallize in the spinel structure with the general formula AX_2O_4, which can be divided into three groups: 1) $M^{II}M_2^{III}O_4$; 2) $M^{IV}M_2^{II}O_4$ and 3) $M^{VI}M_2^{I}O_4$. We shall describe in more detail the first group.

The unit cell of the spinel $M^{II}M_2^{III}O_4$ comprises 32 oxygen ions (8 $M^{II}M_2^{III}O_4$) in cubic close packing (Figure 52). As follows from the formulae given on page 23 the unit cell contains 32 octahedral and 64 tetrahedral holes, in which 8 M^{II} ions and 16 M^{III} ions are arranged. When 8 equivalent tetrahedral holes out out of 64 are occupied by M^{II} ions, and 16 octahedral holes out of 32 by M^{III} ions, the spinel is called normal. Other modes of distribution are also encountered. The 8 tetrahedral holes may be occupied by M^{III} ions, the remaining 8 M^{III} ions and all 8 M^{II} ions occupying the 16 octahedral holes. Such a spinel which may be represented by the formula $M^{III}M^{II}M^{III}O_4$ is called inverse. In some spinels both M^{II} and M^{III} ions are randomly distributed in tetrahedral and octahedral holes. It is convenient to describe different modes of distribution by a parameter η, which gives the fraction of divalent metal ions in octahedral coordination. Thus for the normal spinel $\eta = 0$ and for the inversed one $\eta = 1$. In the case of a random distribution $\eta = 0.67$. Values of the parameter η found experimentally in different spinels are given in Table 9 [54].

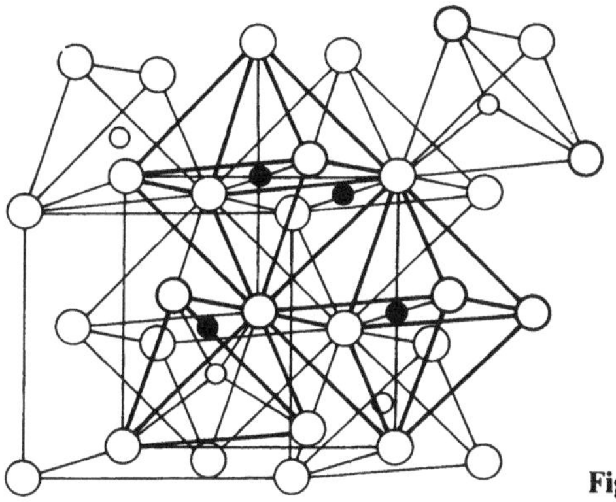

Figure 52. The spinel structure

Table 9. Distribution of cations in octahedral and tetrahedral positions in the spinel lattice (values of parameter η) [54]

Cation	Mg^{2+}	Mn^{2+}	Fe^{2+}	Co^{2+}	Ni^{2+}	Cu^{2+}	Zn^{2+}
Al^{3+}	0	0	0	0	0.76		0
Cr^{3+}	0	0	0	0	0	0.1	0
Fe^{3+}	0.9	0.2	1	1	1	1	0
Mn^{3+}	0	0			1	0	0
Co^{3+}				0			0

Calculation of the lattice energy of a spinel formed only by main group metal ions, e.g. $MgAl_2O_4$, indicates that a normal distribution of cations is the most stable. When however the spinel is composed of transition metal cations, the ligand field stabilization energy, discussed on page 36, must be taken into account in assesment of the stability of different distributions. The values of this energy for first row transition metal ions in octahedral and tetrahedral coordination are summarized in Table 10. The last column gives the difference between the octahedral and tetrahedral stabilization energies. It is sometimes called the site-preference energy. The larger this difference the greater is the tendency of the given ion to choose the octahedral position. The data given in Table 10 indicate that the greatest tendency to occupy the octahedral position is for Cr^{+3} and Mn^{+3} ions. Accordingly, all chromites and manganites except $NiMn_2O_4$ are normal spinels. Converely, the Fe^{+3} ion is not at all stabilized in the ligands field, similarly the Zn^{+2} ion. Thus, zinc ferrite is a normal spinel because only the lattice energy determines the distribution, whereas ferrites of Cu^{+2}, Ni^{+2}, Co^{+2} and Fe^{+2} are inverse, because the high ligand field stabilization energy of these ions in octahedral coordination is the decisive factor, Fe^{+3} ions being displaced to tetrahedral positions.

There are many ways in which defects can form in a spinel structure [55]. The most commonly encountered are vacancies in the cation sublattices: tetrahedral and octahedral. On the other hand, the spinel structure is fairly empty because only 1/8 of the tetrahedral and 1/2 of the octahedral sites are occupied. Thus solubility of the component oxides in the spinel phase is observed, resulting in the appearance of excess cations in the interstitial tetrahedral and octahedral sites. The given cations in the spinel structure can be substituted by many other cations. In this way multicomponent systems can be obtained, their properties being purposefully regulated by proper selection of the component cations and appropriate adjustment of their composition. This also opens good possibilities for studying various factors which influence the catalytic activity of oxide systems [56].

Table 10. Stabilization energy in the ligand field of oxide ions of octahedral and tetrahedral symmetry (in kJ mol^{-1}) [97]

Ion	Stabilization energy		Difference
	octahedral	tetrahedral	
Mn^{2+}	0	0	0
Fe^{2+}	49.9	33.1	16.8
Co^{2+}	93.0	62.0	31.0
Ni^{2+}	122.3	36.0	86.3
Cu^{2+}	90.5	26.8	63.7
Ti^{3+}	87.6	58.7	28.9
V^{3+}	160.5	106.8	53.6
Cr^{3+}	225.0	67.0	158.0
Mn^{3+}	135.8	40.2	95.5
Fe^{3+}	0	0	0

The second structure which will be described here is that of perovskite. It is adopted by compounds in which one of the metal ions has a size similar to that of oxygen ion. In the case of a binary compound composed of A and O ions with $r_A/r_O \approx 1$, a close-packed arrangement of these ions in which the positive ions A are not in contact with each other is possible only at the O/A ratio of at least 3/1, that is for compounds of the composition AO_3. The ions may be arranged in two ways shown in Figure 53 in which single layers of these arrangements are shown. The crystal structures are formed by superposition of such layers, the ions A having a dodecahedral coordination. When metal ions A are mono or divalent, the excess negative charge must be compensated by a second type of cation, characterized by a small radius and occupying the holes in the close-packed structure. Complex oxides, halides etc. are formed in this way. By superposition of layers shown in Figure 53 and the distribution of cations in octahedral holes, a perovskite structure of the composition AXO_3 is formed. It is illustrated in Figure 54. This requires that X be a cation with a preference for octahedral coordination. For stoichiometric AXO_3 compounds there are many combinations of A^{+a} and X^{+x} that satisfy the electroneutrality condition $a + x = 6$, and this structure is adopted by many complex oxides of Group IV, V and VI transition metal ions with alkali or alkali earth metal cations, e.g. $BaTiO_3$, $KNbO_3$, $PbTiO_3$, $SrZrO_3$, $LaMnO_3$, $LaCrO_3$, $KTaO_3$, $BaHfO_3$ etc. It is interesting to note that the perovskite structure is closely related to that of ReO_3.

The possible combinations of A and X ions are restricted by the requirement that A and X must by stable in dodecahedral and octahedral coordina-

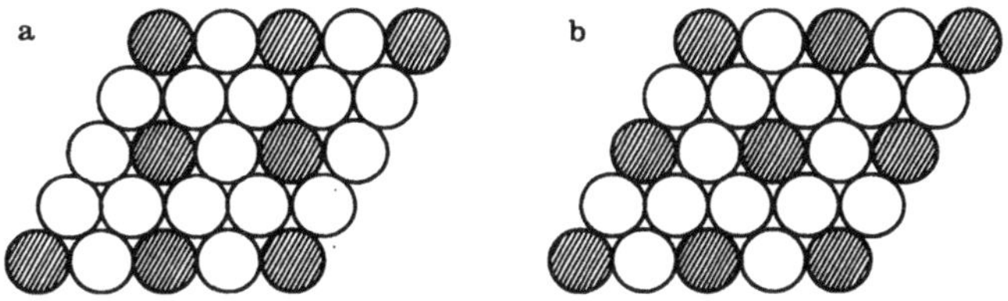

Figure 53. Close-packed layers of ions A and O with the same radius in the structure of AO_3 compound

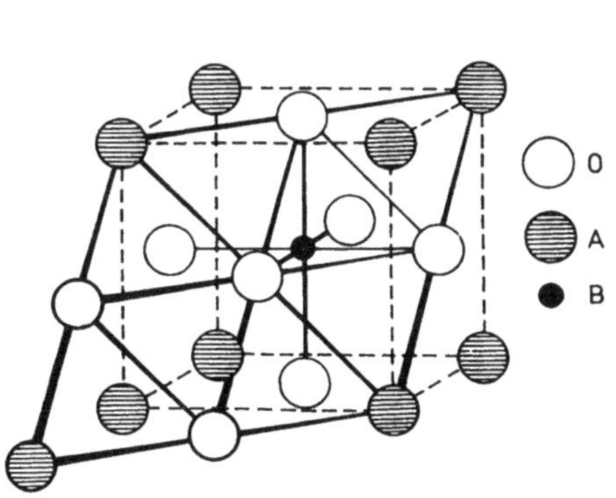

Figure 54. The perovskite structure

tions respectively. There are also conditions imposed by the geometry [57]: radii should have the values $r_A > 0.90$ Å, $r_X > 0.51$ Å, and $0.75 < (r_A + r_O)/\sqrt{2}\,(r_X + r_O) < 1.00$.

Both ions A and X can be substituted by other ions to form a whole variety of solid solutions. Ordering of the solute ions may also take place resulting in the formation of definite compounds.

The perovskite phases are usually nonstoichiometric due to the presence of vacancies in the sublattice of cations A, as well as vacancies in the oxygen sublattice. This is manifested by the existence of tungsten bronzes Na_xWO_3 with a very broad range of the variation of x. When the number of A is much smaller then that of X, the A ions may be either distributed randomly in the dodecahedral sites as it is the case in tungsten bronzes, or may form an ordered superstructure, as in $La_{2/3}TiO_3$. Nonstoichiometry resulting from the presence of oxygen vacancies is also very frequent among the perovskite phases [57].

Of particular importance for the use of perovskites as catalysts is the ability of these structures to stabilize the noble metals. Compounds of Pt, Ir, Os, Re, Rh and Ru may exist in the perovskite structure, the noble metal atoms occupying the X sites. As an example Figure 55 shows the activity of noble metal substituted perovskites as catalysts for the car mufflers [58]. The perovskites are also used as catalysts in fuel cells.

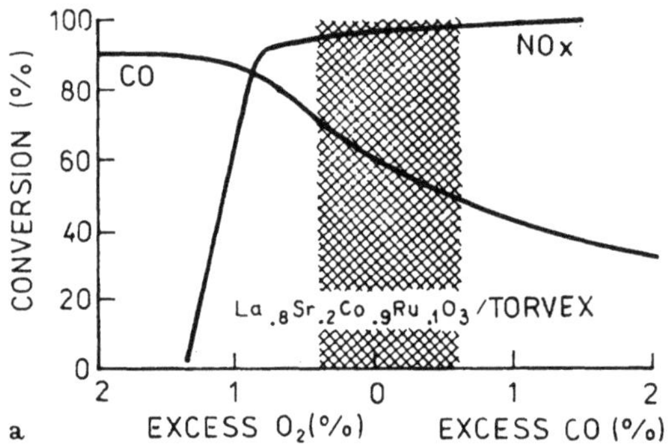

a

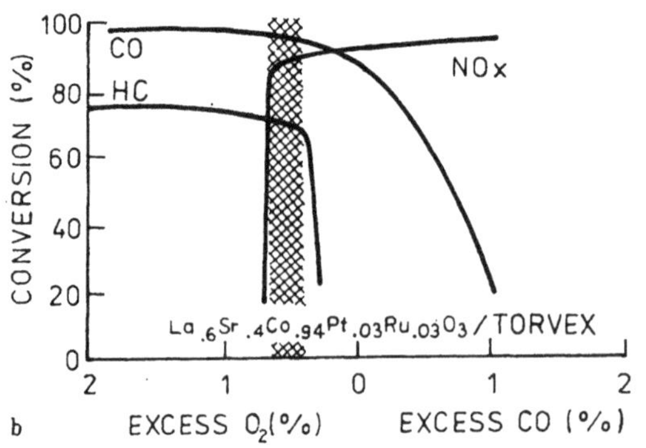

b

Figure 55. Catalytic activity of noble-metal substituted perovskites in the removal of CO and NO_x from exhaust gases [58]

2. Oxysalts of Group V and VI Transition Elements

The characteristic property of Group V and VI transition elements is their ability to form oxy-salts and oxy-acids, both simple as well as complex, in which condensed oxy-ions composed of MO_4 tetrahedra and MO_6 octahedra are the building units. They are described as isopolyoxysalts. The MO_6 octahedra may also form a compact grouping around a central atom of some other element. Such compounds are called heteropoly oxysalts. They may be divided into two classes according to whether the central atom is surrounded by four (tetrahedral) or six (octahedral) oxygen atoms. From the catalytic point of view the most important are the phospho-polyoxysalts.

As an example Figure 56 shows the structure of the phosphomolybdate anion. The central P atom is located at the center of an PO_4 tetrahedron which shares oxygen atoms with 4 groups of three edge-sharing MoO_6 octahedra. Figure 57 illustrates the structure of the telluromolybdate anion as a representative of compounds in which the central ion (Te) is octahedrally surrounded by MoO_6 octahedra sharing edges.

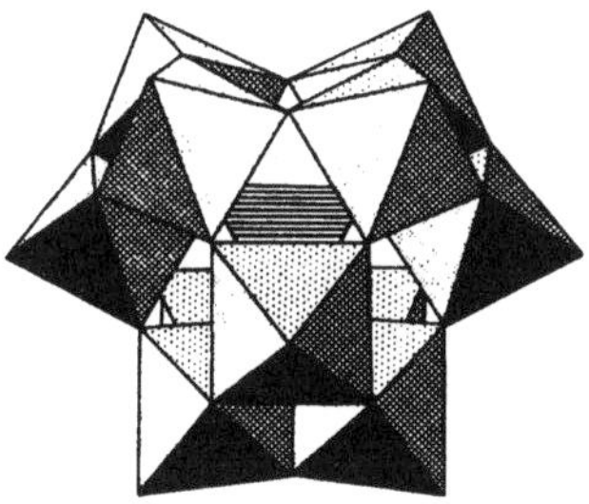

Figure 56. The structure of $PMo_{12}O_{40}^{3-}$ anion

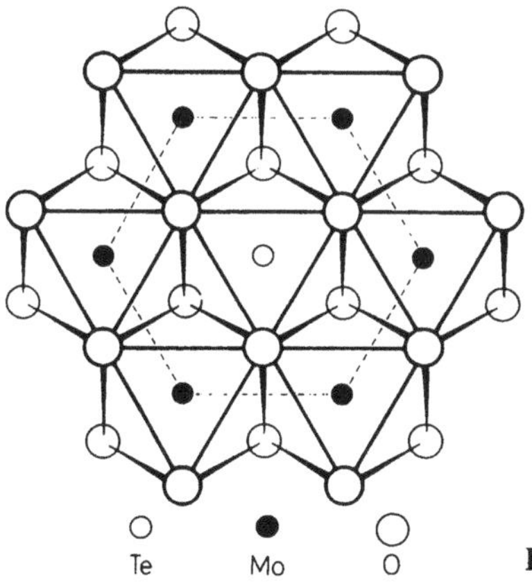

Figure 57. The structure of $TeMo_6O_{24}^{6-}$ anion

5. Oxides of Main Group Elements

A. Acid-Base Catalysts

As already mentioned a large group of catalytic reactions such as cracking, isomerization, dehydration, alkylation etc. proceeds through steps in which the reactant molecules form heterolytic bonds with the catalyst. These bonds may be formed either by using the free electron pair of the catalyst which then plays the role of a base, or that of the reactant molecule, the catalyst manifesting then its acid properties. Detailed description of solids which are commonly used as acidic and basic catalysts is given in [59].

B. Aluminas

Aluminium oxide forms a number of different phases depending on the conditions of the preparation. They are all based on a more or less close-packed oxygen lattice with aluminium ions in the octahedral and tetrahedral sites, and differ by the sequence in which the close-packed oxygen layers are superposed. Three series of aluminas may be distinguished [60]:

1) α-series with hexagonal close-packed lattices, characterized be the sequence ABAB
2) β-series with altenating close-packed lattices, characterized by various sequence of the type ABACABAC, ABAC—CABA etc.
3) γ-series with cubic close-packed lattices, characterized by the sequence ABCABC.

The representative of the first series is α-Al_2O_3 with the corundum structure, described on page 27. The β-series consists of χ- and ϰ-alumina, which are products of the decomposition of gibbsite. Members of the γ-series are obtained on decomposition of the hydroxides bayerite, nordstrandite and boehmite, and can be devided into a low temperature group (γ and η-alumina) and the high temperature group (δ and θ-alumina). The dehydration processes in air can be represented as follows:

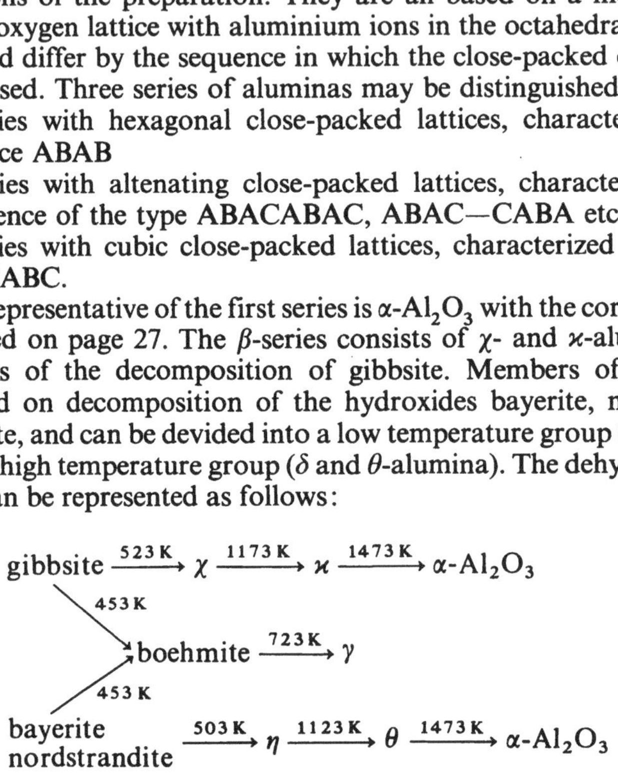

In vacuum the three trihydroxides decompose at low temperatures into an almost completely amorphous product (ϱ-alumina), which at higher temperatures changes into γ- or η-alumina, and further into θ-alumina.

γ-alumina has the structure of a defected spinel. It can be visualized as the structure, in which 1/3 of tetrahedral sites normally occupied in the spinel remain empty (Figure 58) and described by the formula

$$(Al_{0.67}V_{0.33})_{tetr}\,(Al_2)_{oct}\,O_4$$

Active alumina is not pure Al_2O_3 but contains a certain amount of water in form of surface hydroxyl groups [61, 62]. At low temperatures a complete monolayer of OH^- ions can be formed as shown in Figure 58b. During dehydration adjacent OH^- ions are assumed to combine at random, but

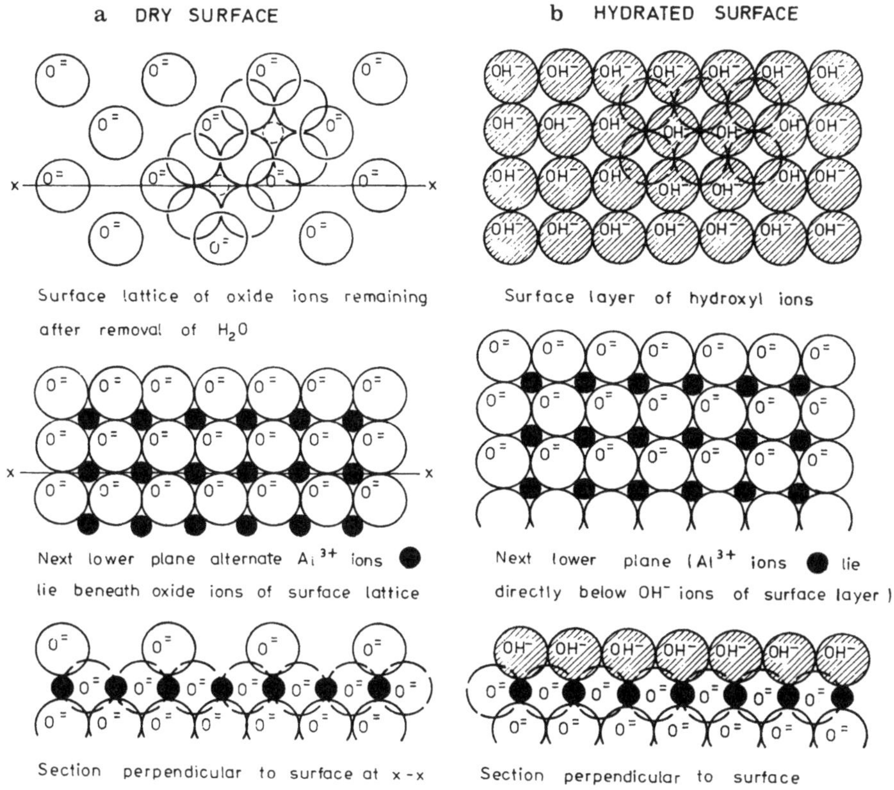

Figure 58. The structure of γ-Al_2O_3

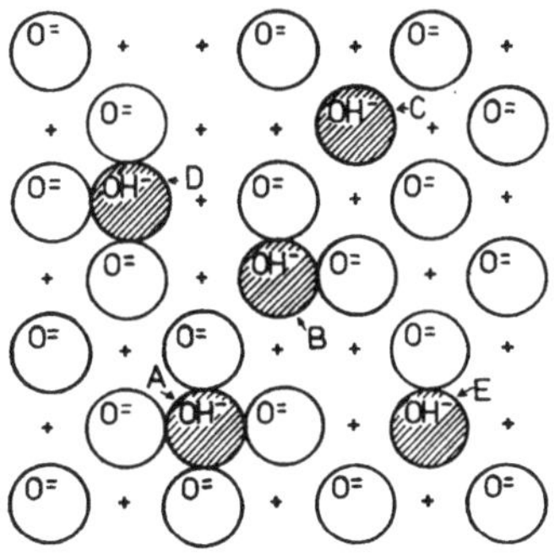

Figure 59. Model of the surface of γ-Al_2O_3

only 2/3 of them can be removed without disturbing local order of an idealized surface of dry alumina exposing the (100) plane, as shown in Figure 58a. Further dehydration entails the formation of disorder. Random removal of all hydroxyl pairs leaves the surface as illustrated in Figure 59 [61]. The remaining isolated hydroxyl groups, covering about 10 percent of the surface are found on five types of sites, having from zero to four nearest oxide neighbours. These five types of hydroxyl ions vary in chemical properties, the A-site ion being the most basic and the C-site ion the most acidic. Thus, the surface of alumina exhibits a heterogeneity. Upon dehydration at increasing temperatures the Brönsted acid sites, numerous at high water content, are gradually converted into Lewis acid sites of different strength. Even after heat treatment in vacuum at 1070 K about 2 percent of the total hydroxyl content is still retained on the surface. It must be also remembered that faces other than the (100)-face may be also exposed. This is certainly true of the frequently used δ-Al_2O_3 consisting of microspheres of 50–300 Å in diameter. Its surface has been approximated by a rhombooctahedron of twenty six sides, made by (100), (110) and (111) planes of the alumina spinel structure [63]. The surface of this material should then contain both octahedral and tetrahedral Al cations.

C. Silica

It follows from the Pauling rules that in the structures of silicates each vertex of the SiO_4 tetrahedron is common only for two tetrahedra, and two tetrahedra have only one vertex in common, that is, they never share edges or faces.

When SiO_4 tetrahedra have no elements in common and are linked only through polyhedra of other cations, orthosilicates are formed, such as olivine $(Mg, Fe)_2SiO_4$, granates $(Mg, Fe, Mn, Ca)_3(Cr, Al, Fe)_2(SiO_4)_3$ etc. When each tetrahedron shares only one vertex with another tetrahedron, the silicate is composed of $[Si_2O_7]^{6-}$ as structural units. Tetrahedra which share two vertices with other tetrahedra form structures composed either of ionic chains $^1_\infty[SiO_3]^{2-}$ (number 1 in the left superscript indicates a one dimensional chain stretching infinitely as marked by the left subscript ∞), or of cyclic ions $(Si_3O_9)^{6-}$ (three-members rings) and $[Si_6O_{18}]^{12-}$ (six-members rings). The Si/O ratio in these compounds amounts to 1/3. A simple silicate chain is shown in Figure 60a, its cross-section in Figure 60b, and in Figure 60c

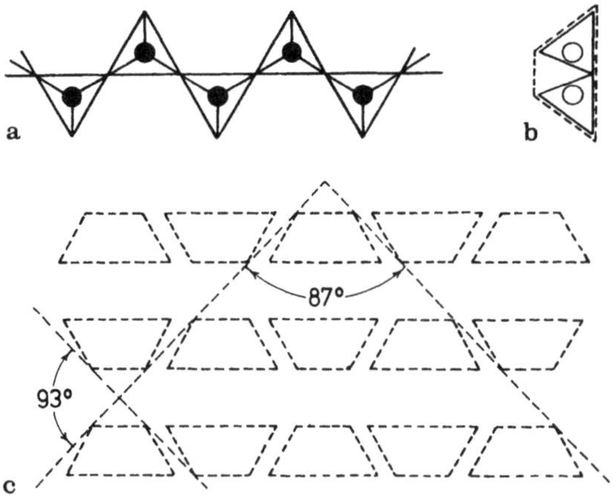

Figure 60. The arrangement of SiO_4 tetrahedra in the chain silicate: **a** simple chain, **b** cross-section of the chain, **c** array of chains in the crystal. Dotted lines represent the clevage planes

the cross-section of the array of parallel chains in the crystal. The cleavage planes are marked by dotted lines. The way in which SiO_4 tetrahedra are linked together to form three- and six-membered rings is illustrated in Figure 61. Simple ionic chains may link together through free vertices to form double chains of composition ${}^{1}_{\infty}[Si_9O_{11}]^{6-}$, in which the Si/O ratio amounts to 1/2.75 (Figure 62).

Linkage of a large number of chains through vertices lying in one plane results in the formation of a layer structure, shown in Figure 63a. In this structure each tetrahedron shares three vertices with other tetrahedra, the Si/O ratio being thus 1/2.5. The structure may be thus described as composed of ${}^{2}_{\infty}[Si_2O_5]^{2-}$ units, the number 2 in the left superscript indicating the bidimensional sheet. Three main classes of layer structures may be distinguished according to the structure of the layer. The first class comprises structures composed of single layers, in which the free vertices of tetrahedra in alternate

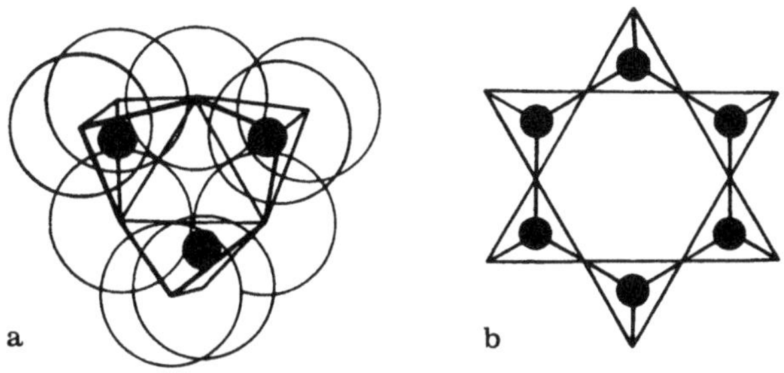

Figure 61. The arrangement of SiO_4 tetrahedra in rings: **a** three-membered, **b** six-membered

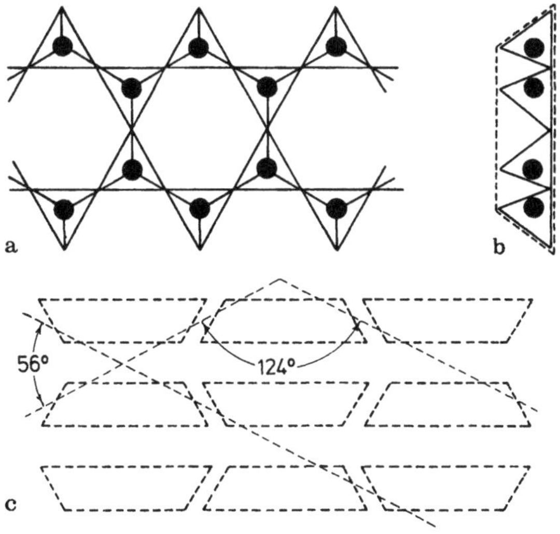

Figure 62. The arrangement of SiO_4 tetrahedra in the double-chain silicate: **a** double-chain, **b** cross-section of the double chain, **c** array of double chains in the crystal. Dotted lines represent the cleavage planes

rings point to opposite sides of the layer, the cations lying between the layers. In the structures of the second class represented by many aluminosilicates the layers are built of tetrahedra pointing with free vertices to one side and linked together in pairs to form double layers. If part of Si^{+4} ions are replaced by Al^{+3} ions, the net negative charge appears on the aluminosilicate layer and is compensated by interlayer cations, which hold together the layers in the crystal (Figure 63b). The third class of structures are composed of sheets of SiO_4 tetrahedra with vertices pointing to one side and linked to sheets of octahedra of cations such as Mg^{+2}, Ca^{+2}, Al^{+3} etc., coordinated by oxygen ions or hydroxyl groups to form a composite layer. The octahedral sheet may also be linked on both sides to a tetrahedral sheet and form a three-sheet composite layer. Such composite layers are neutral and are held together

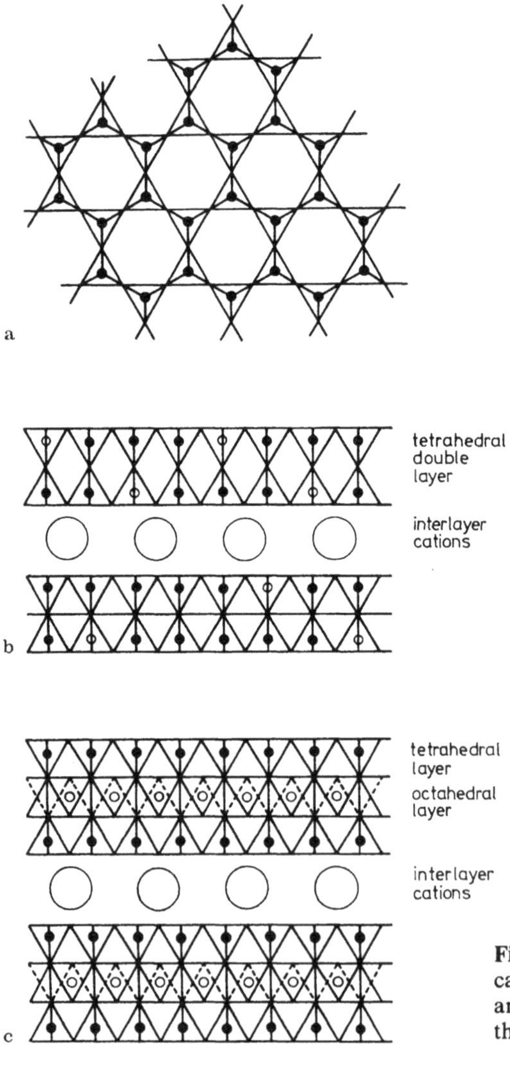

Figure 63. The structure of the layer silicate: **a** single layer, **b** cross-section of the array of double-layers, **c** cross-section of the array of three-sheet composite layers

in the crystal by weak van der Waals forces. This is the reason for the softness of such structures which are present, for instance, in talc. If a certain number of ions, for instance Al^{+3} ions in the octahedral sheet, is substituted by divalent ions the composite layer acquires a charge which must be compensated by additional cations incorporated between the layers (Figure 63c). In some structures the charged composite layers are interleaved with hydrated ions, as in montmorillonite, $[Mg_{0.33}Al_{1.67}Si_4O_{10}(OH)_2]^{-0.33}Na_{0.33}$ shown in Figure 64.

When all four oxygen atoms of the SiO_4 tetrahedron are shared by two tetrahedra, a 3-dimensional silicon-oxygen framework is formed, as in all polymorphic modifications of SiO_2 and such minerals as felspars and zeolites. In all these structures except SiO_2 some of the tetrahedral Si^{+4} ions are substituted by Al^{+3} ions so that other positive ions must be present to neutralize the negative charge of the framework.

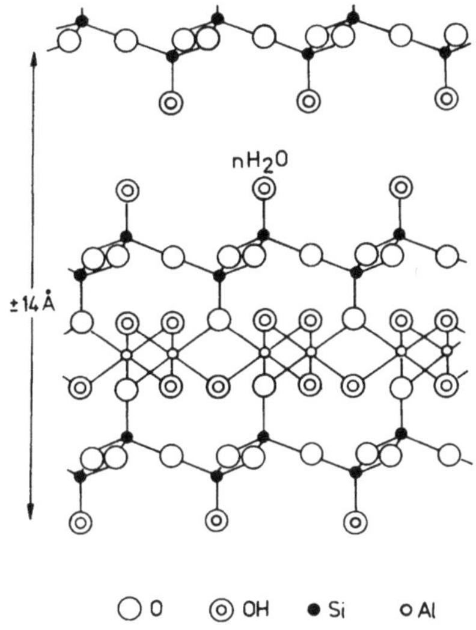

Figure 64. The structure of montmorillonite

Table 11. Structures of silicates

Name	Type of complex anion	Number of shared oxygen ions	O/Si ratio
insular silicates	$[SiO_4]^{4-}$	0	4/1
pyrosilicates	$[Si_2O_7]^{6-}$	1	3.5/1
sixfold (threefold) rings	$[Si_6O_{18}]^{12-}$ $[Si_3O_9]^{6-}$	2	3/1
chain silicates	${}^1_\infty[SiO_3]^{2-}$	2	3/1
band silicates	${}^1_\infty[Si_4O_{11}]^{6-}$	$2\frac{1}{2}$	2.75/1
layer silicates	${}^2_\infty[Si_2O_5]^{2-}$	3	2.5/1
skeleton silicates	${}^3_\infty[SiO_2]$	4	2/1

The substitution of $Si^{+4}O_4$ tetrahedra by $Al^{+3}O_4$ tetrahedra is the fundamental factor responsible for the catalytic properties of aluminosilicates, because it generates acid centers at their surfaces. An Al atom has only 3 valence electrons as compared to 4 electrons in silicon, and on substituting the latter in the tetrahedron it must attract an additional electron, assuming a negative charge. This can be illustrated by a following scheme of the formation of the aluminosilicate lattice:

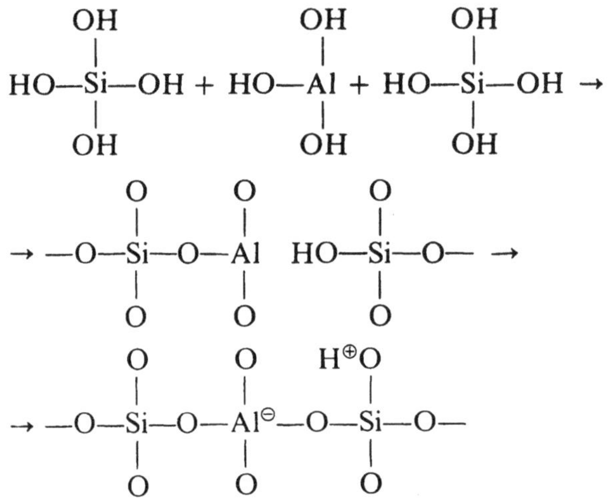

(A)

A center containing a loosely bonded proton is thus formed, playing the role of a Brönsted acid center. On outgassing at higher temperatures the surface becomes dehydroxylated and the Brönsted center is transformed into a Lewis center.

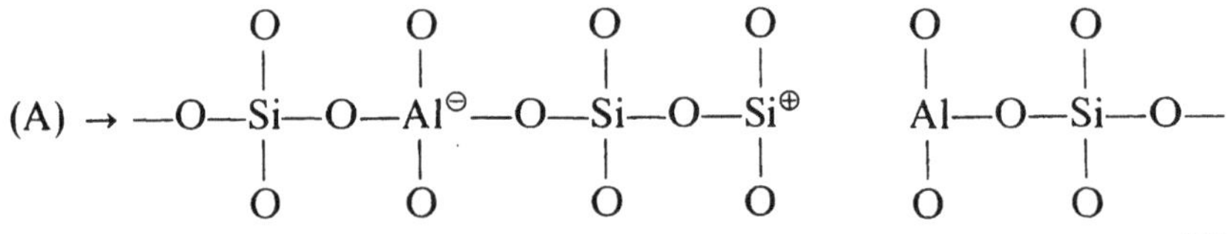

(B)

characterized by a high deficiency of electrons.

All the described structures of silicates are summarized in Table 11.

D. Zeolites

In crystalline aluminosilicates built of SiO_4 tetrahedra partially substituted by AlO_4 tetrahedra and linked through all four vertices into a 3-dimensional framework, one can distinguish definite building units such as chains, rings and regular polyhedra [65], some examples of which are shown in Figure 65. These building units, listed in Table 12, are the basis of the classification of zeolites [66, 67]. From the catalytic point of view two families of zeolites, sodalite and chabazite groups, have attracted most interest. The sodalite group of zeolites is built by linking together in different ways the cubocta-

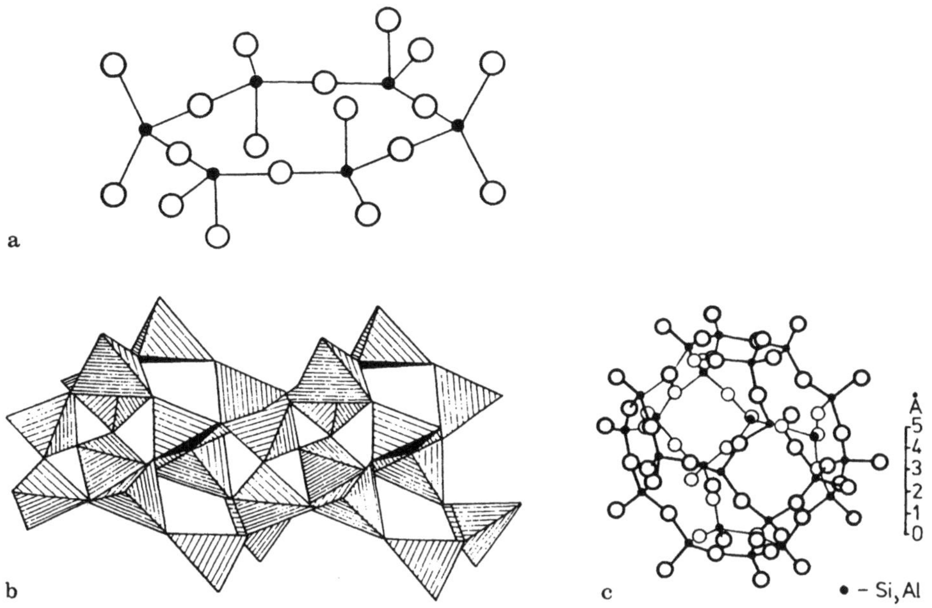

Figure 65. Building units of zeolites: **a** six membered ring; **b** Mordenite chain; **c** cubo octahedron

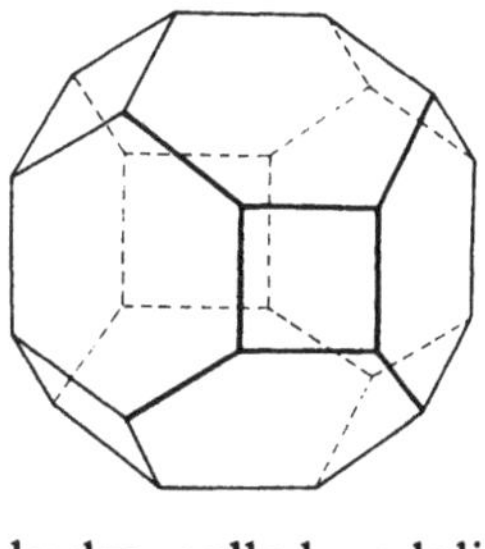

Figure 66. The sodalite unit

hedra called sodalite units. These are truncated octahedra, built of SiO_4 tetrahedra as primary units (Figure 65c). They can be visualized as obtained by superposition of an octahedron and a cube, and cutting away all the vertices of the octahedron (Figure 66). If the sodalite units are linked together through vertices of tetrahedra sticking out of square planes of cubooctahedra in a simple cubic manner, the structure of zeolite A is formed (Figure 67a). This produces an internal cavity (supercage) of the shape of a 26 sided truncated cubooctahedron, having a diameter of 11.4 Å. Access to this supercage is secured through six eight-membered windows of a diameter of about 4.2 Å. Thus, a zeolite A has a three dimensional networt of interconnected supercages, enabling sorption of molecules and their catalytic conversion.

When the sodalite units are linked together by vertices of the tetrahedra protruding from the hexagonal planes in a tetrahedral coordination (diamond-type structure), the structure of the mineral faujasite and its synthetic analogues, zeolite X and Y are produced (Figure 67b). It can be visualized as

Table 12. Some characteristic structural units in zeolites

Structural units	Description
S4R	single four-membered ring
S6R	single six-membered ring
S8R	single eight-membered ring
D4R	double four-membered ring, 6 hedron
D6R	double six-membered ring, 8-hedron
4–1	four-membered ring with one protrusion: five membered assemblage of the natrolite group
5–1	five-membered ring with one protrusion: six-membered assemblage of the mordenite group
4–4–1	nine-membered assemblage of the heulandite group
10–H (I)	10-hedron, type (I), of paulingite
10-H (II)	10-hedron, type (II), of gismondite
11-H	11-hedron, of cancrinite
14-H (I)	14-hedron, type (I), cubo-octahedron of sodalite
14-H (II)	14-hedron, type (II), of gmelinite
17-H	17-hedron of levynite
18-H	18-hedron of paulingite and ZK-5
20-H	20-hedron of chabazite
23-H	23-hedron of erionite
26-H (I)	26-hedron, type (I), truncated cubo-octahedron of A and ZK-5
26-H (II)	26-hedron, type (II), of faujasite

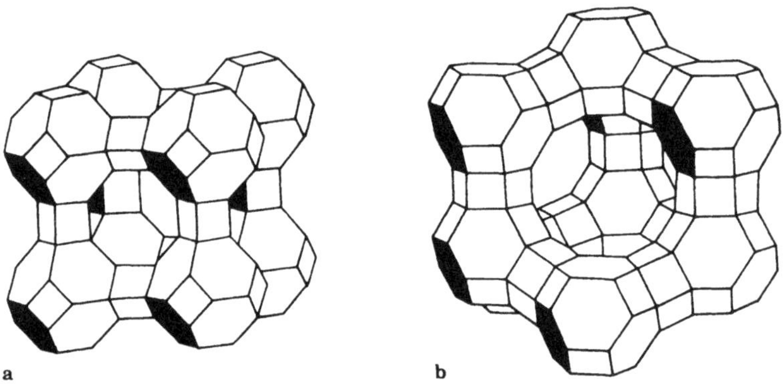

a b

Figure 67. The structure of **a** zeolite Linde A, **b** faujasite

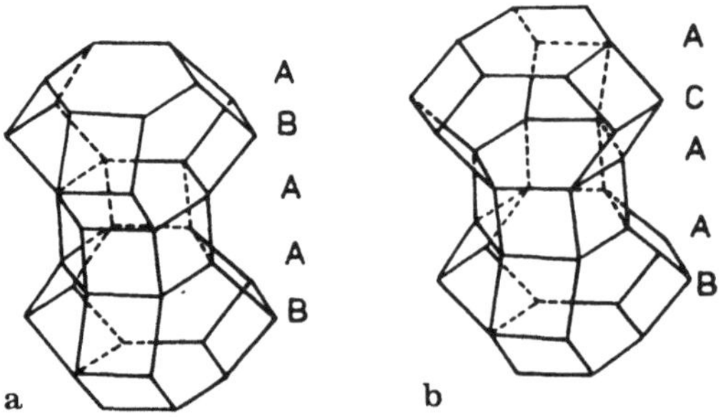

a b

Figure 68. Linkage of polyhedra in **a** offretite, **b** erionite [80] composed of cubooctahedra linked through hexagonal prisms. The resulting large, almost spherical, supercages have a diameter of about 12 Å, and are interconnected through distorted twelve-membered windows of diameter 8 to 9 Å.

The zeolites of chabasite can be considered as constructed from six-membered rings as basic building units, linked into hexagonal sheets, stacked in different sequences. If these six membered rings are denoted by A, B and C the zeolites of this group can be represented by the following stacking sequences (Figure 68):

AAB	offretite
AABB	guelinite
AABBCC	chabasite
AABAAC	erionite

The resulting structures have a one-dimensional system of chanels in the direction of the crystallographic c-axis (Figure 69).

The number of $Al^{+3}O_4$ tetrahedra which can be incorporated into the silicate framework varies in a wide range, and consequently so does the composition of the zeolites, expressed usually in terms of the Si/Al ratio. The highest possible amount of Al^{+3} ions is limited by Löwenstein's empirical rule that AlO_4 tetrahedra can only be linked to SiO_4 tetrahedra and not to each other. This places a lower limit of 1 on the Si/Al ratio, this value being encountered in Linde A zeolites. It varies from 1 in type X to 3 in type Y zeolites and amounts to 4.5–5.0 in mordenite. The upper limit appears to be very high, and recently synthesized zeolites of the ZSM type show the Si/Al ratio as high as 30 or more.

The negative charge introduced on the framework by each incorporated $Al^{+3}O_4$ tetrahedron must be compensated by a cation present within the framework. Thus, the number of these cations divided by their valence must be equal to the number of the $Al^{+3}O_4$ tetrahedra.

The compositions of the unit cells of some typical zeolites are:

Linde 4A	$Na_{12}Al_{12}Si_{12}O_{48}$, 28 H_2O
Linde 5A	$Na_6Ca_3Al_{12}Si_{12}O_{48}$, 30 H_2O
Faujasite	$(Na_2, Ca)_{32}Al_{64}Si_{128}O_{384}$, 25 H_2O
Si/Al = 2	
Mordenite	$Na_8Al_8Si_{40}O_{96}$, 24 H_2O

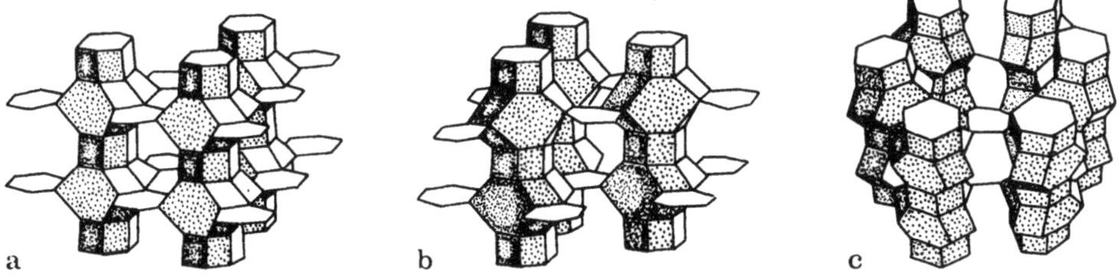

Figure 69. Framework of **a** offretite, **b** erionite, **c** Linde L [80]

The cations are exchangable and by properly selecting the type of cation replacing the Na ion, and the degree of exchange, one can modify the adsorptive and catalytic properties of the zeolite [68, 69]. These cations are not randomly distributed in the empty spaces of the aluminosilicate framework, but occupy certain defined positions [70]. As an example Figure 70 shows a sodalite unit with three hexagonal prisms forming part of the faujasite structure. There are four types of oxygen ions (oxygen ions are at the middle of the edges of the polyhedra) designated by 1, 2, 3 and 4. The cation sites are marked by Roman numerals. Site I is in the center of the hexagonal prism, site II lies opposite the center of the hexagonal face of the sodalite unit in the supercage (in the figure sites II are marked for the three hexagonal faces lying opposite the prisms, i.e. behind the plane of the figure). Inside the sodalite unit, proceeding towards its center from position I, site I′ is reached, and from II, site II′. Radially with II′ and II, site II^x is situated inside the supercage. These sites offer different coordinations to the metal cations and therefore they are occupied in different orders of preference depending on the number and type of cation and its affinity for a given coordination. Site I is generally the most stable, but the cation occupying it is practically inaccessible for reacting molecules and therefore shows no influence on catalytic activity. The most exposed for interaction with the reacting molecules are cations in positions II and II^x, being thus mainly responsible for the modification of catalytic properties of the zeolites.

There are two types of structural hydroxyl groups. The first one is formed by the hydrogen atom H1 linked to oxygen O1 towards the supercage (ir absorption band at 3650 cm^{-1}), while the second is composed of hydrogen H2 attached to oxygen O3 and situated inside the hexagonal prism (ir absorption at 3550 cm^{-1}, indicating a weaker bond and stronger acidity).

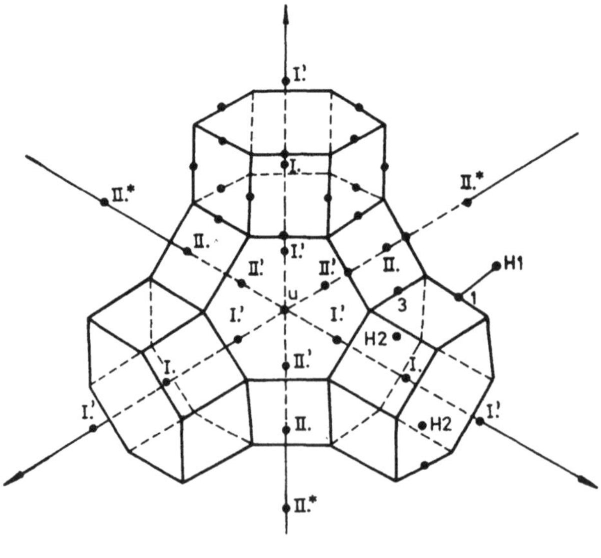

Figure 70. Positions of exchangable cations in faujasite

The properties of zeolites can also be modified by decationation, dehydroxylation and dealumination. Decationation consists in removal of cations from the zeolite by exchange with ammonium ions and their subsequent decomposition by thermal treatment

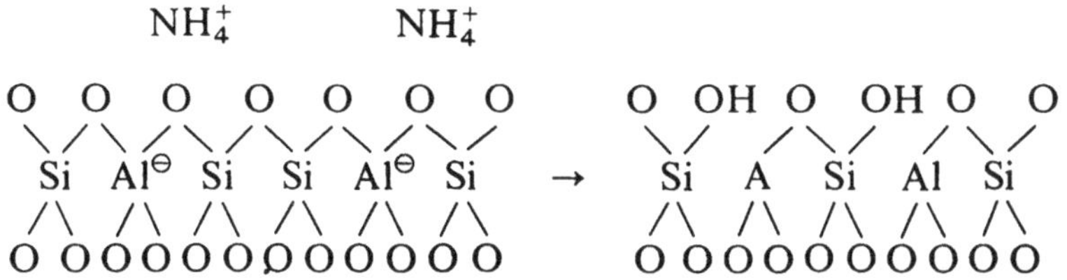

resulting in the formation of a Lewis acid center and a Brönsted acid center. Further heating of such hydrogen zeolite results in dehydroxylation, as shown by scheme (B).

Aluminium can be extracted from the framework of the zeolite by treating them with strong acids or aluminium complexing agents. This process can be visualized as substitution of an Al^{+3} ion by three protons and formation of neutral structural groups composed of 4 hydroxyl groups instead of polar $[Al^{+3}O_4]^{-}Me^{+}$ groups:

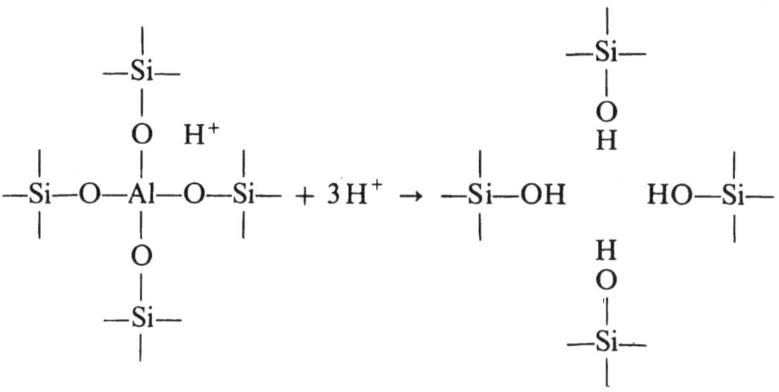

Further heat treatment leads to dehydroxylation of the nests of four hydroxyls and formation of new Si—O—Si bonds, accompanied by some disordering of the framework. The products have increased thermal stability.

Recently a new type of zeolites, called ZSM, was synthesized [71], of which the most important for catalysis is ZSM-5. It is built of five membered rings, eight of which form a building unit shown in Figure 71a. These units are linked through edges to form chains as shown in Figure 71b. The chains are connected into sheets, which then link to form the three dimensional framework. The sheet parallel to the (100) plane is shown in Figure 72 and the channel structure produced by stacking these sheets is presented schematically in Figure 73. This structure contains two intersecting channel systems, one sinusoidal and the other straight, which are believed to be responsible for shape selective catalysis. The unit cell of ZSM-5 has the formula $Na_nAl_nSi_{96-n}O_{192} \cdot 16\,H_2O$, where n is about 3. It has been shown [72, 73]

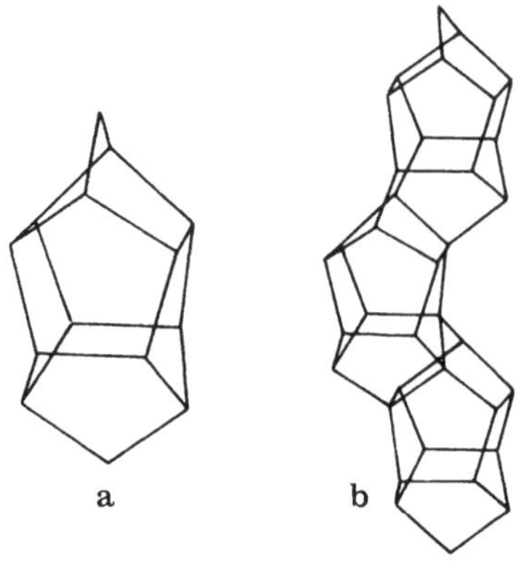

Figure 71. **a** Building unit and **b** its linkage into a chain in ZSM-5 zeolite

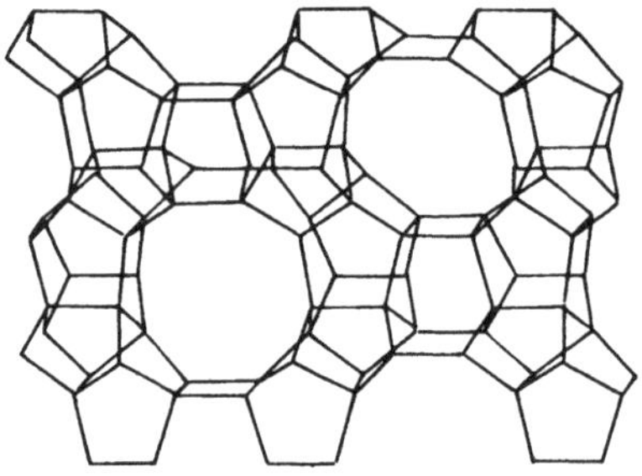

Figure 72. The (100) sheet of the ZSM-5

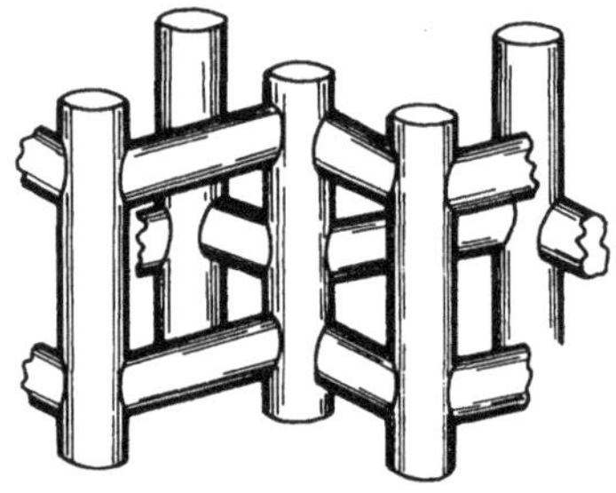

Figure 73. The channel structure in ZSM-5

that zeolite ZSM-5 has outstanding catalytic properties in selective conversion of methanol into high octane-number gasoline, as well as in other catalytic syntheses, isomerization and disproportionation reactions.

E. Phosphates

Last decade has brought increased interest in the application of phosphates as catalysts [74]. This is due to several reasons, one of them being the fact that the structural chemistry of phosphates, based on PO_4 tetrahedra, has much in common with that of silicates, based on SiO_4 tetrahedra. Similar to the way in which SiO_4 tetrahedra link together through vertices in different ways to form a whole variety of structures, PO_4 tetrahedra can link together to form chain and ring ions. However, at variance with the silicates, they cannot form a structure in which each PO_4 tetrahedron would share three vertices with other PO_4 tetrahedra, because it would be electrically neutral. Layer structures analogous to the layer silicates are therefore not possible. Chain or ring ions must be thus interconnected through polyhedra of other elements to form three-dimensional structures.

Of particular interest are phosphates of B, Al, Fe, Ga, and Mn, in which the trivalent metal atoms are also tetrahedrally coordinated. The mean valence of the metal and phosphorus being +4 the tetrahedra are joined together in an alternate way to form a continuous three-dimensional structure analogous to the polymorphic modifications of SiO_2.

$CePO_4$ and $BiPO_4$ appear in two modifications; hexagonal and monolinic. In the hexagonal structures the coordination of the metal is cubic and the oxygen atoms form channels running parallel to the hexagonal axis. This may be one of the reasons for their particular catalytic properties in reactions of dehydrogenation and dehydrocyclisation [75].

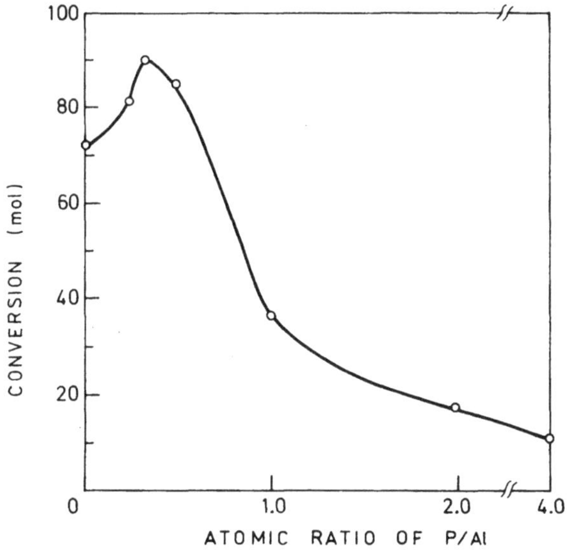

Figure 74. Conversion of *n*-butanol on aluminium phosphate catalysts as function of the P/Al ratio [75]

One of the features of metal phosphates which makes them specially interesting as prospective catalysts is the fact that their surface possesses a variety of different acid and basic sites. This was demonstrated by ir spectroscopic studies of the interaction of water with a boron phosphate surface, which revealed the existence of a number of different OH groups [76]:

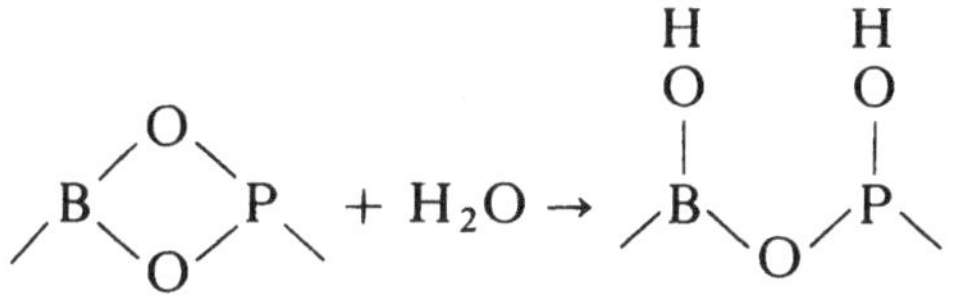

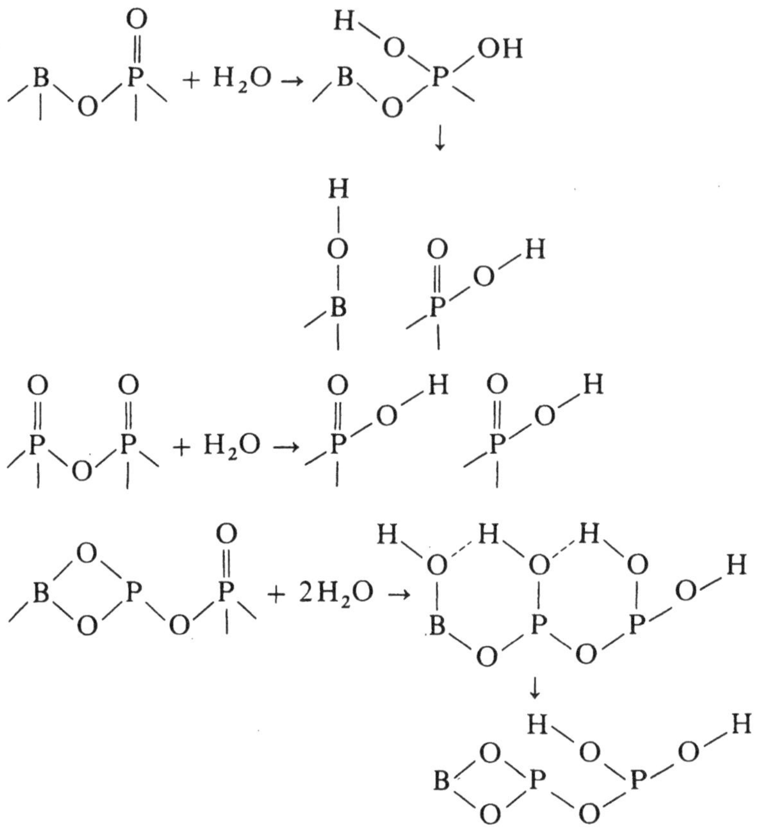

In most catalytic studies, phosphate catalysts with a P/M ratio different from the stoichiometric one turned out to be most active. Unfortunately, practically no information is available as to the structural basis of the non-stoichiometry in the metal oxide-phosphorus oxide systems. As an example Figure 74 shows the conversion of *n*-butanol in the reaction of dehydration on aluminium phosphate catalysts as function of the P/Al ratio [77]. It was recently pointed out [78] that an important role in elimination reactions may be played by basic centers. It is obvious that the ratio of acid to basic centers and their strength will strongly depend on the composition of the surface.

6. Surfaces Phases

A. Surface vs Bulk Composition

In all catalytic studies two fundamental questions have to be answered [79]:
(1) what is the true composition of the surface of a solid catalyst and how it may differ from the bulk. When trying to correlate catalytic activity with other physico-chemical properties some assumption has to be made

as to how the measured properties of the bulk such as x-ray diffraction pattern, chemical composition, electronic structure etc. reflect the state of the surface;

(2) how the reacting medium influences the structure of the catalyst surface and how the resulting changes influence the catalytic reaction itself. For a long time it has been realized that reactants and the catalyst constitute one dynamic system, and that it is expedient to measure the properties of the catalyst in the course of the catalytic reaction.

In recent years additional possibilities to answer these questions have opened, as the last decade has equipped surface chemistry with a number of new techniques [80–83] which yield detailed information on both the atomic and electronic structure of solid surfaces and their interaction with the adsorbed species.

It has been pointed out few years ago [84, 85] that the surface composition of alloys used in catalysis is in general different from the composition of the bulk. For instance, the so called "cherry like" model of an alloy crystallite has been formulated, in which the crystallite is envisaged as composed of the bulk of given composition, enveloped by a thin surface layer of different composition due to the contribution of the surface free energy to the condition of thermodynamic equilibrium.

In a two-component system, the difference Γ_1 between surface and bulk composition of component 1 is given by Gibbs equation:

$$(\partial\gamma/\partial n_1)_{T,V,A_s,n_2} = -\Gamma_1(\partial\mu_1/\partial n_1)_{T,V,A_s,n_2} \quad (44)$$

where γ is the surface tension; n_1 and n_2 the concentrations of components 1 and 2 respectively; μ_1 the chemical potential of component 1; and A_s the surface area. Thus, if the surface tension decreases, Γ_1 will increase with increasing concentration n_1, because $\partial\mu_1/\partial n_1$ is always positive. In the case of an ideal solution and at a temperature high enough for equilibrium to be attained

$$\mu_1 = \mu_1^0 + RT \ln x_1 \quad (45)$$

$$\gamma \approx x_1\gamma_1 + x_2\gamma_2 \quad (46)$$

where x_1 and x_2 are the concentrations, and γ_1 and γ_2 the surface tensions of components 1 and 2 respectively. From these equations it follows that

$$\Gamma_1 = (x_1/RT)(\gamma_2 - \gamma_1)\, aN \quad (47)$$

where a is the average surface area of the components, and N is Avogadro's number. In the case of limited miscibility of the components, segregation of a new surface phase may take place when the surface concentration surpasses a given value.

The problem of surface enrichment is of particular interest in the case of highly dispersed binary alloy catalysts, composed of microclusters of metal supported on suitable carriers. There are several possibilities for the microstructure of such systems. When the two constituents are immisible, separate microclusters of A and B on the carrier may be formed (Figure 75a).

The other limiting case involves constituents of complete miscibility, when microcrystals of single phase solid solution are expected (Figure 75b). There are then two possible microstructures with one component segregated to the surface: enrichment of one component in the surface layer with a nearly homogeneous alloy at the center of the microcluster (Figure 75c), and separation of the crystal into two concentric phases of different composition, one on the inside and one on the outside (Figure 75d).

Experimental data, obtained in recent years by ESCA, AES and work function measurements, supplemented by selective chemisorption, has already confirmed several general conclusions [15, 84]:

— in one phase alloys the surface tends to be enriched by the component with the lower sublimation heat (lower surface energy);
— the surface composition is a function of the composition of the gas phase in contact with the given alloy. The component with the higher heat of adsorption of gases present in the atmosphere tends to accumulate in the surface;
— when an alloy contains two phases in equilibrium, the alloy with the lower sublimation energy tends to form the outer surface.

As an example, Figure 76 shows results of the analysis of the surface composition of Cu—Ni alloy, obtained by Auger spectroscopy [86]. It may be seen that the difference between surface and bulk composition is so great that the first surface atomic layer of a 50 percent copper-nickel alloy contains almost 95 percent of copper. The enrichment rapidly decreases with depth and the 4th layer has already a composition practically identical with the bulk composition.

In the case of oxide catalysts it may also be expected that the composition of the surface will be different from that of the bulk [87]. Let us consider a nonstoichiometric transition metal oxide in which cation vacancies exist as the main type of imperfections. The equilibrium concentration of vacancies in this crystal is given by equation 4.

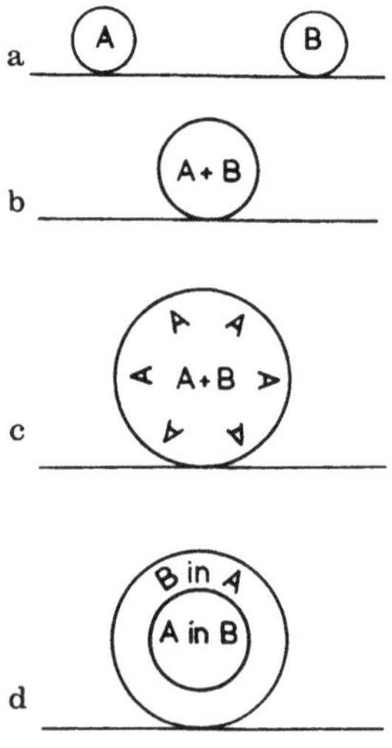

Figure 75. Possible microstructure of a highly dispersed binary alloy on a substrate: **a** separate microcrystals of A and B; **b** single phase mixed microcrystals of A + B; **c** microcrystal of single phase alloy A + B with surface enriched in A; **d** phase separated microcrystal of B in A and A in B solid solutions

On calculating the equilibrium concentration of vacancies in the surface layer, an additional term contributing to the value of ΔG must be taken into account due to the change of the surface free energy on the formation of vacancies:

$$\Delta G^{(s)} = G^{(s)}_{real} - G^{(s)}_{id} = n\,\Delta H - T(\Delta S_k + n\,\Delta S_{osc}) + n\left(\frac{\mathrm{d}\gamma}{\mathrm{d}n}\right) = 0 \quad (48)$$

Obviously, formation of a vacancy at the surface decreases the surface tension due to the relaxation of the lattice, the vacancies will thus accumulate at the surface and their concentration will be higher than in the bulk.

The second factor influencing the concentration of defects in the surface layer is the surface charge resulting from the interruption of the crystal periodicity at the surface (giving rise to Tamm levels), as well as the appearance of various local electron levels due to adsorbed species. They give rise to a planar surface charge compensated by a diffuse space charge inside the crystal resulting in an electric field. The concentration of defects in the crystal varies with the potential Φ of the electric field according to:

$$n = n_\infty \exp(-q\Phi/kT) \quad (49)$$

where n and n_∞ represent the concentration of defects at the surface and inside the crystal respectively, and Φ the electric potential. $\Phi = 0$ inside the crystal.

It may be thus concluded that due to the contribution of surface energy and surface charge the concentration of defects at the surface in the equilibrium conditions is different from the concentration in the bulk. However, equilibration of the gas phase with the bulk of crystallites takes place only on annealing at high temperatures, when the diffusion of defects becomes enough rapid. When an oxide which has been equilibriated at a high temperature in oxygen at a given pressure is then heated under a different pressure at a low temperature at which the diffusion of defects in the lattice is slow, the new equilibrium comprises only the surface layer. An additional gradient of the defect concentration appears, contributing also to the difference between the composition of the surface and the bulk. This contribution depends on the initial conditions of preparation and the conditions of subsequent treatment.

As an example let us discuss the situation in the case of cobalt oxides [87, 88]. Studies of the surface composition of different cobalt oxide phases by photoelectron spectroscopy and work function have shown that samples of CoO as well as Co_3O_4, depending on oxygen pressure in the gas phase, contain at the surface either Co^{+2} or Co^{+3} ions. This may be easily visualized on the basis of a structural model. Co_3O_4 contains Co^{+3} ions in one half of the octahedral sites and Co^{+2} ions in one eighth of the tetrahedral sites. On removal of oxygen by outgassing, the Co^{+3} ions are reduced and the Co^{+2} ions are shifted filling all octahedral sites and forming a surface structure analogous to CoO. The reverse process takes place on annealing in oxygen. It may be thus concluded that irrespective of the composition of the bulk a surface layer exists characterized by a composition which depends on the oxygen pressure in the gas phase.

Let us now discuss the situation in mixed oxides in which one of the components is in excess, for instance, transition metal oxides doped with altervalent ions. Depending on the properties of the dopant it may either accumulate at the surface or escape from the surface into the bulk. When the two oxides are capable of reacting, surface compounds may be formed which determine the adsorptive and catalytic properties of the solid. Figure 77 shows the values of the activation energy for surface diffusion of cations in CoO doped with Cr_2O_3, as calculated from measurements of the changes of work function [89]. It may be concluded that the surface behaves as that of CoO in samples containing less than about 0.4 at percent of chromium. After reaching this concentration a surface transformation takes place resulting in the formation of a surface layer of the $CoCr_2O_4$ spinel.

A similar phenomenon of surface reconstruction may also take place in the case when two component oxides are capable of forming several phases of different structure and composition.

In the Bi_2O_3—MoO_3 system three different phases, Bi_2MoO_6, $Bi_2Mo_2O_9$ and $Bi_2(MoO_4)_3$ are known to be active and selective catalysts for partial oxidation of olefins, but considerable differences of opinion exist as to the sequence of activities of these molybdates. The question may be raised as

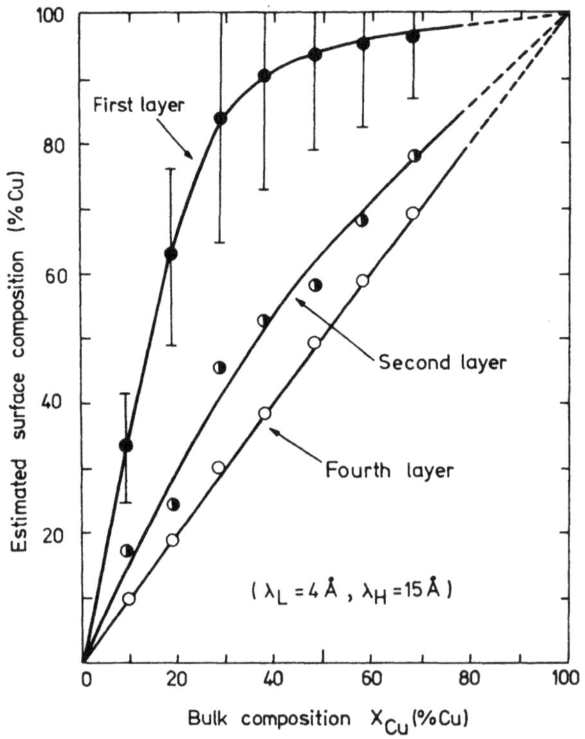

Figure 76. Surface composition of crystallites of Cu—Ni alloy as function of the bulk composition [86]

to whether it is appropriate to discuss the differences in catalytic properties of the three phases in terms of different bulk compositions which may not reflect the true composition of their surfaces. Photoelectron spectroscopy has been exploited to answer this question and studies of the spectra of bismuth molybdate catalysts have been carried out, and the changes occuring on vacuum-heat treatment as well as on contacting the catalyst with reactants in the oxidation reaction have been followed [90]. Figure 78 shows the relative intensities of the Bi($4f_{7/2}$) and Mo($3d_{5/2}$) peaks with respect to the O($1s$) peak, as a function of the Bi/Mo atomic ratio in the molybdates for: fresh samples; those outgassed at 393 K for 3 hr at 10^{-6} Pa (curves IA and IIA); samples which were heated at 743 K in vacuum for 10 hr (curve IB and IIB); and samples treated at 713 K for 5 hr in a flow reactor with the reaction mixture ($C_3H_6:O_2:N_2$ = 24:21:55 vol. percent).

Annealing in vacuum at high temperature leads to a considerable decrease in Bi intensity (curve IB) and to an increase in Mo intensity (curve IIB). This effect is most pronounced in the case of the Bi-rich phase Bi_2MoO_6. It may be thus concluded that on annealing in vacuum the surface layer of grains becomes enriched in molybdenum, whereas in the oxidizing atmosphere of the reaction mixture the composition of the surface is the same as that of a fresh sample. It may be also concluded that the surface composition of the three bismuth molybdates strongly depends on the redox properties of the gas phase. Two important comments emerge from these results. Firstly, as the surface composition depends on the composition of the gas phase, also the catalytic properties of the surface may depend on that parameter, that is, on the conditions of experiment. Secondly, as the various physical properties of the surface are usually determined in a reducing atmosphere of high vacuum, whereas adsorptive and catalytic activity is very often measured in an oxidizing atmosphere, the two groups of measurements refer to different

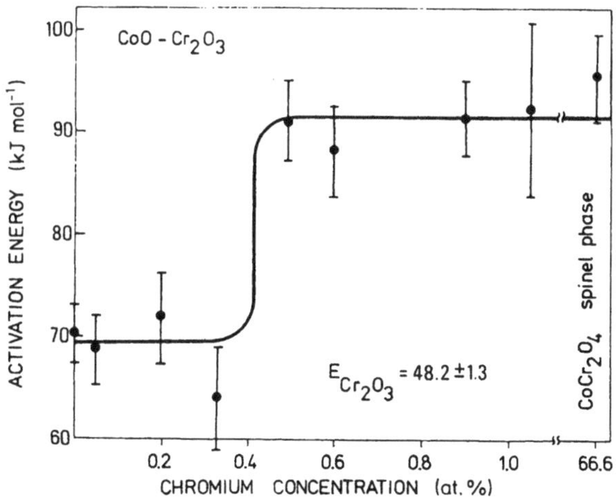

Figure 77. Activation energy of the surface diffusion of cations in the solid solutions of Cr_2O_3 in CoO as function of Cr_2O_3 [89]

states of the surface and correlation of the results should be treated with great caution.

This may be an important factor determining the catalytic properties of the molybdate catalysts as indicated by more recent results [91], in which it was shown that strictly stoichiometric molybdates are inactive in the oxidation of propylene to acrolein and the presence of excess MoO_3 is necessary to render them active.

B. Dynamics of Surface Reconstruction

From the discussion of the preceding paragraph it may be thus concluded that in the conditions of the catalytic reaction, the surface layer of an oxide catalyst will assume the composition which will depend on the composition of the reaction mixture, and will change when the composition of the mixture is changed. This may result in a number of different phenomena such as ordering of defects at the surface, surface reconstruction, and precipitation of a new bidimensional surface phase. They may entail the appearance of new types of active centers at the surface of the catalyst, directing the catalytic reaction into a new path and thus profoundly influencing the selectivity.

In order to demonstrate this effect we shall describe results of the oxidation of propylene at the surface of copper molybdate catalysts.

Figure 79 shows the results of an experiment [92], in which a mixture of propylene and oxygen of composition 1/1 was at first pulsed over the

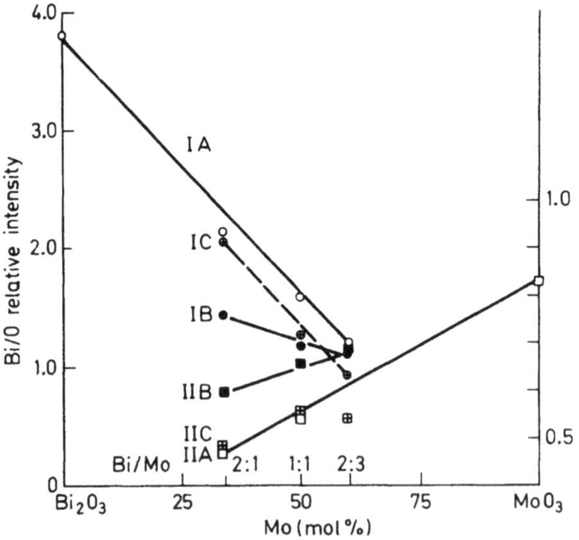

Figure 78. Relative intensities of Bi 4*f* peak (I) and Mo 3*d* peak (II) with respect to O 1*s* peak for bismuth molybdates; A initial samples, B after heating in vacuum at 743 K, C after treatment with reaction mixture ($C_3H_6 + O_2 + N_2$) at 713 K

$Cu^1_2Mo_3O_{10}$ catalyst. Total oxidation was observed with a conversion of about 25 percent. Then pulses of oxygen started to be introduced between the pulses of the reaction mixture. This resulted in a strong increase of the selectivity to acrolein, which after a certain number of pulses attained a new high level. When the pulses of oxygen were stopped, the selectivity to acrolein rapidly dropped back to a very low level. Such cycles have been repeated many times. It may be concluded that on exposing the catalyst to pulses of oxygen, reversible oxidation of the surface takes place resulting in the formation of clusters of new surface phases containing active centers responsible for the selctive oxidation. In order to check this conclusion, changes of the surface composition were followed by photoelectron spectroscopy [93]. The presence of a single peak at the binding energy of 932 eV in the spectrum of the initial sample (curve 1 in Figure 80) indicates that only Cu^{+1} ions in the environment of the $Cu_2Mo_3O_{10}$ lattice are present at the surface. After the oxidation pulses, that is in the state when a plateau at 20 percent selectivity to acrolein is attained, the photoelectron spectrum (curve 2) shows that now the surface is composed of two types of Cu ions, the ones being in a coordination similar to that in $CuMoO_4$, and the others to that in Cu_2O. Comparison with spectra 3 and 4, registered after subsequent reducing and oxidizing cycles, indicates that the reconstruction of the surface is reversible. It may be thus concluded that the surface transformation resulting in the drastic change of selectivity consists in the formation of surface clusters of Cu^{+1} and Cu^{+2} ions in new environments, the appearance of the clusters of $CuMoO_4$ being responsible for the insertion of oxygen and formation of acrolein. Simultaneously, the adsorption of oxygen and its activation leading to total oxidation is depressed.

The examples described here lead to a general conslusion [94], that depending on the properties of the mixture of reactants of the catalytic reaction,

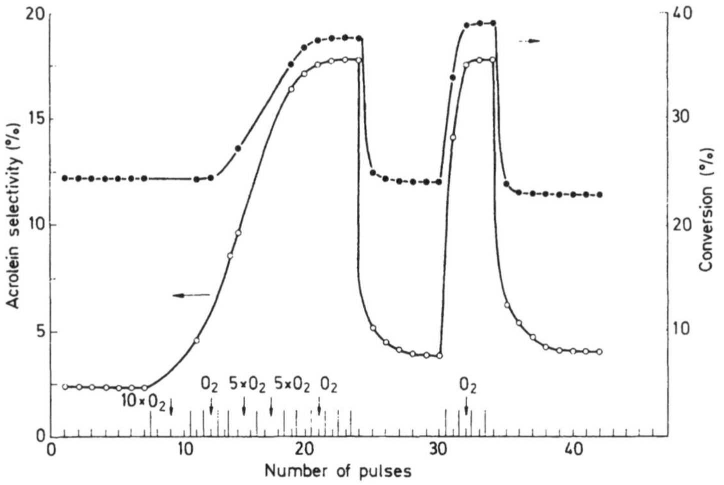

Figure 79. Conversion and selectivity of the oxidation of propylene to acrolein on $Cu_2Mo_3O_{10}$ catalyst as function of the number of $C_3H_6 + O_2$ and O_2 pulses. Reaction temperature 623 K

different surface phases are formed at the surface of the catalyst, directing the reaction along different reaction paths. The surface of the catalyst has a dynamic character and reconstructs depending on the properties of reactants of the catalytic reaction until it attains the structure and composition corresponding to the given steady-state conditions of the reaction. When these conditions are changed, the structure and composition of the catalyst surface changes also, modifying the activity and selectivity of the catalyst itself. This means that in the equation describing the rate of the reaction:

$$r = k\mathrm{f}(p_1 \ldots p_i) \tag{50}$$

it is not only the function f which depends on the pressures of the reactants, but also the rate constant k,

$$k = k(p_1 \ldots p_i)\,. \tag{51}$$

Thus, the conditions of the steady-state influence the catalytic reaction not only directly through the kinetic parameters, but also by modifying the properties of the catalyst.

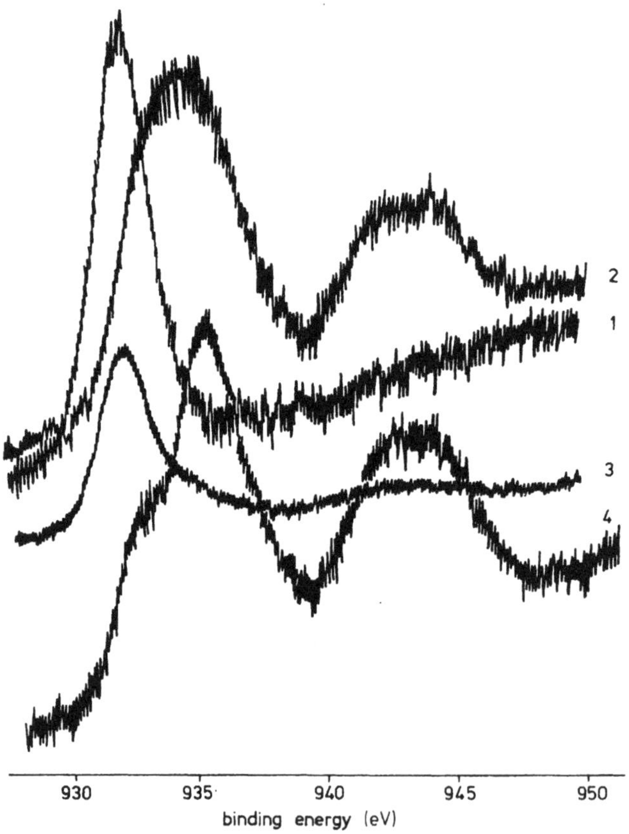

Figure 80. Photoelectron spectra of Cu 2p electrons in $Cu_2Mo_3O_{10}$ catalyst: **1** initial sample, **2** after exposure to O_2 pulses, **3** after outgassing at 623 K, **4** after reoxidation

References

1. Wells, A. F.: Structural Inorganic Chemistry, Oxford: Clarendon Press
2. Clark, G. M.: The Structures of Non-molecular Solids, London: Applied Science Publ. Ltd. 1972
3. Krebs, H.: Grundzüge der Anorganischen Kristallchemie, Stuttgart: Enke 1968
4. Naray-Szabo, I.: Inorganic Crystal Chemistry, Budapest: Academiai Kiado 1969
5. Haber, J.: Entwicklung neuer Vorstellungen in der Katalyse, Lecture to the Anniversary of W. Ostwald, Wiss. Hefte Univers. Leipzig 1980
6. Haber, J., Witko, M.: J. Molec. Catal. (in print)
7. Ozin, G. A.: Catal. Rev. **16**, 191 (1977)
8. Catalyst Handbook, Wolfe Scientific Books, London 1970
9. Mukhlyonov, I. et al.: Catalyst Technology, Moscow: MIR Publ. 1976
10. Preparation of Catalysts, (ed.) Delmon, B., Jacobs, P. A., Poncelet, G.: Elsevier 1976
11. Rogiński, S. Z.: Problemy Kinet. Katal., **8**, 110 (1955)
12. Boreskov, G. K.: Proc. 3rd Intern. Congr. Catalysis, Amsterdam 1964, Amsterdam: North-Holland Publ. 1965, vol. 1 p. 163.
13. Mady, T. E. et al.: Catalysis by Solid Surfaces, In: Treatise on Solid State Chemistry (ed. N. B. Hannay), Plenum Press, vol. 6B., 1976
14. Sinfelt, J. H.: Advan. Catal. **23**, 91 (1973)
15. Ponec, V.: Catal. Rev. **11**, 41 (1975)
16. Sachtler, W. M. H.: Catal. Rev., **14**, 193 (1976)
17. Sinfelt, J. H., Cusumano, J. A.: In: Adv. Materials in Catalysis, London, New York: Academic Press 1977, p. 1
18. a) Palczewska, W.: Advan. Catal. **24**, 245 (1974), b) Lisitchkin, G., Semienienko, K. H.: Izv. AN USSR, Nieorg. Mat. **14**, 1585 (1978.
19. Müller, J. M., Gault, F. G.: Bull. Soc. Chim. Fr., **2**, 416 (1970)
20. Levy, R., Boudart, M.: Science **181**, 547 (1973)
21. Levy, R.: In: Adv. Materials in Catalysis, London, New York: Academic Press 1977 p. 101
22. Venables, J. A., Price, G. L.: In: Epitaxial Growth, London, New York: Academic Press 1975, p. 381
23. Allpress, J. G., Sanders, J. V.: Austr. J. Phys. **23**, 23 (1970)
24. Burton, J. J.: Catal. Rev. **9**, 209 (1974)
25. Kimoto, K., Nishida, I.: J. Phys. Soc. Jap. **22**, 940 (1967)
26. Ogawa, S., Ino, S.: J. Cryst. Growth 13/14, 48 (1972)
27. Srinirasen, S., Wróblowa, H., Bockris, J. O'M.: Advan. Catal. **17**, (1967)
28. a) Anderson, R. B.: In: Catalysis (ed. Emmett, P. H.) Vol. 4, p. 1, Van Nostrand-Reinhold 1956; b) Cusumano, J. A., Dalla-Betta, R. A., Levy, R. B.: Catalysis in Coal Conversion, London, New York: Academic Press 1978.
29. Poraj-Koshitz, M. A., Atovmian, L. O.: Crystalchemistry and Stereochemistry of Coordination Compounds of Molybdenum (in russian), Moscow: Izd. Nauka 1974
30. Kepert, D. L.: The Early Transition Metals. London, New York: Academic Press 1972
31. Orgel, L. E.: Chemistry of Transition Metal Ions, Methuen 1960
32. Schmidke, S.: In: Physical Methods in Advanced Inorganic Chemistry, London 1968, p. 107
33. Figgis, B. N.: Introduction to Ligand Fields, New York: Interscience Publ. 1966
34. Boreskov, G. K.: Discuss. Faraday Soc. **41**, 263 (1966)
35. Greenwood, N. N.: Ionic Crystals, Lattice defects and Nonstoichiometry, London: Butterworths 1968
36. Kröger, F. A.: Chemistry of Imperfect Crystals, vol. 1—3, Amsterdam: North Holland/American Elsevier, 1973
37. Henderson, B. and Hughes, A. E. (ed.): Defects and Their Structure in Nonmetallic Solids, Plenum Press 1976
38. Jarzębski, Z. M.: Oxide Semiconductors, London: Pergamon Press 1973
39. Suchet, J. P.: Crystal Chemistry and Semiconduction, London, New York: Academic Press 1971

40. Brouwer, G.: Philips Res. Rep. 9, 366 (1954)
41. Bielański, A., Dereń, J.: In: Electronic Phenomena in Chemisorption and Catalysis on Semiconductors, Berlin: de Gruyter 1969, p. 149
42. Stone, F. S.: J. Solid State Chem. **12**, 271 (1975)
43. Cimino, A.: Chim. Ind. **56**, 27 (1974)
44. Dyrek, K.: Bull. Acad. Pol. Sci., Ser. Sci. Chim. **21**, 675 (1973)
45. Dyrek, K., Nowotny, J.: unpublished
46. Anderson, J. S.: Surface and Defect Properties of Solids (ed. Roberts, M. W., Thomas, J. M.): (Specialist Periodical Rep.) The Chemical Society, London 1972, vol. 1, p. 1
47. Anderson, J. S., Tilley, R. J. D.: ibid. vol. 3, p. 1
48. Magneli, A.: In: Chemistry of Extended Defects in Non-metallic Solids, Amsterdam: North Holland Publ. 1970, p. 148
49. Dziembaj, R.: J. Solid State Chem. **26**, 159 (1978)
50. Bielański, A., Haber, J.: Catal. Rev. **19**, 1 (1979)
51. Reed, T. B.: In: Chemistry of Extend defects in non-metallic solids, Amsterdam: North Holland Publ. 1970, p. 21
52. Haber, J.: In: Chemistry and Uses of Molybdenum, Proc. IIIrd Conf., Ann Arbor 1979
53. Haber, J., Janas, J., Schiavello, M., Tilley, R. J. D.: J. Catal. (in print)
54. Dunitz, J. D., Orgel, L. E.: Advan. Inorg. Chem. Radiochem. **2**, 1 (1960)
55. Trietiakov, J. D.: Chemistry of Nonstoichiometric oxides (in Russian), Moscow: Izd. MGU 1974
56. Krylov, O. V.: Catalysis by Non-metals, London, New York: Academic Press, 1970
57. Goodenough, J. B., Longo, J. M.: In: Landolt-Börnstein New Series, vol. 4, part a, p. 126–315. Berlin, Heidelberg, New York: Springer 1970
58. Voorhoeve, R. J. H.: In: Advanced Materials in Catalysis, London, New York: Academic Press 1977, p. 129
59. Tanabe, K.: Solid Acids and Bases, London, New York: Academic Press 1970
60. Lippens, B. C., Steggerda, J. J.: In: Physical and Chemical Aspects of Adsorbents and Catalysts, London, New York: Academic Press 1970, p. 171
61. Peri, J. B.: J. Phys. Chem. **69**, 211, 220, 231 (1965)
62. Knözinger, H.: Advan. Catal. **25**, 184 (1976)
63. Parkyns, N. D.: Proc. Intern. Congr. Catal. 5th, 1, 255 (1973)
64. Görlich, E.: Chemia Krzemianów, Warszawa: Wyd. Geologiczne 1957
65. Barrer, R. M.: Chem. Ind. (London). 1203 (1968)
66. Meier, W. M.: Molecular Sieves, Soc. Chem. Ind., London 1968, p. 10
67. Breck, D. W.: Zeolite Molecular Sieves, New York: J. Wiley 1974
68. Venuto, P. B., Landis, P. S.: Advan. Catal. **18**, 259 (1968)
69. Haynes, H. W. Jr.: Catal. Rev. **17**, 273 (1978)
70. Smith, J. V.: In: Molecular Sieve Zeolites, Advan. Chem. Ser. 101, p. 172
71. Kokotailo, G. T. et al.: Nature **272**, 437 (1978)
72. Meisel, S. L. et al.: Chem. Tech. (Berlin) **6**, 86 (1976)
73. Derouane, E. G. et al.: J. Catal. **53**, 40 (1978).
74. Moffat, J. B.: Catal. Rev. **18**, 199 (1978)
75. Sakamoto, T., Egashira, M., Seiyama, T.: J. Catal. **16**, 407 (1970).
76. Moffat, J., Necleman, J. F.: J. Catal. **34**, 376 (1974).
77. Tada, A., Yoshida, M., Hirai, M.: Nippon Kagaku Kaishi, 1379 (1973)
78. Noller, H., Kladnig, W.: Catal. Rev. **13**, 149 (1976)
79. Haber, J.: Proc. IV Intern. Symp. Heterogeneous Catalysis, Varna 1979
80. Methods of surface analysis (ed. Czanderna, A. W.) Amsterdam: Elsevier, 1975
81. Surface properties of materials (ed. Levenson, L. L.): Amsterdam: North Holland Publ. 1975
82. Characterization of Solid Surfaces (ed. Kane, P. F., Larrabee, G. B.): New York: Plenum Press, 1974
83. Adsorption at Solid surfaces, (ed. Brundle, C. R., Todd, C. J.): Surface Sci. **53** (1975)
84. Sachtler, W. M. H., van Santen, R. A.: Advan. Catal. **26**, 69 (1977)
85. Menon, P. G., Prasada-Rao, T. S. R.: Catal. Rev. **20**, 97 (1979)
86. Watanabe, R., Hashiba, M., Yamashira, T.: Surface Sci. **61**, 483 (1976)

87. Haber, J.: Contribution of modern physical techniques to understanding of oxidation catalysis, in Proc. Summer School, Theoretical and Practical Aspects of the oxidation catalysis, Lyon 1978
88. Haber, J.: Z. Chem. **16**, 421 (1976)
89. Dyrek, K., Nowotny, J.: in print
90. Grzybowska, B., et al.: J. Catal. **42**, 327 (1976)
91. Mazurkiewicz, A., Grzybowska, B.: unpublished
92. Haber, J., Wiltowski, T.: Bull. Acad. Pol. Sci., Ser. Sci. Chim. **27**, 785 (1979)
93. Haber, J., Stoch, J., Wiltowski, J.: React. Kinet. Catal. Letters **13**, 161 (1980)
94. Haber, J.: Kinet. Katal. **21**, 1 (1980)
95. Hauffe, K.: Reaktionen in und an festen Stoffen, Berlin, Heidelberg, New York: Springer 1966
96. Wolkenstein, Th.: The Electronic Theory of Catalysis on Semiconductors, Oxford: Pergamon Press 1963
97. Dunitz, J. D., Orgel, L. E.: Advan. Inorg. Chem. Radiochem. **2**, 1 (1960)
98. Mainwaring, D. E.: Proc. Roy. Aust. Chem. Inst. **40**, 163 (1973)

Chapter 3

Catalytic Kinetics: Modelling

G. F. Froment and L. H. Hosten

Laboratorium voor Petrochemische Techniek
Rijksuniversiteit Gent
Krijgslaan 271, B-9000 Gent, Belgium

Contents

1. Introduction and Scope

Kinetic studies are a powerful tool for a better understanding of catalysis. They also form the basis for a more accurate design of chemical reactors and for a better insight into the behavior of an existing reactor.

In kinetic investigations the form of the equation is rarely known a priori, although physico-chemical laws and several formalisms limit the spectrum of possible models. The selection on the basis of experimental data of the best rate equation among a set of rivals also called "model discrimination" will be one of the topics of this chapter. Also unknown, of course, are the values of the parameters. The problem of parameter estimation and of statistical testing of both the model and the parameter estimates will receive considerable attention in this chapter. Too often, indeed, experimental data obtained after painstaking efforts are analyzed by means of elementary if not naive methods, which do not make the most out of the data. On the other hand, powerful methods of data analysis will never compensate for the shortcomings of an experimental program. This is why the design of such a program for optimal discrimination and estimation is extensively dealt with in this chapter.

Even if the approach is rigorous and heavily relies upon computers, the emphasis is not so much on the mathematics as on the possibilities and performance of the various methods. These are illustrated by means of examples generally drawn from practical cases. The fast growing literature on the subject and the rather wide spectrum of techniques to be introduced made it impossible to be exhaustive. For further reading recent reviews on the subject by Bard and Lapidus [1], Kittrell [2], Mezaki and Happel [3], Reilly and Blau [4] and Froment [5, 6] are recommended.

2. Derivation of Rate Equations

In this text, the Hougen-Watson formalism, based upon the Langmuir adsorption isotherm, is adopted for the quantitative formulation of the kinetics of solid catalyzed chemical reactions. Although the underlying assumptions may not always be completely fulfilled it is generally accepted that this approach is the most suitable and reliable way of rationalizing observed catalytic rate data.

Consider the reversible catalytic dehydrogenation of ethanol into acetaldehyde and water:

$$A \rightleftharpoons R + S$$

This reaction may be decomposed into the following elementary steps, describing the interaction of the chemical species A, R, S and the active catalytic sites l.

Adsorption of reactant A:

$$1.\ A + l \rightleftharpoons Al \qquad K_1 = K_A = C_{Al}/p_A \cdot C_l \tag{1}$$

Surface reaction on dual sites:

$$2.\ Al + l \rightleftharpoons Rl + Sl \quad K_2 = C_{Rl} \cdot C_{Sl}/C_{Al} \cdot C_l \tag{2}$$

Desorption of products R and S:

$$3.\ Rl \rightleftharpoons R + l \qquad K_3 = 1/K_R = p_R \cdot C_l/C_{Rl} \tag{3}$$

$$4.\ Sl \rightleftharpoons S + l \qquad K_4 = 1/K_S = p_S \cdot C_l/C_{Sl} \tag{4}$$

Assuming mass action kinetics, the rates can be written:

$$r_1 = k_1(p_A \cdot C_l - C_{Al}/K_1) \tag{5}$$

$$r_2 = k_2(C_{Al} \cdot C_l - C_{Rl} \cdot C_{Sl}/K_2) \tag{6}$$

$$r_3 = k_3(C_{Rl} - p_R \cdot C_l/K_3) \tag{7}$$

$$r_4 = k_4(C_{Sl} - p_S \cdot C_l/K_4) \tag{8}$$

At steady state:

$$r_1 = r_2 = r_3 = r_4 = r_A \tag{9}$$

Equation 9 is a system of three equations which, in conjunction with a balance on the sites,

$$C_t = C_l + C_{Al} + C_{Rl} + C_{Sl} \tag{10}$$

may be solved, at least in principle, to yield the inaccessible concentrations C_{Al}, C_{Rl}, C_{Sl} and C_l. Substituting these into one of the equations 5–8 yields the desired rate equation as a function of observable quantities, i.e. the partial pressures in the gas phase.

Since one of the equations 9 is nonlinear in the unknown surface concentrations, solution is cumbersome and the resulting rate equation is complex and difficult to handle. Simplifications are therefore desirable. The two simplifying assumptions that are commonly made are the existence of a *rate determining step* (rds) and of a most *abundant surface intermediate* (masi) [7]. The principle of the rate determining or slowest step assumes that all driving forces but one in the rate equations 5–8 are almost zero. This implies that thermodynamic equilibrium is nearly established for those reactions and that use can be made of the corresponding equilibrium relations in (1)—(4) to eliminate the concentrations of adsorbed species. If the concentration of the masi is far superior to that of the other adsorbed species the contributions of the latter may be left out of the balance 10 and this again simplifies the rate equation.

As pointed out by Boudart [8] the assumptions of a rds and of a masi must be made with the proper care and it should not be forgotten that both can change with varying process conditions.

Examples of kinetic analyses in which the existence of a rds is not considered are given by Bischoff and Froment [9] and by Bradshaw and Davidson [10]. They also illustrate how difficult it is to detect changes in rds.

Assuming e.g. that the surface reaction 2 is potentially much slower than the adsorption and desorption steps, the overall rate is given by 6. Making use of the equilibrium relations with respect to adsorption and desorption to eliminate the concentrations of adsorbed species, and of the site balance 10 to eliminate the unknown concentration of unoccupied active sites, the following rate equation is obtained:

$$r_A = \frac{kK_A(p_A - p_R p_S/K)}{(1 + K_A p_A + K_R p_R + K_S p_S)^2} \tag{11}$$

K is the equilibrium constant of the overall reaction. K_A, K_R ... represent the adsorption equilibrium constants of species A, R ... and $k = k_2C_t$ is the global rate coefficient.

Similar equations are derived when the adsorption of A, respectively the desorption of R (or S) are rate determining:

$$r_A = \frac{k(p_A - p_R p_S/K)}{1 + K_A p_R p_S/K + K_R p_R + K_S p_S} \tag{12}$$

$$r_A = \frac{kKK_R(p_A - p_R p_S/K)}{p_S + K_A p_A p_S + KK_R p_A + K_S p_S^2} \tag{13}$$

In these equations, the rate coefficient and the adsorption equilibrium constants are unknown. These parameters have to be determined on the basis of experimental data.

3. Methodology of Kinetic Analysis

This section deals with the principles, the techniques, and some of the mathematical tools for estimating the parameters in kinetic models. Kinetic data are frequently collected in flow reactors. When the fractional conversion is very low the reactor is said to be operated in a differential way; if it is high in the integral way. The conservation equation for reactant A in an ideal tubular reactor may be written [11]:

$$F_{A_0} dx_A = r_A dW \tag{14}$$

Consequently the reaction rate is obtained by derivation of the conversion vs space time data:

$$r_A = \frac{dx_A}{d(W/F_{A_0})} \tag{15}$$

When the reactor is operated in a differential manner, conditions are essentially constant throughout the catalyst bed and the differential equation 15 simplifies into an algebraic equation:

$$r_A = \frac{x_A}{W/F_{A_0}} \tag{16}$$

Hence, reaction rates may be calculated in a straightforward way from very low conversion data.

Data obtained in a differential or integral reactor can be processed as shown in Figure 1. The differential method of kinetic analysis is based upon the reaction rates, whereas the integral method of kinetic analysis starts from the gas phase composition directly, as expressed by the conversion (s), partial pressures, etc.

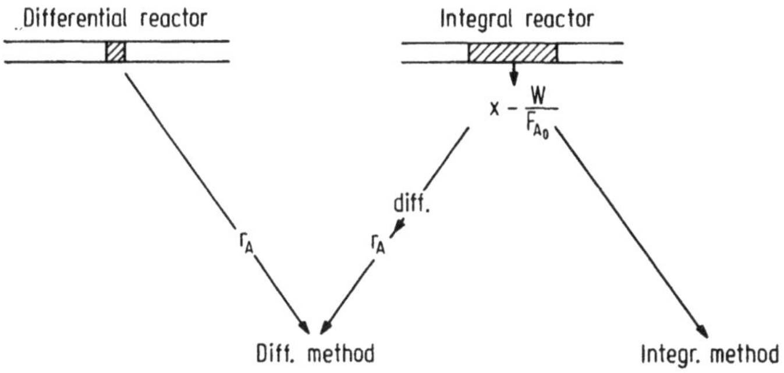

Figure 1. Possible ways of kinetic analysis

A. Single Equation Models

1. Differential Method of Kinetic Analysis

The principles are first outlined by means of the specific reaction $A \rightleftharpoons R + S$ for which rate equations were developed in the previous section. Subsequently these principles are applied to experimental data on the dehydrogenation of ethanol [12].

a) Discrimination and estimation based upon graphical representation of the data

Graphical representation of the data has always been the favorate method of model screening. At extremely low conversion of a feed containing no reaction products ($p_R = p_S = 0$). Equations 12, 11 and 13 reduce to:

$$r_{A_0} = kp_{A_0} \tag{17}$$

$$r_{A_0} = \frac{kK_A p_{A_0}}{(1 + K_A p_{A_0})^2} \tag{18}$$

$$r_{A_0} = k \tag{19}$$

From the graphical representation of these equations shown in Figure 2 it is obvious that observations of the initial rate of the feed component A as a function of the partial pressure is a powerful tool for revealing the rds and hence for the selection of the appropriate rate equation. This is the method of initial rates [13, 14] which appears to be one of the most powerful tools for discrimination between rival rate equations.

To extract quantitative information from plots, the equations often are rearranged into a linear dependence between variables. After rearranging the variables, equation 18 can be written:

$$\sqrt{\frac{p_{A_0}}{r_{A_0}}} = \frac{1}{\sqrt{kK_A}} + \frac{K_A}{\sqrt{kK_A}}\, p_{A_0} \tag{20}$$

Consequently, plotting $\sqrt{\frac{p_{A_0}}{r_{A_0}}}$ versus p_{A_0} should result in a straight line when the reaction rate equation is adequate. Values for $\frac{1}{\sqrt{kK_A}}$ and $\frac{K_A}{\sqrt{kK_A}}$ are determined from the intercept and the slope of the straight line and allow k and K_A to be calculated. The straight line can be drawn by eye. It is preferable, however, to use an objective method, like the *least squares* approach. Thereby the data are confronted with the model by minimizing the sum of the squares of the deviations between observed values and values predicted by the model.

This procedure is commonly called *regression.*

b) Parameter estimation by linear regression

In the same way as equation 18 is modified into equation 20, equation 11 is rearranged into

$$\sqrt{\frac{p_A - p_R \cdot p_S/K}{r_A}} = \frac{1}{\sqrt{kK_A}} + \frac{K_A}{\sqrt{kK_A}}\, p_A + \frac{K_R}{\sqrt{kk_A}}\, p_R + \frac{K_S}{\sqrt{kk_A}}\, p_S \tag{21}$$

Equation 21 is of the form

$$y = a + bp_A + cp_R + dp_S \tag{22}$$

which is linear in the four unknowns a, b, c and d.

The general linear model containing p parameters β is given by

$$y = \beta_1 x_1 + \beta_2 x_2 + \ldots + \beta_p x_p + \varepsilon \tag{23}$$

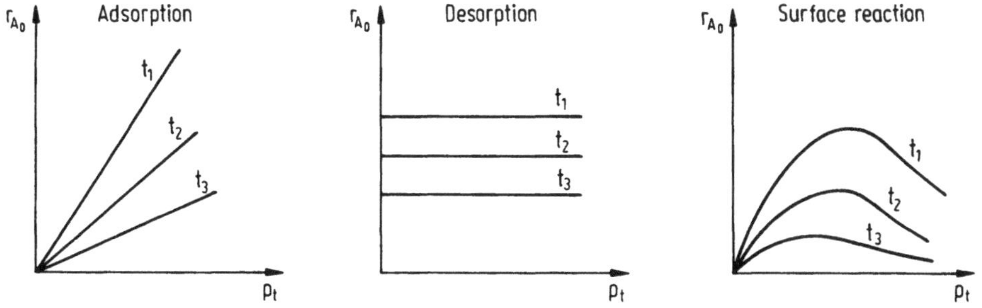

Figure 2. Initial rate behaviour vs total pressure

It is thus assumed that the independent variables, x, are set accurately at each desired level, and hence are known exactly, whereas the dependent variable y is subject to experimental error and hence belongs to the category of observable random or stochastic variables.

To obtain some idea of the magnitude of these parameters, observations of the dependent variable are made for a number of chosen settings of the independent variables.

When n observations of the dependent variable y have been taken, the following set of *observation equations* can be written:

$$\begin{aligned} y_1 &= \beta_1 x_{11} + \beta_2 x_{12} + \dots + \beta_p x_{1p} + \varepsilon_1 = \eta_1 + \varepsilon_1 \\ y_2 &= \beta_1 x_{21} + \beta_2 x_{22} + \dots + \beta_p x_{2p} + \varepsilon_2 = \eta_2 + \varepsilon_2 \\ &\vdots \\ y_n &= \beta_1 x_{n1} + \beta_2 x_{n2} + \dots + \beta_p x_{np} + \varepsilon_n = \eta_n + \varepsilon_n \end{aligned} \tag{24}$$

i. Estimation procedure

Excellent introductions to parameter estimation are given e.g. in the books by Draper and Smith [15], by Graybill [16] and by Himmelblau [17]. The least squares principle minimizes the sum of squares of the unobservable errors with respect to the parameter:

$$S(\boldsymbol{\beta}) = \sum_{i=1}^{n} \varepsilon_i^2 = \sum_{i=1}^{n} (y_i - \eta_i)^2 \xrightarrow{\beta_1, \beta_2, \dots, \beta_p} \text{Min} \tag{25}$$

For a concise notation, the following vectors and matrices are defined:

$$\mathbf{y} = \begin{bmatrix} y_1 \\ y_2 \\ \vdots \\ y_n \end{bmatrix} \qquad \mathbf{X} = \begin{bmatrix} x_{11} & x_{12} \cdots & x_{1p} \\ x_{21} & x_{22} \cdots & x_{2p} \\ \vdots & \vdots & \vdots \\ x_{n1} & x_{n2} & x_{np} \end{bmatrix} \tag{26}$$

$$\boldsymbol{\beta} = \begin{bmatrix} \beta_1 \\ \beta_2 \\ \vdots \\ \beta_p \end{bmatrix} \qquad \mathbf{b} = \begin{bmatrix} b_1 \\ b_2 \\ \vdots \\ b_p \end{bmatrix} \qquad \boldsymbol{\varepsilon} = \begin{bmatrix} \varepsilon_1 \\ \varepsilon_2 \\ \vdots \\ \varepsilon_n \end{bmatrix}$$

The observation equations 24 then read

$$\mathbf{y} = \mathbf{X}\boldsymbol{\beta} + \boldsymbol{\varepsilon} \tag{27}$$

and the least squares criterion

$$\boldsymbol{\varepsilon}^{\mathrm{T}}\boldsymbol{\varepsilon} \xrightarrow{\boldsymbol{\beta}} \mathrm{Min} \tag{28}$$

where T stands for matrix transpose. The solution of 28 is given by

$$\mathbf{b} = (\mathbf{X}^{\mathrm{T}}\mathbf{X})^{-1}\,\mathbf{X}^{\mathrm{T}}\mathbf{y} \tag{29}$$

To stress the fact that criterion 28 never yields the true parameter values $\boldsymbol{\beta}$ but estimates only, the symbol $\boldsymbol{\beta}$ has been replaced by **b**.

ii. Statistical properties and hypothesis testing

Not less important than the mathematical estimation problem itself is the statistical testing of the estimates. This involves the statistical properties of the experimental errors, $\boldsymbol{\varepsilon}$. Readers less familiar with statistical distribution theory, in particular with the normal, standardized normal, χ^2, t and F distributions will find excellent chapters on these topics in books by Bowker and Liebermann [18], Walpole and Myers [19], Lindgren and McElrath [20], Volk [21], among others, or in books dealing with statistics for experimentalists like Cooper [22], Green and Margerison [23] or some of the comprehensive summaries like the series by Andersen [24] in CE refresher or the chapter on statistical distributions in Ullman's Encyclopädie [25].

When the experimental error has zero mean, constant variance σ^2, and is independently distributed, it can be shown that **b** is an unbiased estimate of $\boldsymbol{\beta}$ and that the variance-covariance matrix of the estimates **b** is given by:

$$\mathbf{V}(\mathbf{b}) = (\mathbf{X}^{\mathrm{T}}\mathbf{X})^{-1}\,\sigma^2 \tag{30}$$

Further, when the model is adequate i.e. when there is no *lack of fit*, an unbiased estimate of the experimental error variance is given by the minimum residual sum of squares divided by its corresponding number of degrees of freedom:

$$s^2 = \frac{\sum_{i=1}^{n} (y_i - \hat{y}_i)^2}{n - p} \tag{31}$$

An estimate of **V**(**b**) is obtained by replacing σ^2 in 30 by s^2:

$$\hat{\mathbf{V}}(\mathbf{b}) = (\mathbf{X}^{\mathrm{T}}\mathbf{X})^{-1}\,s^2 \tag{32}$$

Error model

The theory for hypothesis testing is extensively developed for errors which are distributed according to the *Gaussian or normal distribution* only, so that attention will be restricted here to this distribution. Let the errors be normally distributed with zero mean and constant variance σ^2. Assume also that the errors are statistically independent. These properties are written:

$$\boldsymbol{\varepsilon} \sim N(\mathbf{0}, \mathbf{I}\sigma^2); \text{ i.e. } \quad \begin{aligned} E(\boldsymbol{\varepsilon}) &= \mathbf{0} \\ \mathbf{V}(\boldsymbol{\varepsilon}) &= \mathbf{I}\sigma^2 \end{aligned} \tag{33}$$

The likelihood function of the parameters, given the data, then reads:

$$L(\boldsymbol{\beta}|\mathbf{y}) = \frac{1}{(\sqrt{2\pi})^n \sigma^n} \exp\left\{-\frac{1}{2\sigma^2}\sum_{i=1}^{n}(y_i - \eta_i)^2\right\} \tag{34}$$

From equations 25 and 34 it follows that the maximum likelihood criterion and the least squares criterion are identical when the experimental errors obey 33.

When the errors are normally distributed with zero mean but do not possess constant variance and/or are interdependent, i.e.

$$\boldsymbol{\varepsilon} \sim N(\mathbf{0}, \mathbf{V}\sigma^2) \tag{35}$$

unweighted least squares is not appropriate and has to be replaced by weighted least squares:

$$S(\boldsymbol{\beta}) = \sum_{j=1}^{n}\sum_{i=1}^{n} v^{ij}(y_i - \eta_i)(y_j - \eta_j) \xrightarrow{\boldsymbol{\beta}} \text{Min} \tag{36}$$

in which:

$$\mathbf{V}^{-1} = \{v^{ij}\}$$

The maximum likelihood estimates are then given by

$$\mathbf{b} = (\mathbf{X}^T\mathbf{V}^{-1}\mathbf{X})^{-1}\mathbf{X}^T\mathbf{V}^{-1}\mathbf{y} \tag{37}$$

and the covariance matrix by

$$\mathbf{V}(\mathbf{b}) = (\mathbf{X}^T\mathbf{V}^{-1}\mathbf{X})^{-1}\sigma^2 \tag{38}$$

An estimate of the unknown proportionality factor σ^2 is obtained from

$$s^2 = \frac{S(\mathbf{b})}{n-p} = \frac{(\mathbf{y} - \mathbf{Xb})^T\mathbf{V}^{-1}(\mathbf{y} - \mathbf{Xb})}{n-p} \tag{39}$$

Statistical test of model adequacy

To test the statistical adequacy or lack of fit of the model, a number of replicated experiments, n_e say, under at least one set of experimental conditions has to be performed. An estimate of pure error variance is then calculated according to

$$s_e^2 = \frac{\text{Pure error sum of squares}}{\text{number of degrees of freedom}} = \frac{\sum_{j=1}^{n_e}(y_j - \bar{y})^2}{n_e - 1} \tag{40}$$

$\bar{y}$ represents the arithmetic mean of the observations of the n_e replicates. Figure 3 shows the partitioning of sums of squares upon which the adequacy test is based. The lack of fit sum of squares and the pure error sum of squares, divided by their respective number of degrees of freedom, are used in an F-test for lack of fit: if

$$F_c = \frac{\dfrac{\text{Lack of fit sum of squares}}{n - p - n_e + 1}}{\dfrac{\text{Pure error sum of squares}}{n_e - 1}} > F(n - p - n_e + 1, n_e - 1, 1 - \alpha) \tag{41}$$

there is a chance of $1 - \alpha$ that the model is inadequate. The model is rejected because of so-called lack of fit. $F(n - p - n_e + 1, n_e - 1, 1 - \alpha)$ is the α-percentage point of the F-distribution with $n - p - n_e + 1$ and $n_e - 1$ degrees of freedom. Common values for α are 0.05 and 0.01. When the calculated ratio does not exceed the tabulated F-value, no lack of fit is detected and the model is retained.

It is very important to recall that the tests described in what follows are meaningful only when the model exhibits no lack of fit.

Individual t-tests on estimates b_i

Under the assumptions 33, the random variable

$$\frac{|b_i - \beta_i|}{s(b_i)} \sim t(n - p) \tag{42}$$

with

$$s(b_i) = \sqrt{\hat{V}(\mathbf{b})_{ii}} \tag{43}$$

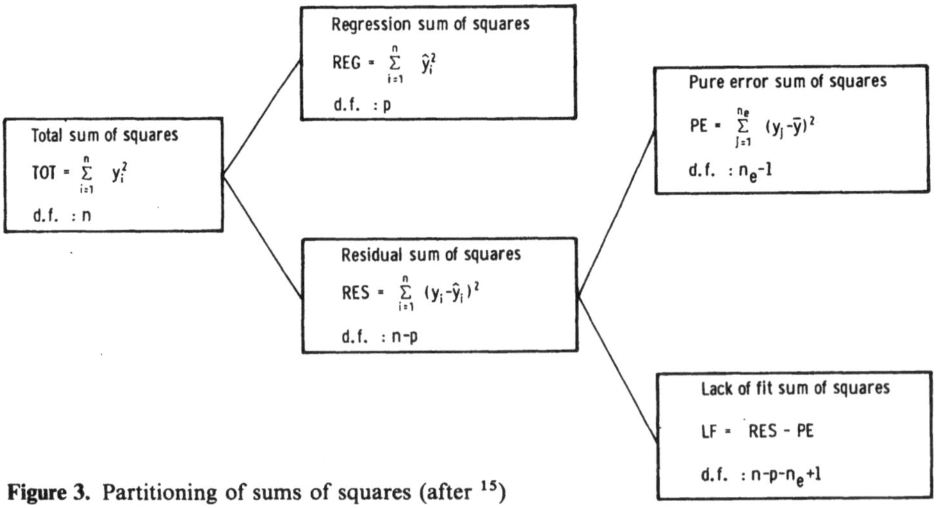

Figure 3. Partitioning of sums of squares (after [15])

is distributed like t with n — p degrees of freedom. This property is used to test the hypothesis

$$H_o : \beta_i = \bar{b}_i$$

provided all other parameters are kept constant at their optimal estimate. $\bar{b}_i$ is a some chosen reference value.

If

$$t_{calc} = \frac{|b_i - \bar{b}_i|}{s(b_i)} > t(n - p, 1 - \alpha/2) \qquad (44)$$

where $t(n - p, 1 - \alpha/2)$ is the $\alpha/2$-percentage point of the t-distribution with n — p degrees of freedom, the hypothesis H_o has to be rejected. Since equation 44 is a two-sided test, the chance of having rejected a correct hypothesis is α. In kinetic modeling, the reference value $\bar{b}_i$ is usually chosen to be zero. Common values for α are 0.05 and 0.01.

Individual confidence limits for the estimates b_i

Of greater interest than these tests against one single reference value $\bar{b}_i$ are the confidence limits. These delimit at once the whole collection of values $\bar{b}_i$ which are not significantly different from the optimal estimate b_i at the selected probability level $1 - \alpha$, provided all other parameters β_j, $j \neq i$, are kept constant upon their optimal estimate b_j. These intervals are defined by

$$b_i - t(n - p, 1 - \alpha/2) \cdot s(b_i) \leq \beta_i \leq b_i + t(n - p, 1 - \alpha/2) \cdot s(b_i) \qquad (45)$$

and are symmetric with respect to the optimal point estimate b_i.

Joint confidence region of the estimates b_i

The joint confidence region of the estimates b_i delimits the region of joint parameter uncertainty, accounting for the simultaneous variation of all the parameters. It is defined by

$$S(\boldsymbol{\beta}) \leq S(\mathbf{b})\,[1 + ps^2\, F(p, n - p, 1 - \alpha)] \qquad (46)$$

where $S(\mathbf{b})$ is the minimum residual sum of squares and $F(p, n - p, 1 - \alpha)$ the α-percentage point of the F distribution with p and n — p degrees of freedom. It can be proved that the boundary of the joint confidence region is defined by all combinations $\boldsymbol{\beta}$ which satisfy

$$(\mathbf{b} - \boldsymbol{\beta})^T\, \mathbf{X}^T\mathbf{X}(\mathbf{b} - \boldsymbol{\beta}) = ps^2\, F(p, n - p, 1 - \alpha) \qquad (47)$$

Equation 47 represents a p-dimensional hyperellipsoid in parameter space, centered at **b**. All parameter combinations enclosed by the ellipsoidal surface do not deviate significantly from the maximum likelihood estimates **b** at the probability level of $1 - \alpha$.

Example (3.A.1.b): *Dehydrogenation of ethanol*

Franckaerts and Froment [12] studied the dehydrogenation of ethanol in an integral plug flow reactor. They report the conversion of ethanol x_A

as a function of W/F_{A0}, at various total pressures and temperatures. A sample of such data is shown in Figure 4. The reaction rates are obtained from the derivatives of the x_A vs W/F_{A0} data in accordance with equation 15. The derivatives were computed numerically means of Newton's interpolation formula.

An r_{A0}/p_{A0} plot is shown in Figure 5. It suggests that the surface reaction on dual sites is the rate determining step. This is confirmed by plotting $\sqrt{p_{A0}/r_{A0}}$ vs. p_{A0} as in Figure 6. Straight lines are obtained so that the initial rate is adequately described by equation 18.

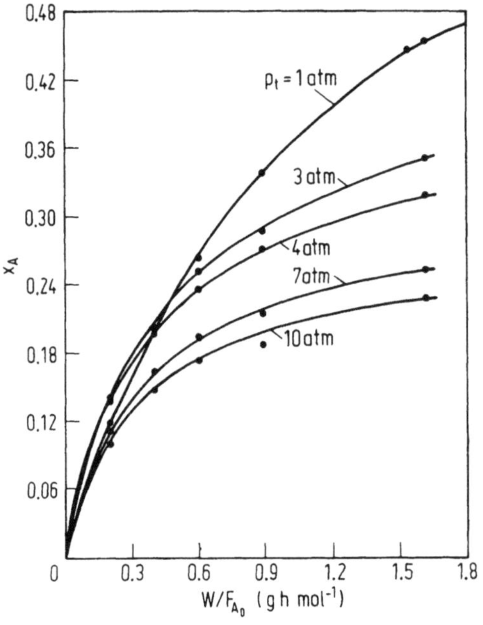

Figure 4. Ethanol dehydrogenation. Sample of data

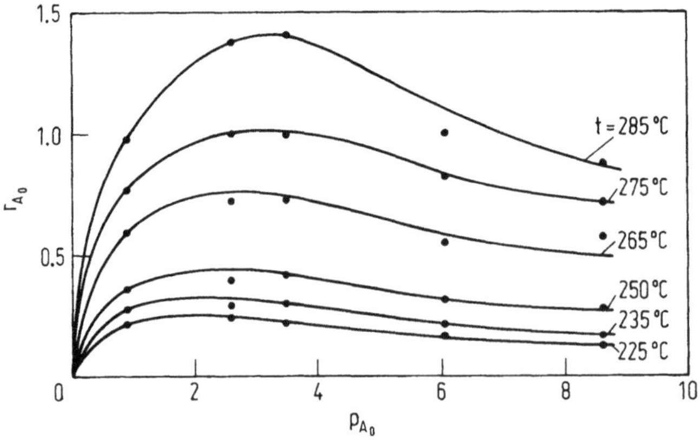

Figure 5. Ethanol dehydrogenation: initial rate vs p_{A_0} behaviour

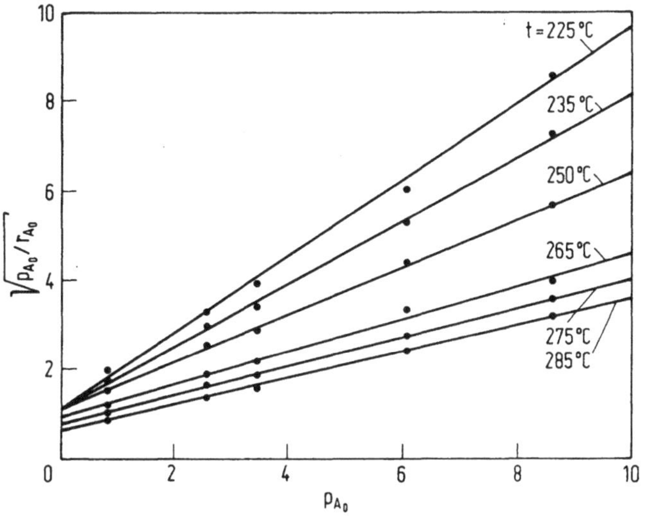

Figure 6. Ethanol dehydrogenation: $\sqrt{r_{A_0}/p_{A_0}}$ vs p_{A_0} plot

Since the ethanol feed was the binary azeotrope with water, a term $K_W p_W$, expressing the eventual adsorption of water on the catalyst has to be included in the denominator of equation 11. The equivalent of equation 21 now contains a fifth term, ep_W, with $e = K_W/\sqrt{kK_A}$. Before proceeding to the estimation of the parameters in equation 21, care should be taken to check the non-singularity of the matrix $\mathbf{X}^T\mathbf{X}$, in other words to check if the rank of **X** is p. This implies that **X** has no linearly dependent columns and that there are at least p linearly independent rows. The second condition means that at least as many experiments at different conditions have to be performed as there are parameters to be estimated and is rather trivial. The first condition may be easily overlooked, however. For the reaction $A \rightleftharpoons R + S$, e.g. p_R will always equal p_S when only pure A (or the azeotropic mixture ethanol-water) is fed to the reactor, and the third and fourth columns in **X** will be identical. This only allows the sum $K_R + K_S$ to be estimated and not the constants individually.

In the present case and neglecting for a moment the term $K_W p_W$, an inspection of the reaction stoichiometry and of the rate equation easily reveals the

Table 1. Ethanol dehydrogenation. Potential for parameter estimation of various possible data sets

	Pure Feed (A)	Mixed Feed (A + R)
Data at $W/F_{A_0} = 0$ only	k, K_A	k, K_A, K_R
Data including $W/F_{A_0} \neq 0$	k, K_A, $K_R + K_S$	k, K_A, K_R, K_S

potential for parameter estimation of various possible data sets. These are shown in Table 1.

If the molar ratio acetaldehyde/ethanol in the feed is represented by μ and the molar ratio water/ethanol by γ (= 0.155 for the azeotropic mixture), the partial pressures of the reaction partners can be expressed in terms of the conversion x_A by means of the following relations:

$$p_A = \frac{1 - x_A}{1 + \mu + x_A + \gamma}; \quad p_R = \frac{x_A + \mu}{1 + \mu + x_A + \gamma};$$
$$p_S = \frac{x_A}{1 + \mu + x_A + \gamma}; \quad p_W = \frac{\gamma}{1 + \mu + x_A + \gamma} \tag{48}$$

Each experimental data point therefore yields a set of values of the independent variables p_A, p_R, p_S and p_W and a value of the "observed" dependent variable $\sqrt{\frac{p_A - p_R \cdot p_S/K}{r_A}}$.

For a feed containing various amounts of A, water and the reaction product R the estimation based upon initial rate data was achieved by means of linear regression:

$$\sum\left(\sqrt{\frac{p_{A0}}{r_{A0}}} - \widehat{\sqrt{\frac{p_{A0}}{r_{A0}}}}\right)^2 \xrightarrow{a,\,b,\,c,\,e} \text{Min}$$

Estimates for a, b, c and e and their 95% confidence limits are given in the upper part of Table 2. Clearly, in none of the cases the parameter e is significantly different from zero, at the 95% probab:lity level. The physical significance of this phenomenon is probably that water is not significantly adsorbed on the catalyst surface so that the corresponding term $K_W p_W$ in the denominator may be deleted without loosing a contributing parameter. After deletion of one of the parameters those remaining in the simplified rate equation have to be re-estimated. Deletion of a non significant parameter should not significantly modify the remaining parameters, however.

Table 2. Ethanol dehydrogenation. Parameter estimation in equation 21 by means of the differential method of kinetic analysis and linear regression

T (°C)	a ± t.s (a)	b ± t.s (b)	c ± t.s (c)	e ± t.s (e)	d ± t.s (d)
225	1.287 ± 0.40	1.239 ± 1.18	6.433 ± 4.17	−2.808 ± 7.59	
250	1.037 ± 0.2	0.617 ± 0.58	2.953 ± 2.06	−0.529 ± 3.75	
275	0.759 ± 0.09	0.504 ± 0.25	1.82 ± 0.89	−1.191 ± 1.62	
225	1.167 ± 0.55	0.856 ± 0.14	8.501 ± 4.53		0.891 ± 4.88
250	1.223 ± 0.21	0.488 ± 0.06	2.745 ± 1.85		0.598 ± 1.77
275	0.798 ± 0.08	0.291 ± 0.02	1.88 ± 0.76		0.309 ± 0.73

This is shown in the second part of Table 2 which presents the results based on

$$\sum\left[\sqrt{\frac{p_A - p_R p_S/K}{r_A}} - \widehat{\sqrt{\frac{p_A - p_R p_S/K}{r_A}}}\right]^2 \xrightarrow{a,b,c,d} \text{Min} \tag{49}$$

on all data, after deletion of the $K_W p_W$ term.

c) Parameter estimation by non-linear regression

i. Estimation procedure

The procedure discussed and illustrated above estimates the parameters in a rather simple way, at the expense, however, of mathematical and particularly statistical rigour. The parameters obtained in this way cannot be claimed to be the best possible ones, guaranteeing that the predicted rates match as closely as possible the observed rates. Indeed, they do not minimize the sum of squares of deviations associated with the true reaction rate, but instead the deviations with respect to some composite function of both dependent and independent variables. Further, the rearrangement which turns the true reaction rate equation 11 into 21 does not preserve the error properties and this is an even more serious objection.

Hence, even when the experimental error on the rate meets the statistical requirements 33, this will not be so for the error with respect to the group $\sqrt{\frac{p_A - p_R \cdot p_S/K}{r_A}}$.

The statistical calculations are then no longer valid, since these are explicitly based on the properties summarized in 33. The shortcomings may be overcome by minimizing the deviations between observed and predicted values of the reaction rate itself:

$$\sum_{i=1}^{n} (r_{A_i} - \hat{r}_{A_i})^2 \rightarrow \text{Min} \tag{50}$$

Since the rate equation 11 is a non linear function of the parameters, minimization of the objective function 50 is an iterative process. This then requires an efficient optimization strategy, like the methods of Newton-Raphson, Rosenbrock, Fletcher-Powell, the Simplex method of Nelder & Mead, Newton-Gauss and its modification by Marquardt, etc. An excellent introduction to non linear regression analysis is found in the books by Draper and Smith [15], by Bard [26] and by Himmelblau [17].

Consider the general non linear model

$$y_i = f(\mathbf{x}_i; \boldsymbol{\beta}) + \varepsilon_i \tag{51}$$

and the corresponding least squares criterion for the estimation of the parameters

$$S(\boldsymbol{\beta}) = \sum_{i=1}^{n} [y_i - f(\mathbf{x}_i, \boldsymbol{\beta})]^2 \xrightarrow{\boldsymbol{\beta}} \text{Min} \tag{52}$$

The Newton-Gauss technique has been found to be very effective for the minimization of sums of squares. It is based on linearization of the model equation in Taylor series around an initial guess $\mathbf{b}_0$ for $\boldsymbol{\beta}$, thereby neglecting all partial derivatives of second and higher order. It involves repetitive evaluation of

$$\Delta\mathbf{b}_{i+1} = (\mathbf{J}_i^T\mathbf{J}_i)^{-1}\,\mathbf{J}_i\mathbf{r}_i \tag{53}$$

$$\mathbf{b}_{i+1} = \mathbf{b}_i + \Delta\mathbf{b}_{i+1} \tag{54}$$

until convergence is achieved. The symbols have the following meaning:

$$\mathbf{J}_i = \{J_{ikl}\} = \left\{\frac{\partial f(\mathbf{x}_k, \boldsymbol{\beta})}{\partial \beta_l}\bigg|_{\boldsymbol{\beta}=\mathbf{b}_i}\right\} \tag{55}$$

$$\mathbf{r}_i = \mathbf{y} - \mathbf{f}(\mathbf{x}, \mathbf{b}_i) \tag{56}$$

To prevent possible divergence of the Newton-Gauss procedure and retain at the same time its desirable properties, like quadratic convergence, Levenberg [27] and later Marquardt [28] developed a constrained Newton-Gauss procedure, which turns out to be a compromise between the methods of steepest descent and of Newton-Gauss. In this compromise, the parameter improvements are calculated by means of

$$\Delta\mathbf{b}_{i+1} = (\mathbf{J}_i^T\mathbf{J}_i + \lambda_i\mathbf{I})^{-1}\,\mathbf{J}_i^T\mathbf{r}_i \tag{57}$$

where $\mathbf{I}$ is a $p \times p$ identity matrix and λ_i is a Lagrange multiplier which is continuously adapted and controls both direction and length of the step. The flow diagram of Figure 7 outlines the procedure. Optimization techniques are described in full detail in books by Wilde [29], Beveridge and Schechter [30], Hoffmann and Hofmann [31], Bard [26], among others.

ii. Statistical testing

The tests described under paragraph III.A.1.b.ii. rely explicitly upon the linear character of the model. Although not rigorously valid for non linear models, they do provide tentative information about the parameter estimates, and ought to be considered as an integral part of any data analysis. The basis for the reliability analysis is the expansion of the non linear model in Taylor series around the converged parameter estimates and the application of the linear theory to the truncated series. It is assumed that the truncated series closely approximates the true non linear model throughout the range delimited by the confidence intervals.

An estimate of the approximate covariance matrix of the estimates $\mathbf{b}$ is given by the analog of equation 32:

$$\hat{\mathbf{V}}(\mathbf{b}) = (\mathbf{J}_*^T\mathbf{J}_*)^{-1}\,s^2 \tag{58}$$

where s^2 is defined in 31, and where $\mathbf{J}_*$ is evaluated at the converged $\mathbf{b}$ values. From here onwards, no distinction will be made between $\mathbf{X}$ and $\mathbf{J}_*$ and only the symbol $\mathbf{X}$ will be used, even when dealing with non linear models.

Approximate individual confidence intervals are computed from 45. An exact joint confidence region is still given by 46. The probability level of the

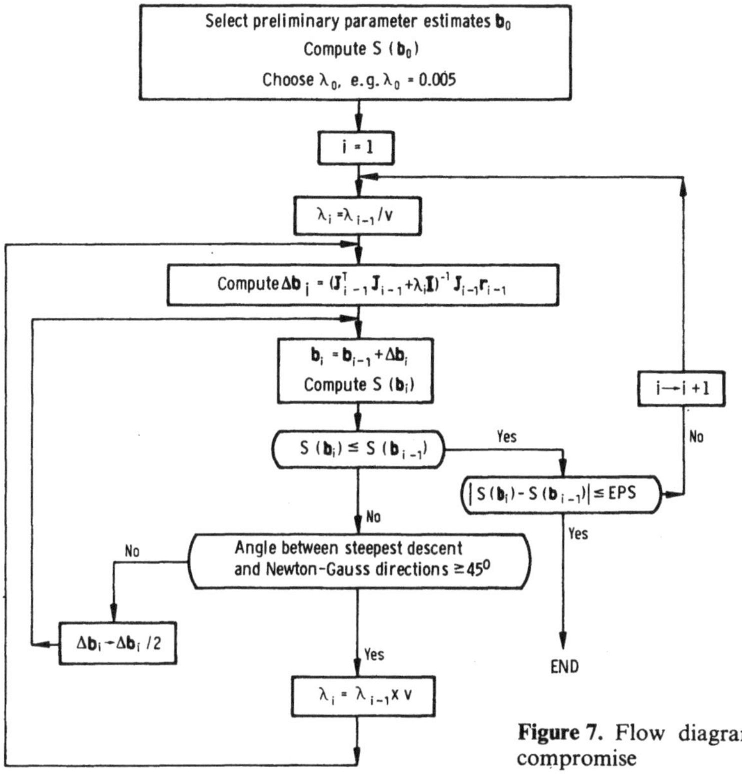

Figure 7. Flow diagram of Marquardt's compromise

contour, 1 — α, is only approximate, however. This region is not ellipsoidal, but generally banana-shaped. An approximate ellipsoidal joint confidence region at a correct probability level 1 — α is given by 47. Calculated values of the objective function for sets of parameter values on the ellipsoidal boundary locus are not identical, however. The magnitude of the deviations as well as the degree of deviation of the banana-shaped contour from the ellipsoidal surface depends on the degree of nonlinearity of the model. For further details on nonlinearity tests, the reader is referred to Beale [32] and Guttman and Meeter [33].

Example (3.A.1.c): *Disproportionation of propylene*
Recently, Hattikudur and Thodos [34] investigated the disproportionation of propylene over a tungsten oxide-silica catalyst. In the first part of their study, low reactant conversions, ranging from about 0.02 to 0.10, were observed in a differential reactor. Experiments were performed at five temperature levels. Corresponding initial rates were calculated from equation 16.

Application of the Yang-Hougen technique for discrimination on the basis of initial rates led the authors to adopt a dual site mechanism with the surface reaction rate controlling. For initial conditions:

$$r_0 = \frac{Cp_3^2}{(1 + K_3p_3)^2}$$

Estimates for C and K_3 were subsequently obtained by non linear regression, using

$$\sum\left[r_0 - \frac{Cp_3^2}{(1 + K_3p_3)^2}\right]^2 \xrightarrow{C,\,K_3} \text{Min}$$

Table 3 presents their results, including the approximate 95% individual confidence intervals on the parameter estimates. This case study will be taken up again in the section on the integral method of kinetic analysis.

2. *Integral Method of Kinetic Analysis*

Reaction rates are seldom directly measured. When integral reactors are used, the rates have to be determined by graphical or numerical derivation of observed conversion vs. space time curves.

Taking derivatives of experimental data is known to be an error magnifying rather than an error smoothing procedure. In addition, the distribution characteristics of the error on this derived dependent variable are always completely unknown, so that statistical calculations, based on assumptions 33 lack a sound basis. The appropriate way of data analysis is to use the data as they have been obtained. The conversion and the reaction rate of a component are related to one another by the continuity equations 14 and 15 for that component, with the initial condition:

$$x_A(0) = x_{A_0} \quad \text{when} \quad W/F_{A_0} = 0 \tag{59}$$

Equation (14) can be formally integrated:

$$\int \frac{dx_A}{r_A} = \int \frac{dx_A}{f(x_A)} = \int d\left(\frac{W}{F_{A_0}}\right) \tag{60}$$

With rate equations of the Hougen-Watson type and isothermal conditions, the quadrature in the left hand side of 60 can generally be carried out analytically, yielding an expression:

$$\frac{W}{F_{A_0}} = \varphi(x_A, p_t; \text{rate parameters}) \tag{61}$$

Discrimination based on graphical representation of some test quantities, like intrinsic parameters, is practically out of the question because of the intricacy of the formulae.

Table 3. Disproportionation of propylene. Hattikudur and Thodos' non linear least squares analysis per temperature based on initial rate data

Temp. (°F)	$C \pm t.s(C)$	$K_3 \pm t.s(K_3)$
650	0.276 ± 0.024	0.203 ± 0.010
700	0.744 ± 0.111	0.294 ± 0.025
750	2.098 ± 0.237	0.478 ± 0.031
800	5.105 ± 0.535	0.660 ± 0.038
850	18.06 ± 1.55	1.150 ± 0.052

In order to distinguish between the models, one is almost forced to proceed to the stage of parameter estimation. The residuals could be based on the space time, resulting in the criterion:

$$\sum_{i=1}^{n}\left(\frac{W}{F_{A0_i}}-\frac{\hat{W}}{F_{A0_i}}\right)^2 \xrightarrow{\text{rate parameters}} \text{Min} \tag{62}$$

In some cases, expression 61 is linear in groups of the parameters, so that only linear regression is involved (see e.g. Hosten and Froment [35]). In many situations, catalyst weight W and feed molar flow rate F_{A_0} are known with great precision, however, so that W/F_{A_0} may not be a proper regression variable. The parameters then have to be estimated by means of

$$\sum_{i=1}^{n}(x_{A_i}-\hat{x}_{A_i})^2 \rightarrow \text{Min} \tag{63}$$

or some criterion based on a variable related to the conversion. This again implies non linear regression. There is an additional difficulty, moreover, since equation 61 is implicit with respect to the conversion. The equation therefore has to be solved at the current values of the parameters for $\hat{x}_{A_i}$ for each experiment and the result substituted in 63.

When the quadrature in 60 cannot be performed analytically, the continuity equation 15 with its initial condition 59 is kept in the form of a differential equation, and numerically integrated up to the observation nodes $(W/F_{A_0})_i$, $i = 1, 2, ..., n$ by means of a routine like the Runge-Kutta (-Gill/Merson) or Adams-Moulton predictor-corrector methods [36].

In a study of the data analysis in pentane isomerization Froment and Mezaki [37] compared both criteria 62 and 63 and found an excellent agreement between both sets of estimates.

Example (3.A.2.α): *Ethanol dehydrogenation*

Peterson and Lapidus [38] performed an extensive integral kinetic analysis on Franckaerts and Froment's data on ethanol dehydrogenation, based on numerical integration of the conservation equation of the reactant, ethanol. Unfortunately, their results are subject to distrust because of a serious error in their continuity equation. Moreover, numerical integration of that equation is not necessary here, since it can be achieved analytically, as is shown in what follows.

Franckaerts and Froment's data at 275 °C are now reanalyzed by means of the integral method of kinetic analysis. Substitution of the expressions 48 in the rate equation 11 e.g. followed by analytical integration of 60 yields

$$\frac{W}{F_{A_0}} = \frac{1}{C}(a_1\tau_1 + b_1\tau_2 + c_1\tau_3) \tag{64}$$

in which τ_1, τ_2 and τ_3 are functions of the conversion and process conditions:

$$\tau_1 = \frac{1}{\sqrt{D}} \log \left| \frac{(2c_2x + b_2 - \sqrt{D})(b_2 + \sqrt{D})}{(2c_2x + b_2 + \sqrt{D})(b_2 - \sqrt{D})} \right|$$

$$\tau_2 = \frac{1}{2c_2}\left(\log \frac{a_2 + b_2x + c_2x^2}{a_2} - b_2 \cdot \tau_1\right) \tag{65}$$

$$\tau_3 = \frac{1}{c_2}\left(x - \frac{b_2}{2c_2} \log \frac{a_2 + b_2x + c_2x^2}{a_2} + \frac{b_2^2 - 2a_2c_2}{2c_2} \cdot \tau_1\right)$$

$$a_2 = (1 + \mu + \gamma)\, p_t$$

$$b_2 = -\left(\gamma + \mu\left(1 + \frac{p_t}{K}\right)\right) p_t$$

$$c_2 = -\left(1 + \frac{p_t}{K}\right) p_t$$

$$D = b_2^2 - 4a_2c_2$$

C, a_1, b_1 and c_1 are functions of the process conditions and rate parameters:

$$C = kK_A$$

$$a_1 = [1 + \gamma + \mu + (K_A + K_R\mu)\, p_t]^2 \tag{66}$$

$$c_1 = [1 - (K_A - K_R - K_S)\, p_t]^2$$

$$b_1 = 2\sqrt{a_1c_1}$$

Proceeding in the same way with the adsorption and desorption rate equations, 12 and 13, leads to equations identical to 64. The functional dependence of C and a_1, b_1, c_1 on the process conditions and parameters is different, however:

Adsorption rate controlling

$$C = k$$

$$a_1 = (1 + \mu + \gamma)(1 + \mu + \gamma + K_Rp_t\mu)$$

$$b_1 = [2 + (K_R + K_S)\, p_t]\,(1 + \mu + \gamma) + \left(\frac{K_A}{K} p_t + K_R\right) \mu p_t \tag{67}$$

$$c_1 = 1 + \left(\frac{K_A}{K} p_t + K_R + K_S\right) p_t$$

Desorption rate controlling

$$
\begin{aligned}
C &= kKK_R \\
a_1 &= KK_R(1 + \mu + \gamma)\, p_t \\
b_1 &= [1 + \mu + \gamma + K_A p_t - KK_R(\mu + \gamma)]\, p_t \qquad (68) \\
c_1 &= (1 - K_A p_t - KK_R + K_S p_t)\, p_t
\end{aligned}
$$

Clearly, 64 in conjunction with 65 to 68 do not allow discrimination by mere visual inspection, as was possible in the differential method of kinetic analysis.

The parameters have to be estimated in all three models by means of criterion 63. The implicit equation 64 is solved by the fast and powerful "Memory" method, developed by Shacham and Kehat [39]. Marquardt's algorithm is used to optimize the objective function 63. Table 4 summarizes the results.

The adsorption and desorption models may be discarded on the basis of significantly negative parameter estimates and/or obviously worse fits than the surface reaction model. Note that in the latter model K_S is not significantly different from zero.

In the present example the numerical evaluation of the partial derivatives $\frac{\partial x_A}{\partial k}, \frac{\partial x_A}{\partial K_A}, \ldots$ required by the Marquardt procedure, caused severe convergence problems so that analytical derivatives are preferred. Whereas analytical derivation presents no problems in explicit equations, total differentials are involved for models of the type considered here. Equation 60 is differentiated and all differentials but the one related to the parameter for which the partial derivative is sought are set equal to zero.

Table 4. Ethanol dehydrogenation. Integral method of kinetic analysis on data at 275 °C and comparison with DMKALR for the surface reaction model

	$k \pm t.s(k)$	$K_A \pm t.s(K_A)$	$K_R \pm t.s(K_R)$	$K_S \pm t.s(K_S)$	Residual sum of squares of deviations
Surface reaction	4.4005 ± 0.541	0.4223 ± 0.05	3.1048 ± 0.603	− 0.241 ± 0.447	0.00313
Adsorption	0.6389 ± 0.15	13.053 ± 2.295	1.2066 ± 1.604	− 5.597 ± 1.487	0.01111
Desorption	0.5584 ± 0.96	− 0.07215 ± 0.78	0.6836 ± 0.96	0.5165 ± 4.16	0.0362
Surface reaction DMKALR	4.298 ± 0.422	0.365 ± 0.058	2.35 ± 1.145	0.386 ± 0.883	

For $\frac{\partial x_A}{\partial k}$ e.g., the result is given by

$$\left.\frac{\partial x_A}{\partial k}\right|_{K_A, K_R, K_S; W/F_{A0}} = -r_A \int_0^{x_A^*} \left(\frac{\partial}{\partial k}\frac{1}{r_A}\right) dx_A \tag{69}$$

For the surface reaction model e.g., r_A is given by 11. Again for Hougen-Watson rate equations, the quadrature in the right hand side of 69 can generally be carried out analytically. The upper limit of the integral in 69, x_A^*, is the value of the conversion which satisfies equation 64 for the current values of the rate parameters. Using these analytical derivatives and preliminary parameter estimates, e.g. obtained from the differential method of kinetic analysis and linear regression, convergence is usually rapidly achieved.

The last line of Table 4 gives the estimates of k, K_A, K_R and K_S for the surface reaction model, obtained by the *differential method of kinetic analysis and linear regression* (DMKALR). The associated confidence limits were computed according to the rule of propagation of errors:

$$\mathrm{Var}[u(a, b)] = \left(\frac{\partial u}{\partial a}\right)^2 \mathrm{Var}(a) + \left(\frac{\partial u}{\partial b}\right)^2 \mathrm{Var}(b) + 2\frac{\partial u}{\partial a}\frac{\partial u}{\partial b}\mathrm{Cov}(a, b) \tag{70}$$

In 70 u is a random variable, which is a non linear function of two other random variables a and b. For the surface reaction model e.g. $u = k$, $a = 1/\sqrt{kK_A}$, $b = K_A/\sqrt{kK_A}$, so that $u = 1/ab$. The required information on the reliability of a and b, i.e. the variance-covariance matrix of the groups a and b (and c and d) in equation 22 is supplied by the analysis discussed in paragraph 3.A.1.b. The estimate for K_S is positive but nonetheless not significantly different from zero. The agreement between the two sets of estimates is fairly good.

Example (3.A.2.β): *Propylene disproportionation*

The full rate equation developed by Hattikudur and Thodos [34] for the disproportionation of propylene may be written:

$$r = \frac{C(p_3^2 - p_2p_4/K)}{(1 + K_3p_3 + K_2p_2 + K_4p_4)^2} \tag{71}$$

After determining C and K_3 by means of differential reactor data, as outlined in Example (3.A.1.c), $K_2 + K_4$ was subsequently estimated from integral reactor data with pure propylene as the feed. The integrated continuity equation for propylene again reads like 64 in which τ_1, τ_2 and τ_3 are given by 65 and

$$a_1 = \left(\frac{1}{p_t} + K_3\right)^2, \quad c_1 = \left(\frac{K_2 + K_4}{2} - K_3\right)^2, \quad b_1 = 2\sqrt{a_1c_1}$$

$$a_2 = 1, \quad b_2 = -2, \quad c_2 = 1 - \frac{1}{4K}$$

Using the values for C and K_3 obtained in the previous stage (see Example (3.A.1.c)), and introducing the values of x for given p_2, p_3 in 64, point values for $K_2 + K_4$ were computed per experiment and averaged afterwards.

The temperature dependence for the constants, C, K_3 and $K_2 + K_4$ was determined on the basis of the Arrhenius-relation:

$$\ln K_i = \ln A_i - \frac{E_i}{R} \cdot \frac{1}{T} \tag{72}$$

Straight lines are obtained when $\ln K_i$ is plotted vs $\frac{1}{T}$. The straight lines can be drawn by eye or, preferably, by means of a least squares criterion. The regression is linear in this case:

$$\sum \left[\ln K_i - \left(\ln A_i - \frac{E_i}{R} \cdot \frac{1}{T} \right) \right] \xrightarrow{\ln A_i, E_i} \text{Min.}$$

Hattikudur and Thodos' analysis, summarized here and in section 3.A.1.c., violates several assumptions of estimation theory:

a) The initial rates were not observed directly, but computed from experiments at low conversion, according to 16. Consequently, an unweighted least squares analysis, based on the reaction rates, is statistically appropriate only when the variance of x increases quadratically with W/F. This is rather doubtful.

b) The two step procedure in which C and K_3 are first estimated from the differential reactor data and then kept at these values in the subsequent determination of $K_2 + K_4$ based upon the integral method of kinetic analysis is an unorthodox approach yielding sub-optimal parameters values.

c) Averaging the point values of $K_2 + K_4$ yields an estimate to which no properties can be attributed. It must be recognized, however, that point values for a parameter permit testing the adequacy of the model: trends would indicate systematic deviations between the model and the data, i.e. lack of fit.

d) Determining the temperature dependence of the parameters on the basis of Arrhenius-diagrams does not yield the best possible estimates for A_i and E_i, since these do not minimize the deviations between observed conversions and values predicted by the model. Also, confidence limits for $\ln A_i$ and E_i are doubtful, since it is not known if the error on $\ln K_i$ satisfies the properties required by estimation theory.

Hattikudur and Thodos' data will now be analyzed rigorously with the estimation criterion based upon the conversion and by considering both types of data, differential and integral, simultaneously. In a first stage the data at each temperature are analyzed separately. The estimates and the individual confidence limits at the 95 % probability level are given in Table 5.

From a comparison of Tables 3 and 5 it is seen that the difference between the two sets of estimates is negligible from the statistical point of view. The point estimates are indeed located within their respective confidence intervals. The confidence limits obtained in this work, however, are nearly double

Table 5. Disproportionation of propylene. Parameter estimates and 95% individual confidence limits based on integral method of kinetic analysis per temperature

Temp. (°F)	$C \pm t.s(C)$	$K_3 \pm t.s(K_3)$	$K_2 + K_4 \pm t.s(K_2 + K_4)$	
			This work	Hattikudur & Thodos
650	0.293 ± 0.051	0.205 ± 0.033	0.496 ± 0.22	0.444 ± 0.23
750	2.636 ± 0.48	0.541 ± 0.071	0.797 ± 0.31	0.826 ± 0.17
850	14.4 ± 2.84	1.018 ± 0.13	1.487 ± 0.37	1.049 ± 0.18

the original ones. This gives rise to the suspicion that Hattikudur & Thodos calculated standard deviations and not the confidence intervals or else, that their values underestimate these uncertainty intervals. This can be attributed to a different regression equation and, to a lesser extent, to the non-inclusion of $K_2 + K_4$ in the equation for the initial rates.

In a second stage, the temperature dependence of the parameters is determined directly from the simultaneous treatment of the data at all temperatures. For this purpose, all parameters are replaced by the exponential temperature dependence:

$$K_j = A_j \, e^{-\frac{E_j}{RT}}$$

For reasons of convergence, the following reparameterization is recommended [40]:

$$K_j = A_j^* \, e^{-\frac{E_j}{R}\left(\frac{1}{T} - \frac{1}{\bar{T}}\right)} \tag{73}$$

$\bar{T}$ is the mean temperature of all experiments. The A_j^* and E_j are now estimated directly by means of:

$$\sum (x_i - \hat{x}_i)^2 \xrightarrow{A_j^*, E_j;\; j=1,2,3} \text{Min.}$$

The estimates obtained from the analysis per temperature served as preliminary guess for this simultaneous treatment. The analysis presented here includes all of the 93 available data points, presented in Tables II and V of the original publication. It guarantees that the best possible estimates for all six parameters are obtained. Also, the statistical premises for estimation have been respected, so that a reliable statistical analysis for the parameters is obtained. Table 6 compares results.

The agreement between both sets of estimates is remarkable: from the statistical point of view the differences are insignificant. Note that the adsorption enthalpies have the wrong sign. This throws some doubt on the physico-chemical meaning of the model. In addition to Hattikudur & Thodos' tests on the adequacy of the model, an analysis of residuals [15] was carried out.

The residual deviations $x_i - \hat{x}_i$ against the settings of the independent variable, W/F, and the predicted values of the response, $\hat{x}$, are shown in

Table 6. Disproportionation of propylene. Temperature dependence of the parameters. Hattikudur & Thodos: by means of Arrhenius diagrams. This work: by means of an analysis of all data at all temperatures simultaneously

Hattikudur & Thodos	This work
$C = 1.229\ 10^{10}\ e^{-30\,040/RT}$	$C = 3.061\ 10^{10}\ e^{-(31\,054\ \pm\ 2\,030)/RT}$
$K_3 = 4\,082\ e^{-12\,140/RT}$	$K_3 = 6\,799\ e^{-(12\,719\ \pm\ 1\,490)/RT}$
$K_2 + K_4 = 166\ e^{-7\,154/RT}$	$K_2 + K_4 = 5\,608\ e^{-(8\,638\ \pm\ 4\,036)/RT}$

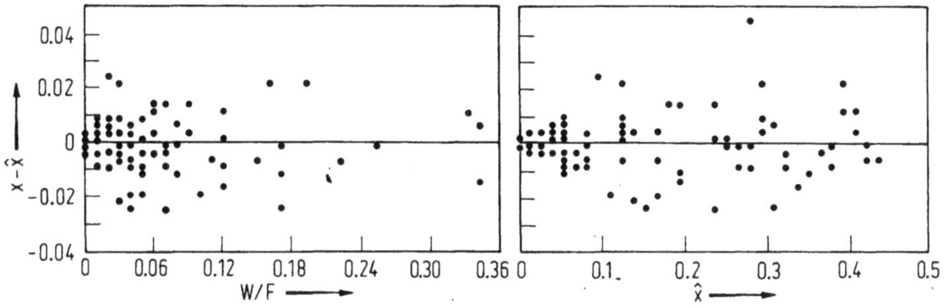

Figure 8. Disproportionation of propylene. Analysis of residuals

Figure 8 for all 93 observations. The uniform bands confirm the adequacy of the rate equation 71 at all temperatures and all conversions.

B. Complex Reactions

The rigorous simulation of a reactor in which a set of reactions is taking place requires kinetic equations for practically all the reactions. There are very few examples of kinetic analysis of a complex reacting network. With the tools now available for chemical analysis and for data treatment there is little justification for limiting the kinetic analysis to simplistic global reaction schemes. The general model for such a structure consisting of v dependent variables may be written:

$$\begin{aligned} y_1 &= f_1(\mathbf{x}, \boldsymbol{\beta}) + \varepsilon_1 = \eta_1 + \varepsilon_1 \\ y_2 &= f_2(\mathbf{x}, \boldsymbol{\beta}) + \varepsilon_2 = \eta_2 + \varepsilon_2 \\ &\quad\vdots \qquad\qquad\quad \vdots \\ y_v &= f_v(\mathbf{x}, \boldsymbol{\beta}) + \varepsilon_v = \eta_v + \varepsilon_v \end{aligned} \tag{74}$$

1. Error Model

Assuming that n observations have been taken of the v response variables, the two dimensional array of n × v errors is organized as follows:

$$\begin{array}{cc|cccccccc}
 & & \multicolumn{8}{c}{\text{Response No.}} \\
 & & 1 & 2 & \dots & j & \dots & v & & \\
\text{Experiment No.} & 1 & \varepsilon_{11} & \varepsilon_{12} & \dots & \varepsilon_{1j} & \dots & \varepsilon_{1v} & \rightarrow & \boldsymbol{\varepsilon}^{(1)} \\
 & 2 & \varepsilon_{21} & \varepsilon_{22} & \dots & \varepsilon_{2j} & \dots & \varepsilon_{2v} & \rightarrow & \boldsymbol{\varepsilon}^{(2)} \\
 & \vdots & \vdots & \vdots & & \vdots & & \vdots & & \\
 & i & \varepsilon_{i1} & \varepsilon_{i2} & \dots & \varepsilon_{ij} & \dots & \varepsilon_{iv} & \rightarrow & \boldsymbol{\varepsilon}^{(i)} \\
 & \vdots & \vdots & \vdots & & \vdots & & \vdots & & \\
 & n & \varepsilon_{n1} & \varepsilon_{n2} & \dots & \varepsilon_{nj} & \dots & \varepsilon_{nv} & \rightarrow & \boldsymbol{\varepsilon}^{(n)} \\
 & & \downarrow & \downarrow & & \downarrow & & \downarrow & & \\
 & & \boldsymbol{\varepsilon}_1 & \boldsymbol{\varepsilon}_2 & & \boldsymbol{\varepsilon}_j & & \boldsymbol{\varepsilon}_v & &
\end{array} \tag{75}$$

The following fairly general error model applies to many practical situations. It assumes that:

1) all errors are normally distributed with zero mean, $E(\boldsymbol{\varepsilon}_j) = \mathbf{0}; j = 1, 2, ..., v$
2) the n errors corresponding to the n observations of the j-th response, $\boldsymbol{\varepsilon}_j$, are statistically independent and have constant variance σ_{jj}
3) the v errors within the i-th experiment, $\boldsymbol{\varepsilon}^{(i)}$, are statistically interdependent with covariance matrix:

$$\boldsymbol{\Sigma} = \{\sigma_{ij}\} = \begin{bmatrix} \sigma_{11} & \sigma_{12} \cdots \sigma_{1v} \\ \sigma_{12} & \sigma_{22} \quad\; \sigma_{2v} \\ \vdots & \\ \sigma_{1v} & \sigma_{2v} \quad\; \sigma_{vv} \end{bmatrix} \tag{76}$$

or $E(\boldsymbol{\varepsilon}_k \boldsymbol{\varepsilon}_l^T) = \mathbf{I}\sigma_{kl}$; k and $l = 1, 2, \dots, v$

When the errors $\boldsymbol{\varepsilon}^{(i)}$ are statistically independent, $\boldsymbol{\Sigma}$ simplifies to a diagonal matrix.

4) this covariance matrix is the same for all experiments.

The likelihood function of the parameters, given the observations, then reads:

$$L(\boldsymbol{\beta}|\mathbf{y}) = \frac{1}{(2\pi)^{\frac{nv}{2}} (\det \boldsymbol{\Sigma})^{\frac{n}{2}}} \exp\left\{-\frac{1}{2} \sum_{i=1}^{v} \sum_{j=1}^{v} \sigma^{ij} \sum_{k=1}^{n} (y_{ki} - \eta_{ki})(y_{kj} - \eta_{kj})\right\} \tag{77}$$

where σ^{ij} are the elements of the inverse of $\boldsymbol{\Sigma}$:

$$\boldsymbol{\Sigma}^{-1} = \{\sigma^{ij}\} \tag{78}$$

The criterion for data analysis depends upon the knowledge of $\boldsymbol{\Sigma}$.

2. Matrix $\boldsymbol{\Sigma}$ known

The maximum likelihood parameter estimates are obtained by maximizing the likelihood with respect to the parameters, or equivalently by

$$S(\boldsymbol{\beta}) = \sum_{i=1}^{v} \sum_{j=1}^{v} \sigma^{ij} \sum_{k=1}^{n} (y_{ki} - \eta_{ki})(y_{kj} - \eta_{kj}) \xrightarrow{\boldsymbol{\beta}} \text{Min.} \tag{79}$$

This is the generalized least squares criterion which reduces to the well known single response criterion 25 when $v = 1$. The criterion is also applicable when $\boldsymbol{\Sigma}$ is known up to a constant only.

a) Model equations linear in the parameters

When all v equations are linear with respect to the parameters the following set of observation equations for the j-th response can be written:

$$\begin{aligned} y_{1j} &= \beta_1 x_{1j1} + \beta_2 x_{1j2} + \dots + \beta_p x_{1jp} + \varepsilon_{1j} \\ y_{2j} &= \beta_1 x_{2j1} + \beta_2 x_{2j2} + \dots + \beta_p x_{2jp} + \varepsilon_{2j} \\ &\vdots \\ y_{nj} &= \beta_1 x_{nj1} + \beta_2 x_{nj2} + \dots + \beta_p x_{njp} + \varepsilon_{nj} \end{aligned} \tag{80}$$

The first subscript of the independent variables refers to the experiment, the second to the response, the third to the variable, or parameter. The second subscript is required because a parameter, β_i say, is not necessarily coupled with the same independent variable in all responses. The following vectors and matrix are defined in a way similar to that already encountered in the single response case:

$$\mathbf{y}_j = \begin{bmatrix} y_{1j} \\ y_{2j} \\ \vdots \\ y_{nj} \end{bmatrix} \quad \boldsymbol{\beta} = \begin{bmatrix} \beta_1 \\ \beta_2 \\ \vdots \\ \beta_p \end{bmatrix} \quad \mathbf{b} = \begin{bmatrix} b_1 \\ b_2 \\ \vdots \\ b_p \end{bmatrix} \quad \boldsymbol{\varepsilon}_j = \begin{bmatrix} \varepsilon_{1j} \\ \varepsilon_{2j} \\ \vdots \\ \varepsilon_{nj} \end{bmatrix}$$
$$\mathbf{X}_j = \begin{bmatrix} x_{1j1} & x_{1j2} \cdots & x_{1jp} \\ x_{2j1} & x_{2j2} \cdots & x_{2jp} \\ \vdots & \vdots & \vdots \\ x_{nj1} & x_{nj2} & x_{njp} \end{bmatrix} \tag{81}$$

The sets of observation equations 80 for all responses are equivalent with

$$\begin{aligned} \mathbf{y}_1 &= \mathbf{X}_1\boldsymbol{\beta} + \boldsymbol{\varepsilon}_1 = \boldsymbol{\eta}_1 + \boldsymbol{\varepsilon}_1 \\ \mathbf{y}_2 &= \mathbf{X}_2\boldsymbol{\beta} + \boldsymbol{\varepsilon}_2 = \boldsymbol{\eta}_2 + \boldsymbol{\varepsilon}_2 \\ &\vdots \\ \mathbf{y}_v &= \mathbf{X}_v\boldsymbol{\beta} + \boldsymbol{\varepsilon}_v = \boldsymbol{\eta}_v + \boldsymbol{\varepsilon}_v \end{aligned} \tag{82}$$

and the generalized least squares criterion with

$$S(\boldsymbol{\beta}) = \sum_{i=1}^{v} \sum_{j=1}^{v} \sigma^{ij}(\mathbf{y}_i - \mathbf{X}_i\boldsymbol{\beta})^T (\mathbf{y}_j - \mathbf{X}_j\boldsymbol{\beta}) \xrightarrow{\boldsymbol{\beta}} \text{Min.} \tag{83}$$

After some manipulations the following results are obtained for the parameter estimates $\mathbf{b}$, their covariance matrix, $\mathbf{V(b)}$, and their joint confidence region Θ:

$$\mathbf{b} = \left(\sum_{i=1}^{v} \sum_{j=1}^{v} \sigma^{ij} \mathbf{X}_i^T \mathbf{X}_j \right)^{-1} \sum_{i=1}^{v} \sum_{j=1}^{v} \sigma^{ij} \mathbf{X}_i^T \mathbf{y}_j \tag{84}$$

$$\mathbf{V}(\mathbf{b}) = \left(\sum_{i=1}^{v}\sum_{j=1}^{v}\sigma^{ij}\mathbf{X}_i^T\mathbf{X}_j\right)^{-1} \tag{85}$$

$$(\mathbf{b}-\boldsymbol{\beta})^T\left(\sum_{i=1}^{v}\sum_{j=1}^{v}\sigma^{ij}\mathbf{X}_i^T\mathbf{X}_j\right)(\mathbf{b}-\boldsymbol{\beta}) = \frac{p}{nv-p}S(\mathbf{b})\,F(p, nv-p, 1-\alpha) \tag{86}$$

or

$$S(\boldsymbol{\beta}) = S(\mathbf{b})\left[1 + \frac{p}{nv-p}F(p, nv-p, 1-\alpha)\right] \tag{87}$$

b) Non-linear models

Multiresponse non linear models are dealt with in a way identical to that followed for single response non linear models. The extended Marquardt algorithm reads:

$$\Delta\mathbf{b}_{s+1} = \left(\sum_{i=1}^{v}\sum_{j=1}^{v}\sigma^{ij}\mathbf{X}_{i_s}^T\mathbf{X}_{j_s} + \lambda_s\mathbf{I}\right)^{-1}\sum_{i=1}^{v}\sum_{j=1}^{v}\sigma^{ij}\mathbf{X}_{i_s}^T\mathbf{r}_{j_s} \tag{88}$$

where

$$\mathbf{X}_{i_s} = \{X_{ikl_s}\} = \left\{\frac{\partial f_i(\mathbf{x}_k;\boldsymbol{\beta})}{\partial\beta_l}\bigg|_{\boldsymbol{\beta}=\mathbf{b}_s}\right\} \tag{89}$$

$$\mathbf{r}_{j_s} = \{r_{jk_s}\} = \{y_{kj} - f_j(\mathbf{x}_k, \mathbf{b}_s)\} \tag{90}$$

3. Matrix $\boldsymbol{\Sigma}$ Unknown

The likelihood 77 is now a function of both the elements of $\boldsymbol{\Sigma}$ and the parameters. First, the maximum likelihood estimate **S** for $\boldsymbol{\Sigma}$ is sought by minimizing the logarithm of the likelihood with respect to all elements of $\boldsymbol{\Sigma}$. This yields:

$$\mathbf{S} = \{S_{kl}\} = \left\{\frac{1}{n}\sum_{i=1}^{n}(y_{ik}-\eta_{ik})(y_{il}-\eta_{il})\right\} \tag{91}$$

The matrix obtained in this way is substituted for $\boldsymbol{\Sigma}$ into the likelihood 77 and the result is minimized with respect to the parameters. After some manipulations, the following result is obtained:

$$\det \mathbf{S} \xrightarrow{\boldsymbol{\beta}} \text{Min} \tag{92}$$

where **S** is defined in 91, or explicitly:

$$\mathbf{S} = \frac{1}{n}\times \begin{bmatrix} \sum_i(y_{i1}-\eta_{i1})^2 & \sum_i(y_{i1}-\eta_{i1})(y_{i2}-\eta_{i2}) & \cdots & \sum_i(y_{i1}-\eta_{i1})(y_{iv}-\eta_{iv}) \\ \sum_i(y_{i1}-\eta_{i1})(y_{i2}-\eta_{i2}) & \sum_i(y_{i2}-\eta_{i2})^2 & \cdots & \sum_i(y_{i2}-\eta_{i2})(y_{iv}-\eta_{iv}) \\ \vdots & \vdots & & \vdots \\ \sum_i(y_{i1}-\eta_{i1})(y_{iv}-\eta_{iv}) & \sum_i(y_{i2}-\eta_{i2})(y_{iv}-\eta_{iv}) & \cdots & \sum_i(y_{iv}-\eta_{iv})^2 \end{bmatrix} \tag{93}$$

Box and Draper [41] derived this criterion on different theoretical grounds. When the error covariance matrix $\boldsymbol{\Sigma}$ is not known, parameter estimation is always a non linear regression problem. Besides, the appropriate criterion 92 is not a sum of squares, so that the Newton-Gauss or Marquardt techniques cannot be used. The techniques which optimize other types of objective functions already mentioned under 3.A.1.c.i have to be resorted to. To avoid pitfalls inherent to the determinant criterion [42, 43], no linear dependences may exist between the columns of $\mathbf{Y} = [\mathbf{y}_1, \mathbf{y}_2 \dots \mathbf{y}_v]$.

The boundary surface of an approximate $100(1 - \alpha)\%$ confidence region for the parameter estimates is defined by all parameter combinations which satisfy [44]

$$|\mathbf{S}| = |\mathbf{S}|^{\min} [\exp \chi^2(v - 1, 1 - \alpha)/n] \tag{94}$$

$|\mathbf{S}|^{\min}$ is the minimum value of the determinant of $\mathbf{S}$ and χ^2 $(v - 1, 1 - \alpha)$ the α-percentage point of the χ^2 distribution with $v - 1$ degrees of freedom.

The use of the determinant criterion 92 has been documented by Mezaki and Butt [45] and by Hosten and Froment [46].

Table 7 summarizes a series of conditions and the appropriate maximum likelihood criteria for parameter estimation.

Example (3.B): *o-Xylene oxidation on a vanadium pentoxide catalyst*
In two recent papers Froment [5, 6] discussed a number of possible ways of multivariate kinetic analysis of the catalytic air oxidation of o-xylene on V_2O_5-catalyst. The mechanism considered for this reaction is as follows:

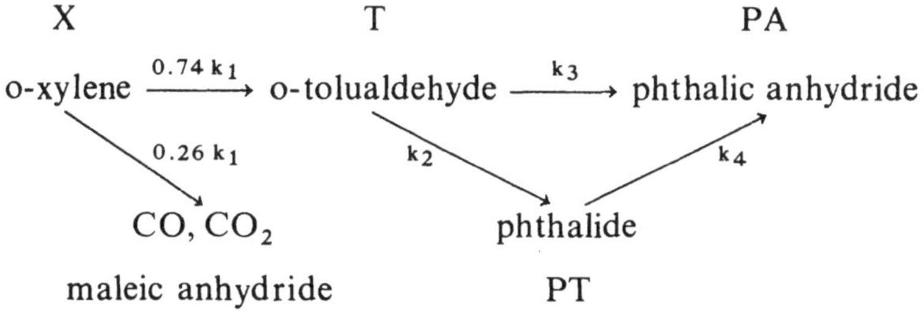

In each step the surface reactions on single sites were assumed to be rate controlling. To reduce the number of parameters, the adsorption equilibrium constants for all the reaction products P were assumed to be identical. The rate equations then take the following form:
rate of disappearance of o-xylene:

$$r_X = \frac{k_1 K_a K_X p_0^{1/2} p_X}{1 + K_X p_X + K_P p_P} \tag{95}$$

net rate of production of o-tolualdehyde:

$$r_T = r_X \frac{0.74 k_1 p_X - (k_2 + k_3) p_T}{k_1 p_X} \tag{96}$$

Table 7. Multivariate maximum likelihood parameter estimation criteria and conditions for application

Criterion	Conditions for application
$\sum_{k=1}^{v} \sum_{l=1}^{v} \sigma^{kl} \sum_{i=1}^{n} (y_{ik} - \eta_{ik})(y_{il} - \eta_{il})$	Experimental errors interdependent with known covariance matrix Σ.
$\sum_{k=1}^{v} \frac{1}{\sigma_{kk}} \sum_{i=1}^{n} (y_{ik} - \eta_{ik})^2$	Experimental errors independent with unequal known variances.
$\sum_{k=1}^{v} \sum_{i=1}^{n} (y_{ik} - \eta_{ik})^2$	Experimental erros independent with equal (known or unknown) variances.
det S, where $S = \{s_{kl}\} = \left\{\sum_{i=1}^{n} (y_{ik} - \eta_{ik})(y_{il} - \eta_{il})\right\}$	Experimental errors interdependent with unknown covariance matrix Σ. No missing observations.
$\det S \xrightarrow{\beta,\, y^*} \text{Min.}$	Experimental errors interdependent with unknown covariance matrix Σ. Few observations, y^*, are missing.
$\sum_{k=1}^{v} \left[\sum_{i=1}^{n_k} (y_{ik} - \eta_{ik})^2\right]^{n_k/2}$	Experimental errors independent with unknown and unequal variances. Many, n_k say, observations of the k-th response are missing.

net rate of production of phthalide:

$$r_{PT} = r_X \frac{k_2 p_T - k_4 p_{PT}}{k_1 p_X} \tag{97}$$

rate of production of phthalic anhydride:

$$r_{PA} = r_X \frac{k_3 p_T - k_4 p_{PT}}{k_1 p_X} \tag{98}$$

The experimental data consist of a set of total conversions of o-xylene and conversions into the intermediate and desired final products taken at a set of W/F_{X_0} values.

i. Differential method of kinetic analysis

The differential method of kinetic analysis makes use of the rates of the reactions. These are obtained by numerical differentiation of the conversion-space time data $x = f_1(W/F_{X_0})$, $u = f_2(W/F_{X_0})$, etc. These data could be analyzed per response, whereby estimates of k_1, K_a, K_X and K_P would be generated from the use of the objective function $\Sigma(r_X - \hat{r}_X)^2$. Estimation based on $\Sigma(r_T - \hat{r}_T)^2$ would in principle yield estimates for the parameters appearing in r_X but also for $k_2 + k_3$. Similarly, estimation based on $\Sigma(r_{PT} - \hat{r}_{PT})^2$ would give estimates for k_1, K_a, K_X, K_P and in addition for k_2 and k_4. The problem would then be to select a consistent set of parameters. In any event, such a procedure is of no practical value, since estimation of four to six parameters from rate data with respect to intermediate products present in relatively low concentrations which vary only slowly along the reactor, would most likely fail because of mathematical singularities.

A true multiresponse analysis is required here. In the absence of evidence of correlation among the errors of the different responses, a simplified form of criterion 79 could be used in which only sums of squares of deviations are considered, each sum being appropriately weighted for the difference in precision with which the various responses are measured (Table 7):

$$w_1 \sum (r_X - \hat{r}_X)^2 + w_2 \sum (r_T - \hat{r}_T)^2 + w_3 \sum (r_{PT} - \hat{r}_{PT})^2 + w_4 \sum (r_{PA} - \hat{r}_{PA})^2 \xrightarrow{k_1, K_a, K_X, K_P, k_2, k_3, k_4} \text{Min} \quad (99)$$

According to 79 the weighting factors have to be inversely proportional to the variance of the errors. This may cause difficulties, since replicated experiments are required to assess the variance of a variable, and replicates are often not available. Alternately, the determinant criterion, based on all four responses, could be used.

ii. Integral method of kinetic analysis

For reasons already outlined above it is preferable, to estimate the parameters on the basis of the experimental data as such. Again, analysis per response is not considered. Substituting the reaction rate equations in the respective continuity equations yields a set of four differential equations:

$$\frac{dx}{dW/F_{X_0}} = r_X = f(x, \text{parameters}) \quad (100)$$

$$\frac{du}{dW/F_{X_0}} = r_T = f(x, u, \text{parameters}) \quad (101)$$

$$\frac{dv}{dW/F_{X_0}} = r_{PT} = f(x, u, v, \text{parameters}) \quad (102)$$

$$\frac{dz}{dW/F_{X_0}} = r_{PA} = f(x, u, v, \text{parameters}) \quad (103)$$

To avoid the frequent, time consuming numerical integrations of the set of equations 100–103 one can proceed through the product distribution equations: 1) Dividing equation 101 by equation 100 yields a differential equation which can be solved analytically for u as an explicit function of x 2) Dividing equation 102 by equation 100 and substituting u by its function of x, obtained in 1), yields another differential equation which can be solved for v as an explicit function of x 3) The same way can be followed with 103. As a result of these integrations, the product distribution relations emerge:

$$u = f_1\left(x; \frac{k_2}{k_1}, \frac{k_3}{k_1}\right) \quad (104)$$

$$v = f_2\left(x; \frac{k_2 + k_3}{k_1}, \frac{k_4}{k_1}\right) \quad (105)$$

$$z = f_3\left(x; \frac{k_2}{k_1}, \frac{k_3}{k_1}, \frac{k_4}{k_1}\right) \quad (106)$$

Estimates of the ratios $\frac{k_2}{k_1}$, $\frac{k_3}{k_1}$ and $\frac{k_4}{k_1}$ can be obtained by means of the determinant criterion, based on the response variables u, v and z, and considering x as an independent variable. To obtain estimates of k_1, K_a, K_X and K_P, 100 is integrated analytically to give a relation:

$$\frac{W}{F_{X_0}} = f_4(x; k_1, K_a, K_X, K_P) \tag{107}$$

which is implicit in x, and the objective function $\Sigma(x - \hat{x})^2$ is minimized with respect to these parameters. From the ratios of rate coefficients determined in the previous stage and knowledge of k_1, the other rate coefficients follow. Although statistically not entirely orthodox, this two step procedure should be fairly straight-forward in its application. The rigorous procedure would consist of minimizing

$$\det\begin{bmatrix} \sum(x-\hat{x})^2 & \sum(x-\hat{x})(u-\hat{u}) & \sum(x-\hat{x})(v-\hat{v}) & \sum(x-\hat{x})(z-\hat{z}) \\ \sum(x-\hat{x})(u-\hat{u}) & \sum(u-\hat{u})^2 & \sum(u-\hat{u})(v-\hat{v}) & \sum(u-\hat{u})(z-\hat{z}) \\ \sum(x-\hat{x})(v-\hat{v}) & \sum(u-\hat{u})(v-\hat{v}) & \sum(v-\hat{v})^2 & \sum(v-\hat{v})(z-\hat{z}) \\ \sum(x-\hat{x})(z-\hat{z}) & \sum(u-\hat{u})(z-\hat{z}) & \sum(v-\hat{v})(z-\hat{z}) & \sum(z-\hat{z})^2 \end{bmatrix} \tag{108}$$

with respect to all seven parameters simultaneously. Calculated values $\hat{x}$, $\hat{u}$, $\hat{v}$ and $\hat{z}$ at the given values of W/F_{X_0} and current values of the parameters are obtained through equations 107 and 104/106. In the absence of accurate preliminary parameter estimates, great difficulties might be encountered, however.

4. Parameter Estimation in Differential Equations

a) Parameter estimation

As soon as the reaction is somewhat complex or leads to rate equations which are non linear in the measured variables, analytical integration of the set of continuity equations is excluded. An important class of models for chemical reaction systems is given by simultaneous first order ordinary (non linear) differential equations

$$\frac{d\mathbf{x}}{dt} = \mathbf{f}(t, \mathbf{x}; \boldsymbol{\beta}) \tag{109}$$

with initial conditions $\mathbf{x}(0) = \mathbf{c}$. The symbol t represents an independent variable like time or reactor length, and $\mathbf{x}$ is a vector of s state variables, which are not necessarily the variables that are experimentally observed. The measured variables $\mathbf{y}$ are related with the state variables through known relations which generally have an algebraic character:

$$\mathbf{y}(t) = \mathbf{h}(t, \mathbf{x}(t)) + \boldsymbol{\varepsilon} \tag{110}$$

Consider for instance a catalytic reaction with considerable mass and heat transfer resistance across the film surrounding the catalyst particle. Surface concentrations and temperature then differ from the values in the bulk of the gas. Only the latter can be measured. In this case surface concentrations and temperature are the state variables **x**, whereas the corresponding bulk quantities are the observed variables **y**.

For an observation of the variables **y** at $t = t_r$:

$$\mathbf{y}(t_r) = \mathbf{h}(t_r, \mathbf{x}(t_r)) + \boldsymbol{\varepsilon}^{(r)} \tag{111}$$

Provided the experimental error is multivariate normally distributed,

$$\boldsymbol{\varepsilon}^{(r)} \sim N(\mathbf{O}, \boldsymbol{\Sigma}) \tag{112}$$

the maximum likelihood parameter estimates are obtained by minimizing

$$S(\boldsymbol{\beta}) = \sum_{r=1}^{n} [\mathbf{y}(t_r) - \mathbf{h}(t_r, \mathbf{x}(t_r))]^T \boldsymbol{\Sigma}^{-1} [\mathbf{y}(t_r) - \mathbf{h}(t_r, \mathbf{x}(t_r))] \tag{113}$$

with respect to the parameters when $\boldsymbol{\Sigma}$ is known or by minimizing det S, where

$$\mathbf{S} = \{S_{ij}\} = \left\{ \sum_{r=1}^{n} [y_{ri} - h_i(t_r, \mathbf{x}(t_r))][y_{rj} - h_j(t_r, \mathbf{x}(t_r))] \right\} \tag{114}$$

when $\boldsymbol{\Sigma}$ is not known.

$S(\boldsymbol{\beta})$ and det **S** contain $\boldsymbol{\beta}$ via **h** which depend on **x**, which in turn are functions of $\boldsymbol{\beta}$ defined by the state equations 109. To compute $S(\boldsymbol{\beta})$ or det S, the state equations 109 have to be integrated numerically for each of the n experiments. Since the relationship $\mathbf{x}|\boldsymbol{\beta}$ is always non linear, minimization of $S(\boldsymbol{\beta})$ or det S is always an iterative process. The solution of this problem requires an appropriate optimization routine to minimize $S(\boldsymbol{\beta})$ or det S, coupled to a differential equation solving routine to integrate the state equations 109. Essentially, estimation of parameters in differential equations is not different from estimation in algebraic equations, except that the calculation of the residuals for each experiment requires the integration of a set of differential equations, instead of simple function calculations. Estimation in differential equations is dealt with in detail by Seinfeld and Lapidus [47]. Clearly, a considerable amount of computational effort is involved. Success is not always guaranteed, especially when only poor preliminary parameter estimates are available. A number of short-cut techniques have been developed in recent years. Some of the most convenient: the reduction of integral reactor data to differential data, Himmelblau-Jones-Bischoff [48], Cresswell-Quon [49], Vanden Bosch-Hellinckx [50] and Cresswell-Paterson [51] techniques, were tested by Hosten [52] who compared their performance with the mathematically rigorous procedure.

b) Reliability analysis of parameter estimates

A lucid development of the theory is presented in Rosenbrock and Storey [53]. Suffice it here to summarize the basic formulae. The covariance matrix is given by

$$\mathbf{V}(\mathbf{b}) = \mathbf{H}^{-1} \tag{115}$$

where:

$$\mathbf{H} = \sum_{r=1}^{n} \mathbf{G}_r^T \mathbf{D}_r^T \mathbf{\Sigma}^{-1} \mathbf{D}_r \mathbf{G}_r \tag{116}$$

$$\mathbf{G}_r = \mathbf{G}(t_r) = \{G_{ij}(t_r)\} = \left\{\frac{\partial h_i(t_r, \mathbf{x}(t_r))}{\partial x_j}\right\}; \quad \begin{array}{l} i = 1, 2, \ldots, v \\ j = 1, 2, \ldots, s \end{array} \tag{117}$$

$$\mathbf{D}_r = \mathbf{D}(t_r) = \{D_{ij}(t_r)\} = \left\{\frac{\partial x_i(t_r)}{\partial \beta_j}\right\}; \quad \begin{array}{l} i = 1, 2, \ldots, s \\ j = 1, 2, \ldots, p \end{array} \tag{118}$$

with

$$\frac{d\mathbf{D}}{dt} = \mathbf{A}(t) \cdot \mathbf{D}(t) + \mathbf{B}(t); \; \mathbf{D}(0) = \mathbf{0} \tag{119}$$

$$\mathbf{A}(t_r) = \{A_{ij}(t_r)\} = \left\{\frac{\partial f_i(t_r, \mathbf{x}(t_r); \mathbf{b})}{\partial x_j}\right\}; \quad i \text{ and } j = 1, 2, \ldots, s \tag{120}$$

$$\mathbf{B}(t_r) = \{B_{ij}(t_r)\} = \left\{\left.\frac{\partial f_i(t_r, \mathbf{x}(t_r); \boldsymbol{\beta})}{\partial \beta_j}\right|_{\boldsymbol{\beta}=\mathbf{b}}\right\}; \quad \begin{array}{l} i = 1, 2, \ldots, s \\ j = 1, 2, \ldots, p \end{array} \tag{121}$$

Equations 119 are the *sensitivity equations.* They describe the sensitivity of the state variables $\mathbf{x}$ with respect to the parameters $\boldsymbol{\beta}$. The procedure for determining the variance-covariance matrix is as follows: once the optimal parameter estimates $\mathbf{b}$ have been obtained, a new integration of the set of state equations 109 and the sensitivity equations 119, is started; at all sampling times t_r as well as at all intermediate times required by the numerical integration technique, the (calculated) values of the state variables $\mathbf{x}(t)$ emerge, so that the matrices $\mathbf{A}(t)$ and $\mathbf{B}(t)$ can be evaluated and introduced into the sensitivity equations; $\mathbf{D}_r$ and $\mathbf{G}_r$ are then calculated from equations 118 and 117; hence the matrix $\mathbf{G}_r^T\mathbf{D}_r^T\mathbf{\Sigma}^{-1}\mathbf{D}_r\mathbf{G}_r$ can be computed for each t_r and added to the running sum; after completion of all sampling points t_r, inversion of the final matrix $\mathbf{H}$ yields the covariance matrix of the parameter estimates. The evaluation of $\mathbf{V}(\mathbf{b})$ thus requires the numerical integration of $s + s \times p$ simultaneous first order ordinary differential equations. Rosenbrock and Storey also derive the equations for the case in which the initial conditions of the state variables are not known with certainty. These are considered as supplementary parameters and are estimated from the data along with the model parameters. This has slight repercussions on the expression for $\mathbf{H}$. When the error covariance matrix $\mathbf{\Sigma}$ is known, an approximate ellipsoidal joint confidence region at the $(1 - \alpha)$ probability level is given by

$$(\mathbf{b} - \boldsymbol{\beta})^T \mathbf{H}(\mathbf{b} - \boldsymbol{\beta}) = S(\mathbf{b}) \frac{p}{nv - p} F(p, nv - p, 1 - \alpha) \tag{122}$$

S(**b**) is the minimum value of the objective function 113. An exact joint confidence region at an approximate probability level of 1 — α is given by all sets of **β** which satisfy

$$S(\boldsymbol{\beta}) = S(\mathbf{b})\left[1 + \frac{p}{nv - p} F(p, nv - p, 1 - \alpha)\right] \tag{123}$$

When the error covariance matrix **Σ** is unknown and the parameters are estimated by criterion 114, the joint confidence region is defined by 94. Further details and documentation on these principles can be found in the work of Heineken et al. [54] and of Emig and Hosten [55].

5. Lumping

In industrial processes the reaction network may become very complicated, even more than in the phthalic anhydride synthesis example dealt with in 3.B.4. Severe convergence problems are likely when the number of parameters to be estimated is large. What is meant by "large" is difficult to say and depends on such factors like the structure of the network, the type of data and their accuracy, the mathematical form of the model.

In addition to the above situation many processes dealing with petroleum fractions like catalytic reforming, catalytic cracking, hydrocracking and hydrodesulphurization involve extremely complex mixtures, so that even a complete analysis of the feed is a problem. Each feed component often leads to a complicated reaction network in itself. Two steps can be taken to keep the complexity within tractable limits: reaction sequences in the network can be lumped or components of the feed can be lumped into a number of groups. Lumping of reaction sequences should not alter the essential characteristics of the network, bur very little has been done that has general significance so far. Lumping feed components or intermediates has to be done in such a way that the evolution of the lump is practically identical to that of the sum of its individual components.

Wei and Kuo [56] presented a very elegant technique to discover appropriate lumps for monomolecular reaction systems, described by the rate equations

$$\frac{d\mathbf{c}}{dt} = \mathbf{K}\mathbf{c}; \qquad \mathbf{c}(0) = \mathbf{c}_0 \tag{124}$$

The matrix **K** has to be known, however, so that this analysis is of little practical interest for the problems considered here, namely estimation of the unknown rate coefficient matrix **K** from experimental data.

An interesting extension of this approach has recently been presented by Bailey [57]. Luss and coworkers [58–62] derived lower and upper bounds for the lump concentration as a function of space time for reaction systems involving many parallel or consecutive first and n-th order reactions. These bounds can be constructed from the initial composition and the first and second moments of the experimental curve of lump concentration vs space time at initial conditions. If these bounds diverge too much the lump considered is inappropriate and further detailing is required. Weekman and coworkers [63] applied lumping to industrial catalytic cracking of gasoil. Liu and Lapidus [64] applied observer theory [65] with respect to lumping in monomolecular reaction systems. Finally, Ozawa [66] and Lee [67, 68] provided some guidance for choosing suitable lumps.

4. Discrimination Between Rival Models

The previous chapter dealt with parameter estimation in a *given* model. Quite frequently, however, the problem is, to select the appropriate mathematical model among a certain number of possible candidates. A few decades ago, valuable techniques were developed for selecting among rival candidates on the basis of already available experimental data. In recent years, however, attention has focused on optimal discrimination between rival models by means of experiments specifically designed for that purpose. The experimental program is then designed to achieve discrimination by means of a minimum number of experiments. Both aspects of model discrimination are discussed in what follows.

A. A Posteriori Techniques

1. Methods of Diagnostic Parameters

A diagnostic parameter is a quantity which somehow measures the behaviour of the model and thus contributes to the diagnosis of its suitability (Hunter and Mezaki [69], Kittrell, Hunter and Mezaki [70]).

a) Method of intrinsic parameters (Kittrell and Mezaki [71])

This method which makes use of parameters which are intrinsically present in the model is best outlined by means of an example [72].

Consider again the reaction $A \rightleftharpoons R + S$ and the rate equation 11 for surface reaction rate controlling. For a feed consisting of pure A:

$$p_A = \frac{1-x}{1+x}\, p_t; \qquad p_R = p_S = \frac{x}{1+x}\, p_t \tag{125}$$

The rate equation can be written in the form:

$$r_A = \frac{(1-x^2)\, p_t - x^2 p_t^2/K}{(C_1 + C_2 x)^2} \tag{126}$$

in which the denominator has been rearranged into two terms, one containing the conversion and one independent of conversion. The quantities C_1 and C_2 are functions of the kinetic parameters and the total pressure only:

$$C_1 = 1/\sqrt{kK_A} + K_A/\sqrt{kK_A}\, p_t \tag{127}$$

$$C_2 = 1/\sqrt{kK_A} + (K_R + K_S - K_A)/\sqrt{kK_A}\, p_t \tag{128}$$

When the model is correct, C_1 and C_2 have to satisfy a linear relationship in p_t. Also, both straight lines need to have the same intercept. To verify this, C_1 and C_2 have to be expressed in terms of the experimental variables. For initial conditions, i.e. $x = 0$, (126) reduces to:

$$C_1 = \sqrt{p_t/r_{A_0}} \tag{129}$$

From equations 127 and 129, it is clear that the test on C_1 entirely corresponds to the method of the initial rates, outlined in Section 3.A.1.a. This is a test on the model for low conversions. It is of course desirable to test the model at high conversions as well. With C_1 given by 129, 126 can be rewritten to give

$$C_2 = \left(\sqrt{\frac{(1 - x^2)\, p_t - x^2 p_t^2/K}{r_A}} - \sqrt{\frac{p_t}{r_{A_0}}}\right)\frac{1}{x} \tag{130}$$

A plot of C_1 and C_2 as a function of p_t for the ethanol dehydrogenation data of Franckaerts and Froment [12] at 275 °C is shown in Figure 9.

The linearity of both correlations confirms that the surface reaction rate model 126 is adequate for the complete range of conversions.

There is no need to apply the high conversion test to the adsorption and desorption models, since these can be rejected on the basis of the test on C_1: the adsorption model requires that p_t/r_{A_0} is independent of total pressure, whereas the desorption model requires that p_t/r_{A_0} is proportional to total pressure. Both conditions are incompatible with the observation that over a wide range $\sqrt{p_t/r_{A_0}}$ is proportional to p_t, as required by the rate expression when the surface reaction is rate determining. The effect of water in the feed has been neglected here, since it does not affect the diagnostics.

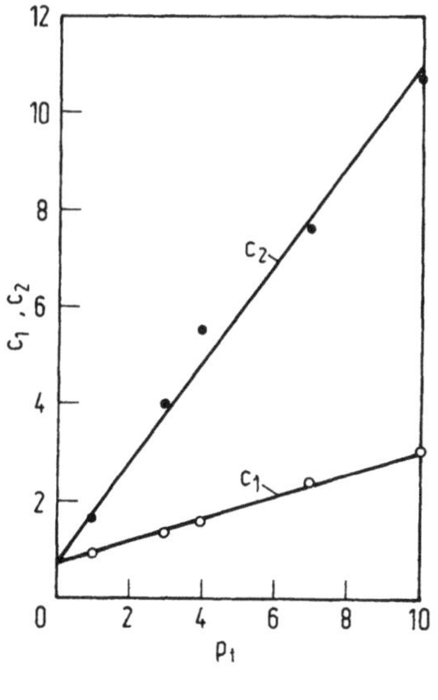

Figure 9. Application of method of intrinsic parameters to ethanol dehydrogenation data

b) Methods of non-intrinsic parameters

A non-intrinsic parameter does not appear in the model, but is introduced for the sole purpose of facilitating the discrimination between rival models.

i. Selection between two models

The method developed by Williams and Kloot [73] and brought to the attention again by Mezaki and Kittrell [74] only requires simple linear regression, an undeniable advantage.

Suppose a selection between two non linear models has to be made:

$$\begin{aligned} E(y) &= f_1(\mathbf{x}, \boldsymbol{\beta}_1) \\ E(y) &= f_2(\mathbf{x}, \boldsymbol{\beta}_2) \end{aligned} \tag{131}$$

A new dependent variable is defined:

$$z = y - \frac{1}{2}(\hat{y}_1 + \hat{y}_2) \tag{132}$$

where $\hat{y}_i$ is the least squares prediction of the dependent variable y under model i.

If Model 1 were correct, $\mathbf{b}_1$ would be close to $\boldsymbol{\beta}_1$ and

$$E(y) = E(\hat{y}_1) = \eta_1 \simeq \hat{y}_1 \tag{133}$$

Hence, when $\hat{y}_1 - \hat{y}_2$ is considered as an independent variable:

$$E(z) = \frac{1}{2}(\hat{y}_1 - \hat{y}_2) \tag{134}$$

If model 2 were correct:

$$E(z) = -\frac{1}{2}(\hat{y}_1 - \hat{y}_2) \tag{135}$$

Both relations 134 and 135 may be condensed into one single equation:

$$E(z) = \lambda(\hat{y}_1 - \hat{y}_2) \tag{136}$$

When model 1 is correct λ takes on the value $+0.5$, when model 2 is correct the value -0.5. Regression of z on $\hat{y}_1 - \hat{y}_2$ will point out which model is best. Since it cannot be expected that the least squares estimation will lead to a value of λ exactly corresponding to one of those two values, the confidence limits have to be evaluated. When this interval includes $+0.5$ and not -0.5, model 1 is the best candidate. In the reverse case, model 2 is preferred.

Example (4.A.1.b.i): *Dehydrogenation of ethanol*

This technique will now be applied to the previously used data on ethanol dehydrogenation at 275 °C. The rival models are the integrated conservation equations corresponding to the three possible rate determining steps, i.e. 64 with the appropriate expressions for C, a_1, b_1 and c_1. The adsorption and desorption equations are each compared with the surface reaction equation.

Equation 132 implies that all three models are first fitted to the data to obtain the fitted values of the dependent variable, the conversion. This fitting was already performed in Example (3.A.2.a). The predicted conversions for all experiments according to all models are computed from equation 64 using the parameter estimates listed in Table 4 and are presented in Table 8 together with the corresponding observed conversions. For a clear graphical representation, the Surface reaction model is labeled model 1 when analyzing the pair Surface reaction-Adsorption, whereas it is labeled model 2 when analyzing the pair Surface reaction-Desorption. Figure 10 shows the values of z vs. the difference in model predictions, i.e.

$$y - \frac{1}{2}(\hat{y}_s + \hat{y}_a) \text{ vs } \hat{y}_s - \hat{y}_a$$

for the Surface reaction-Adsorption pair, and

$$y - \frac{1}{2}(\hat{y}_d + \hat{y}_s) \text{ vs } \hat{y}_d - \hat{y}_s$$

for the Surface reaction-Desorption pair. Least squares straight lines are subsequently fitted through the two sets of data. Information on the slopes is also given in Figure 10. The value +0.5 is nearly included in the confidence interval of the slope, related to the first pair of models indicating that model 1, i.e. surface reaction rate determining is the appropriate model, whereas −0.5 is included in the confidence interval of the slope related to the second

Table 8. Ethanol dehydrogenation. Observed conversions and values predicted by the surface-, adsorption- and desorption models 64

y	$\hat{y}_s$	$\hat{y}_a$	$\hat{y}_d$	y	$\hat{y}_s$	$\hat{y}_a$	$\hat{y}_d$
0.118	0.1287	0.1152	0.0905	0.214	0.227	0.2131	0.2404
0.196	0.2117	0.2114	0.1795	0.254	0.2664	0.2423	0.3163
0.262	0.2728	0.2792	0.238	0.1	0.0927	0.126	0.0969
0.339	0.3371	0.3443	0.3069	0.148	0.1398	0.1528	0.1571
0.446	0.4346	0.4341	0.4154	0.175	0.1696	0.1696	0.1921
0.454	0.4424	0.441	0.4242	0.188	0.1963	0.1858	0.2192
0.524	0.5241	0.5128	0.5166	0.229	0.2253	0.2102	0.2282
0.59	0.5802	0.565	0.579	0.082	0.08	0.0807	0.1176
0.6	0.5869	0.5718	0.5863	0.14	0.1434	0.1463	0.2034
0.14	0.1355	0.1803	0.0982	0.23	0.2391	0.2423	0.3202
0.2	0.2064	0.2301	0.1712	0.17	0.1657	0.1642	0.1929
0.25	0.2534	0.2634	0.2252	0.198	0.2035	0.2033	0.2365
0.286	0.2984	0.2895	0.2794	0.26	0.2674	0.2675	0.3085
0.352	0.3616	0.3342	0.3561	0.108	0.1025	0.0935	0.1105
0.14	0.1282	0.1696	0.098	0.173	0.177	0.1711	0.1937
0.196	0.1933	0.2113	0.1689	0.3	0.2822	0.2794	0.3107
0.235	0.2357	0.2372	0.2195	0.123	0.1152	0.1021	0.1025
0.271	0.2756	0.2622	0.2681	0.193	0.1975	0.1917	0.1852
0.32	0.3294	0.3007	0.3315	0.3	0.3038	0.3042	0.2986
0.112	0.1077	0.1428	0.0975	0.195	0.1853	0.2085	0.1779
0.163	0.1616	0.1743	0.1627	0.28	0.2655	0.2636	0.27
0.194	0.1958	0.194	0.2046				

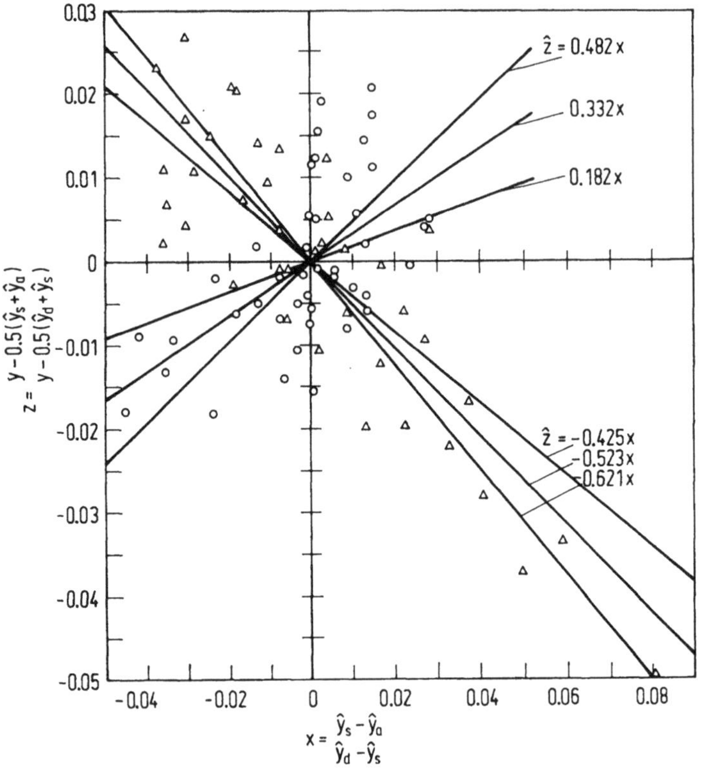

Figure 10. Application of method of non-intrinsic parameter to ethanol dehydrogenation data

pair of models, indicating that model 2, again the surface reaction model, is the best candidate. The surface reaction model is unambiguously selected among the three rival models.

ii. Selection between more than two models

Wilks [75] developed a technique for simultaneously comparing m non-linear models (see also Himmelblau [17]):

$$\begin{aligned} y &= f_1(\mathbf{x}, \boldsymbol{\beta}) + \varepsilon \\ y &= f_2(\mathbf{x}, \boldsymbol{\beta}) + \varepsilon \\ &\vdots \\ y &= f_m(\mathbf{x}, \boldsymbol{\beta}) + \varepsilon \end{aligned} \tag{137}$$

After fitting all these models to the data, the dependent variable y is approximated by a linear combination of the predicted values under the m models:

$$y = \alpha_1 \hat{y}_1 + \alpha_2 \hat{y}_2 + \ldots \alpha_m \hat{y}_m (+\delta) \tag{138}$$

δ is a fictitious error term.

Before estimating the non-intrinsic parameters $\alpha_1, \alpha_2, \ldots, \alpha_m$, these are scaled, so that their sum and hence the sum of their estimates equals 1:

$$\sum_{i=1}^{m} \alpha_1 = \sum_{i=1}^{m} a_i = 1 \tag{139}$$

Estimation of $\boldsymbol{\alpha}$, subject to 139 is a problem of constrained optimization, which is solved by means of the classical Lagrangian multiplier technique.

The criterion is written:

$$S(\boldsymbol{\alpha}) = \boldsymbol{\delta}^T\boldsymbol{\delta} + \lambda(\alpha_1 + \alpha_2 + \ldots + \alpha_m - 1) \xrightarrow{\boldsymbol{\alpha}, \lambda} \text{Min} \tag{140}$$

with λ the unknown Lagrangian multiplier.

Setting the derivatives of $S(\boldsymbol{\alpha})$ with respect to $\boldsymbol{\alpha}$ and λ zero leads to

$$\mathbf{a} = \frac{(\mathbf{Y}^T\mathbf{Y})^{-1}\,\mathbf{1}}{\mathbf{1}^T(\mathbf{Y}^T\mathbf{Y})^{-1}\,\mathbf{1}} \tag{141}$$

in which

$$\mathbf{Y} = \{Y_{ij}\} = \{y_i - \hat{y}_{ij}\}$$

$$\mathbf{Y}^T\mathbf{Y} = \{(\mathbf{Y}^T\mathbf{Y})_{kl}\} = \left\{\sum_{i=1}^{n} (y_i - \hat{y}_{ik})(y_i - \hat{y}_{il})\right\}$$

$$\mathbf{1}^T = [1 \quad 1 \ldots 1]$$

From 141 it is seen that the estimate a_i is the sum of the elements of the i-th row of $(\mathbf{Y}^T\mathbf{Y})^{-1}$, divided by the sum of all its elements.

An estimate of the covariance matrix of the estimates $\mathbf{a}$ is given by

$$\hat{\mathbf{V}}(\mathbf{a}) = \frac{1}{n - m + 1}(\mathbf{Y}^T\mathbf{Y})^{-1}\left[\mathbf{I} - \frac{\mathbf{1}.\mathbf{1}^T(\mathbf{Y}^T\mathbf{Y})^{-1}}{\mathbf{1}^T(\mathbf{Y}^T\mathbf{Y})^{-1}\,\mathbf{1}}\right] \cdot \frac{1}{\mathbf{1}^T(\mathbf{Y}^T\mathbf{Y})^{-1}\,\mathbf{1}} \tag{142}$$

Due to the scaling, the magnitudes of the estimates a_i are a direct measure of the suitability of the model: a_i is the fraction of $\mathbf{y}$ that is accounted for by model i; the closer a_i is to 1, the smaller the contribution of all models, but the i-th is required to approximate $\mathbf{y}$. Further, approximate t-tests against a fixed reference value $\bar{a}_i$ may be carried out:

$$H_0 : a_i = \bar{a}_i \rightarrow \frac{a_i - \bar{a}_i}{s(a_i)} \sim t(n - m + 1) \tag{143}$$

Alternatively two contributing terms a_i and a_j, whereby $a_i > a_j$, may be tested against each other upon significant difference:

$$H_0 : a_i > a_j \rightarrow \frac{a_i - a_j}{s(a_i - a_j)} \sim t(n - m + 1) \tag{144}$$

When calculating the standard deviation $s(a_i - a_j)$ it should be remembered that a_i and a_j are not statistically independent, so that the covariance between a_i and a_j must be taken into account:

$$\mathrm{Var}(a_i - a_j) = \mathrm{Var}(a_i) + \mathrm{Var}(a_j) - 2\,\mathrm{Cov}(a_i, a_j) \tag{145}$$

Due to the null hypothesis formulation, H_0, it is obvious that 144 is a one-sided test, so that the computed value of the test quantitiy must be compared with $t(n - m + 1, 1 - \alpha)$.

Example (4.A.1.b.ii): *Dehydrogenation of ethanol*

With the data given in Table 8, the matrices **Y** and $\mathbf{Y}^T\mathbf{Y}$ are immediately prepared. The estimates of the Wilks parameters and their covariance matrix are computed according to 141 and 142 in a straightforward manner. Estimates and individual t-tests at the 95% probability level against a reference value of zero are presented below.

	a_1	a_2	a_3
estimate	0.8223	0.1712	0.0064
$t = a_i/s(a_i)$	8.01	2.19	0.13

The figures exhibit a very pronounced preference for model 1, i.e. the Surface reaction model. From a comparison of the calculated t-values with the 5-percentage point of the t-distribution with $43 - 3 + 1 = 41$ degrees of freedom, 1.684, it follows that a_1 is strongly and a_2 only moderately significantly different from zero. Hence the Surface reaction model is undoubtedly selected.

2. The Likelihood Ratio as a Discrimination Criterion

Selection between models on the basis of the likelihood ratio [76–79] is discussed here for normally distributed errors only. The joint probability density function of the n experimental errors, having properties summarized in 33, or the likelihood function, is given by

$$p(\boldsymbol{\varepsilon}) = \frac{1}{(\sqrt{2\pi})^n \sigma^n} \exp\left(-\frac{1}{2\sigma^2}\boldsymbol{\varepsilon}^T\boldsymbol{\varepsilon}\right) \tag{146}$$

The maximum likelihood estimate s^2 for σ^2 is proved to be

$$s^2 = \mathbf{e}^T\mathbf{e}/n \tag{147}$$

where **e** is the vector of residuals $\mathbf{y} - \hat{\mathbf{y}}$.

For selection between two models the ratio of the maximum values of both likelihoods is calculated:

$$\frac{(L_1)_{max}}{(L_2)_{max}} = \left(\frac{\mathbf{e}_2^T\mathbf{e}_2}{\mathbf{e}_1^T\mathbf{e}_1}\right)^{n/2} \tag{148}$$

This ratio measures the odds in favor of model 1 above model 2 and is called the likelihood ratio. To come to a practical arrangement, two numbers, A and B, are defined such that:

$$0 < B < 1 < A \tag{149}$$

These numbers are chosen in accordance with the risk of not rejecting a wrong hypothesis [80], i.e. wrong model. These probabilities are given below.

	Hypothesis 1 correct	Hypothesis 2 correct
Accept Hypothesis 1	$1 - \zeta$	α
Accept Hypothesis 2	ζ	$1 - \alpha$

α and ζ are small numbers like 0.05 or 0.1. A and B are then related to α and ζ according to:

$$A = \frac{1 - \zeta}{\alpha}, \qquad B = \frac{\zeta}{1 - \alpha} \tag{150}$$

The strategy is as follows. If

$$\frac{(L_1)_{max}}{(L_2)_{max}} < B \qquad \text{Model 1 is rejected} \tag{151}$$

$$\frac{(L_1)_{max}}{(L_2)_{max}} > A \qquad \text{Model 1 is preferred} \tag{152}$$

$$A < \frac{(L_1)_{max}}{(L_2)_{max}} < B \qquad \text{No decision can be made} \tag{153}$$

When a selection has to be made among more than two models, the model with the smallest residual sum of squares can be taken as reference and all other models compared with this reference. In the above form 148, the likelihood ratio test is strictly applicable only when the models contain the same number of parameters. If not, the biased estimate 147 for σ^2 should be replaced by the unbiased estimate 31. The amended likelihood ratio is then calculated according to

$$\frac{(L_i)_{max}}{(L_j)_{max}} = \left(\frac{n - p_i}{n - p_j}\right)^{n/2} \exp \frac{p_i - p_j}{2} \left(\frac{\mathbf{e}_j^T \mathbf{e}_j}{\mathbf{e}_i^T \mathbf{e}_i}\right)^{n/2} \tag{154}$$

An illustration of the use of the likelihood ratio will be incorporated in Example (IV.B.1)., dealing with a sequential experimental design strategy for discrimination.

3. Physico-Chemical Nature of the Parameters

Parameters in a mechanistic model have a physico-chemical meaning so that they are subject to constraints: rate coefficients and adsorption equilibrium constants e.g. cannot be negative. Hence, when their estimates are significantly negative in the statistical sense the model cannot be accepted (Chou [81]). When the estimate is negative but not significantly, the assumptions giving rise to the model will not be rejected as a whole. Such a parameter is not worse than a parameter which is positive, but not significantly different from 0. What has to be done is to determine it more precisely, or to delete it from the model, if possible.

The reaction rate coefficient increases exponentially with temperature according to the Arrhenius law:

$$k = e^{-\Delta H/RT + \Delta S/R} \tag{155}$$

Adsorption equilibrium constants decrease according to a similar law. Therefore, estimates of activation energies have to be statistically positive and those for adsorption enthalpies negative. In addition to these stringent conditions on ΔH, Boudart et al. [82] developed a few criteria which should be fulfilled by the estimates of ΔS.

4. Goodness of Fit

If more than one model remains in competition after application of the various criteria given above there is not much left than a scrutiny of the deviations between observed values and those predicted by the model. For all competing models, model adequacy (lack of fit) F-tests should be conducted, whenever possible. Plots of residual deviations against the calculated values of the dependent variable and the settings of the independent variables are also recommended [15].

These should show uniform bands: trends point towards model inadequacies. Ultimately the model which fits the data best and/or contains the smallest number of parameters will be preferred among the remaining rivals.

B. Sequential Experimental Designs For Discrimination Between Rival Models

In a sequential procedure, each experiment is designed only after the previous experiment has been performed and all available data have been subjected to a simultaneous analysis. The information thus obtained is then utilized in the design of the next experiment so as to make it optimal for discrimination purposes.

A sequential design for discrimination comprises two criteria: the design criterion, which selects the experimental conditions for the next experiment accounting for the current information, and the model screening criterion, which tests the plausibility of the competing models as the data are gathered. Both criteria are inserted into a loop, which also includes a module for updating the parameters, i.e. by re-estimating the parameters in all rival models, by means of the method of least squares e.g. This cycle is schematically represented in Figure 11.

The most difficult and time consuming step in this sequence of operations is undoubtedly the conduction of the experiment. The three other stages consist of purely mathematical operations. Among these the parameter estimation is the dominant load.

The techniques described in what follows heavily rely on statistical principles. Unless otherwise stated, the experimental error is assumed to possess the properties summarized in 33 while the nonlinear models are considered to behave linearly in the vicinity of the maximum likelihood parameter estimates.

1. The Box-Hill Criterion

Hunter and Reiner [83] were the first to propose a systematic experimental design procedure for discrimination between two rival models. Starting from statistical considerations they arrived at a simple expression for the measure of the divergence or discrimination between the two models:

$$D(\mathbf{x}_k) = [\hat{y}^{(1)}(\mathbf{x}_k) - \hat{y}^{(2)}(\mathbf{x}_k)]^2 \tag{156}$$

in which $\hat{y}^{(i)}(\mathbf{x}_k)$ is the prediction of the dependent variable corresponding to the settings $\mathbf{x}_k$ under model i. Given some preliminary values for the parameters, so that $D(\mathbf{x}_k)$ can be evaluated for any feasible combination $\mathbf{x}_k$, the next experiment has to be conducted at those settings $\mathbf{x}^*$ for which $D(\mathbf{x})$ is maximum.

Box and Hill [84] argued that the divergence is not only determined by the magnitude of the predicted response values but also by their statistical uncertainties: the $\hat{y}^{(i)}$ variables are indeed random and not deterministic quantities. Large differences between the model predictions are not necessarily significant when the associated uncertainty is important. This is illustrated in Fig. 12 which shows the current predictions of two models and the associated confidence limits. An experiment, leading to a result that falls in the region where the confidence limits overlap has less decisive power than an experiment in the vicinity of x_3 e.g., where the difference between predictions is not maximum, but where no overlapping of confidence limits occur. For a linear model the variance of $\hat{y}$ at settings $\mathbf{x}_k$ is given by

$$V(\hat{y}(\mathbf{x}_k)) = \sigma_k^2 = \mathbf{x}_k^T(\mathbf{X}^T\mathbf{X})^{-1}\,\mathbf{x}_k\sigma^2 \tag{157}$$

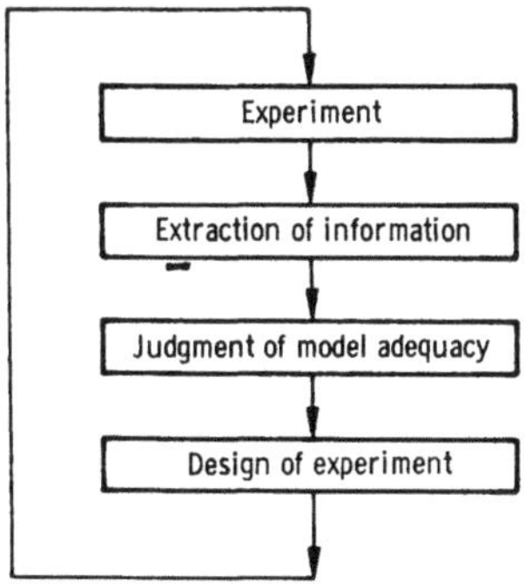

Figure 11. Schematic representation of a sequential discrimination procedure

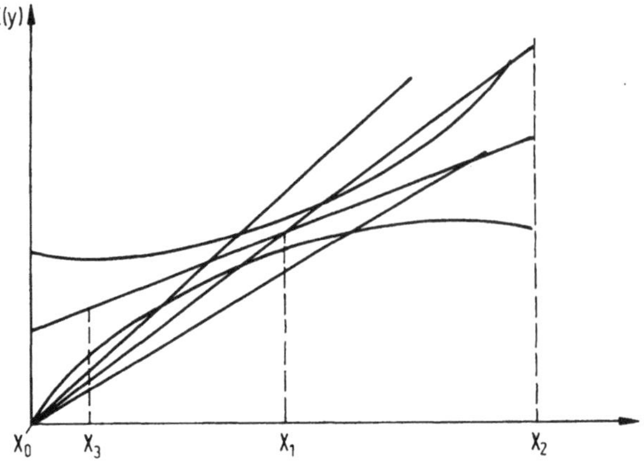

Figure 12. Rival models

where **X** is the matrix of settings of the independent variables with respect to the n experiments, defined in 26. For nonlinear models, the settings have to be replaced by the partial derivatives of the dependent variable with respect to the parameters:

$$V(\hat{y}^{(i)}(\mathbf{x}_k)) = \sigma_{ki}^2 = \mathbf{v}_k^{(i)^T}(\mathbf{X}_i^T\mathbf{X}_i)^{-1}\,\mathbf{v}_k^{(i)}\sigma^2 \tag{158}$$

The superscript in **v** and the subscript in **X** refer to the model. Starting from information theory Box and Hill derived the following expression for the divergence between m rival models:

$$D(\mathbf{x}_k) = \sum_{i=1}^{m-1}\sum_{j=i+1}^{m} \pi_{i,n}\pi_{j,n}\left\{\frac{\sigma_{ki}^2 - \sigma_{kj}^2}{(\sigma^2+\sigma_{ki}^2)(\sigma^2+\sigma_{kj}^2)} - [\hat{y}^{(i)}(\mathbf{x}_k) - \hat{y}^{(j)}(\mathbf{x}_k)]^2 \times \left(\frac{1}{\sigma^2+\sigma_{ki}^2} + \frac{1}{\sigma^2+\sigma_{kj}^2}\right)\right\} \tag{159}$$

After having performed and analyzed n experiments, the (n + 1)-th run has to be conducted at those settings of **x** which maximize D(**x**). $\Pi_{i,n}$ is the probability or degree of adequacy of model i, reached after n experiments. It is called a priori probability with respect to the design of run (n + 1). After the experiment has been performed this probability is updated and called a posteriori probability. In turn this probability will then serve as a priori probability in the design of experiment n + 2.

From an inspection of equation 159 it is clear that

1) experimental conditions giving rise to largely diverging model predictions, $\hat{y}$, strongly contribute to D
2) the divergence is inversely proportional to the variances of the predicted values of the response variable, so that conditions giving rise to predictions which largely diverge but which are highly unreliable at the same time, contribute to a lesser extent
3) since the prior probabilities act as multipliers greater weight is given to models which approximate the data best

After completion of the observation, the models are tested again upon their adequacy. Box and Hill express this degree of adequacy by means of the Bayesian posterior probabilities of the models:

$$\pi_{i,n+1} = \frac{\pi_{i,n} p_i}{\sum_{j=1}^{m} \pi_{j,n} p_j}; \quad i = 1, 2, \ldots, m \tag{160}$$

where p_i is the probability density of the (n + 1)-th observation, y_{n+1}, under model i:

$$p_i = p_i(y_{n+1}) = \frac{1}{\sqrt{2\pi(\sigma^2 + \sigma^2_{n+1,i})}} \exp\left[-\frac{1}{2(\sigma^2 + \sigma^2_{n+1,i})}(y_{n+1} - \hat{y}^{(i)}_{n+1})^2\right] \tag{161}$$

Discrimination is achieved as soon as the posterior model probability of one model closely approaches the value of 1.

Although $D(\mathbf{x})$ is a continuous function of $\mathbf{x}$ and hence could be optimized by some of the optimization techniques mentioned in paragraph 3.A.1.c, grid search is commonly used. The experimental region of interest is delimited and a grid is constructed; each grid point corresponds to a feasible experimental combination; $D(\mathbf{x})$ is then evaluated in each node. Figure 13 is a flow diagram of the Box-Hill sequential experimental design procedure.

Hunter and Hill [85] extended the single response criterion (159) to multi-response cases.

Example (4.B.1): *Catalytic hydrogenation of mixed iso-octenes*
The catalytic hydrogenation of mixed iso-octenes was investigated by Hougen and Watson [86] in a study which has become a classic and which was elaborated further by Blakemore and Hoerl [87].

Using the criteria described above two models were retained out of 17 rivals:

$$r_A = \frac{k p_A p_B}{(1 + K_A \sqrt{p_A} + K_B p_B + K_C p_C)^3} \tag{162}$$

$$r_A = \frac{k p_A p_B}{(1 + K_A p_A + K_B p_B + K_C p_C)^2} \tag{163}$$

where A, B and C refer to hydrogen, **i**-octenes and **i**-octanes respectively. Bard and Lapidus [1] performed a computer simulation of a sequential experimental program for discrimination between these two rate equations. Such a computer simulation technique is a powerful tool for evaluating developments in mathematical modeling. It consists of defining a model and a set of parameter values which are considered to adequately describe the physical phenomena. In practice, these true parameter values are never known, of course. "Observations" are subsequently constructed by adding random deviates to the calculated true response values. These random deviates represent the experimental errors which would have been committed in practice and hence have to be generated in such a way that they possess the properties set forth in estimation theory.

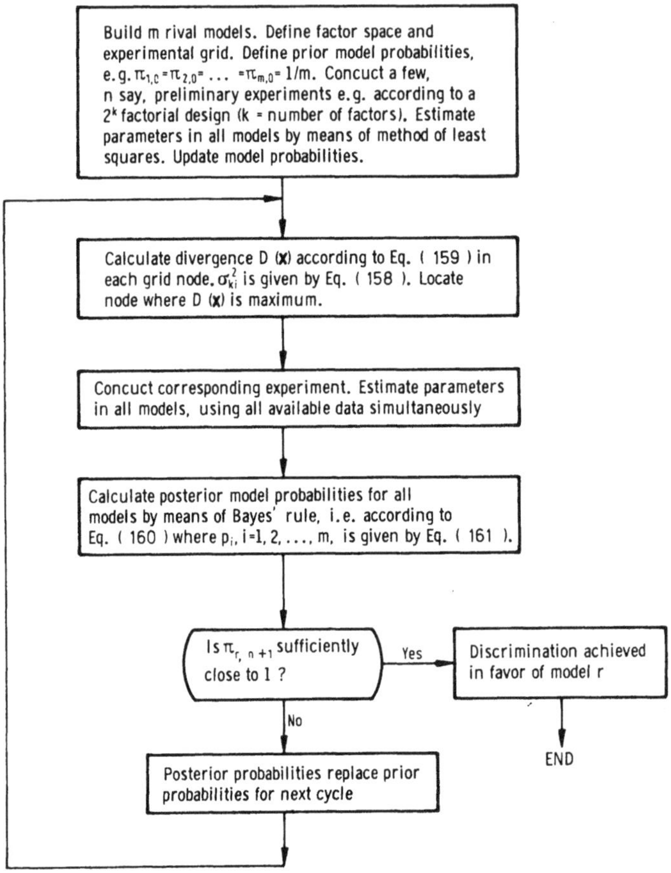

Figure 13. Details of the Box-Hill sequential experimental design for discrimination

Bard and Lapidus generated a series of 24 experiments by means of the Box-Hill criterion, using rate equation 163 as the correct model. To calculate the model predictions and the associated variances featuring in the design criterion 159, some information on the parameters in the models is required. This is why a few preliminary experiments have to be conducted before the sequential procedure can be initiated.

To find out how soon the procedure would overcome misleading results, caused by possible "dishonest" experimental errors e.g., six preliminary experiments were generated by means of the "false" rate equation 162. The settings of the independent variables p_A, p_B and p_C, were based upon an a priori fractional 3^3-factorial design (see e.g. Hicks [88]). These independent variables were confined to the following grid:

p_A: 0.1; 1.3; 2.5

p_B: 0.1; 1.55; 3

p_C: 0.05; 1.375; 2.7.

From the likelihood ratio, comparing the odds in favour of model 163 over model 162, the confidence in preferring model 163 over model 162 can be defined as

$$C = \frac{(L_2)_{max}}{(L_2)_{max} + (L_1)_{max}} \tag{164}$$

This confidence was used to test the adequacy of model 163 after each experiment. The results of the discrimination are presented in Table 9. The C-values indicate that the procedure unambiguously selects the correct model.

To compare the performance of such a sequential directed approach with that of a classical design, Bard and Lapidus also simulated a program of 27 runs according to a full 3^3 factorial design. These are formed by considering all possible combinations of the three independent variables p_A, p_B and p_C at the three levels which constitute the grid mentioned above. This factorial design therefore also includes the six preliminary experiments used in the sequential approach. The last line of Table 9 summarizes the result. The final confidence in model 163, 0.481, indicates that there is complete indifference between the two models after completion of the experimental program. The corresponding confidence of 0.978 for the sequential approach strikingly illustrates the power of the latter.

2. The Hsiang-Reilly Procedure

The posterior model probabilities as defined in 160 do not account for the parameter distributions. This is equivalent with assuming that these distributions are diffuse, or "locally" uniform, i.e. uniform in the "region of interest". Yet, knowledge about the parameter distribution increases as experimentation progresses and information is gathered. Hsiang and Reilly [89] developed another Bayesian technique which explicitly takes these distributions into account.

This procedure is not based on a linearization of the model, and does not require the experimental error to satisfy 33. The joint error density has to be known, but may be arbitrary. Knowledge of distribution parameters, like mean or variance, may even be lacking. These are then treated as supplementary model parameters.

3. Non-Bayesian Discrimination Procedures

The Bayesian developments are based on advanced statistical principles. It is important also to direct the attention on an alternative non-Bayesian attack of the problem [90].

Table 9. Hydrogenation of mixed iso-octenes. Simulation of a sequential design for optimal discrimination and comparison with a factorial approach

Expt.	p_A	p_B	p_C	r	c
1	0.1	1.55	1.375	0.0041	
2	2.5	1.55	1.375	0.07932	
3	1.3	0.1	1.375	0.00508	
4	1.3	3	1.375	0.05633	
5	1.3	1.55	0.05	0.04912	
6	1.3	1.55	2.7	0.04292	5×10^{-5}
7	2.5	3	2.7	0.09769	0.563
8	1.78	0.68	0.05	0.03709	0.391
9	0.58	2.42	0.05	0.02649	0.654
10	1.78	0.39	0.05	0.02243	0.801
11	2.5	1.84	1.375	0.08126	0.737
12	0.34	2.42	1.11	0.01589	0.759
13	1.54	0.68	0.05	0.03363	0.836
14	0.34	2.42	3.15	0.01588	0.874
15	2.02	3	2.17	0.08211	0.901
16	2.5	1.84	1.64	0.08494	0.861
17	1.54	0.68	0.05	0.03364	0.911
18	0.34	2.42	0.05	0.01638	0.924
19	2.02	3	2.435	0.07961	0.853
20	2.5	1.84	1.905	0.08108	0.924
21	1.54	0.68	0.05	0.03411	0.956
22	0.34	2.42	0.05	0.01525	0.975
23	1.54	0.68	0.05	0.03146	0.956
24	0.34	2.42	0.05	0.01627	0.964
25	2.02	3	2.7	0.08174	0.980
26	2.5	1.84	2.17	0.07816	0.987
27	1.54	0.68	0.05	0.03096	0.978
28	2.5	1.84	1.11	0.07775	0.974
29	0.34	2.42	1.11	0.01610	0.978
30	2.5	1.84	2.7	0.08084	0.983
Full 3^3 factorial design = 27 experiments					0.481

a) Model adequacy and design criteria

The fundamental idea behind the model adequacy test is that the minimum residual sum of squares, divided by its number of degrees of freedom, viz 31, is an unbiased estimate of the variance of the experimental error, for the correct model only. For any rival, non-adequate model, this quantity over-estimates the error variance, since it estimates the true error variance plus the systematic deviation of the model from the data, called lack of fit. Hence, experiments are performed until the lack of fit of the non-adequate models is so pronounced that it can be detected by some statistical test. Three cases may be considered, depending upon the degree of knowledge about the error variance σ^2.

—σ^2 known.

The random variable

$$\chi_i^2 = \frac{(n - p_i)\, s_i^2}{\sigma^2} \sim \chi^2(n - p_i) \tag{165}$$

is distributed like χ^2 with $n - p_i$ degrees of freedom provided s_i^2 is an unbiased estimate of σ^2. When σ^2 is precisely known, this property may be used to test upon lack of fit. The calculated values χ_i^2 are compared with the α-percentage points of the χ^2 distribution with appropriate numbers of degrees of freedom, $\chi^2(n - p_i, 1 - \alpha)$. Models for which $\chi_i^2 > \chi^2(n - p_i, 1 - \alpha)$ are rejected on the basis of a significant lack of fit.

—σ^2 unknown, but an independent estimate s_e^2 is available. An independent estimate of the pure error variance may be obtained by replicating n_e times, a given experiment:

$$s_e^2 = \frac{\sum_{i=1}^{n_e} (y_i - \bar{y})^2}{n_e - 1} \tag{166}$$

The random variable

$$F_i = \frac{s_i^2}{s_e^2} \sim F(n - p_i, n_e - 1) \tag{167}$$

is distributed like F with $n - p_i$ and $n_e - 1$ degrees of freedom provided s_i^2 is an unbiased estimate for σ^2. The model adequacy is tested by comparing the calculated F_i values with the corresponding α-percentage points of the F distribution. Models for which $F_i > F(n - p_i, n_e - 1, 1 - \alpha)$ are discarded.

— No information on σ^2.

The m estimates s_i^2 of error variance plus eventual lack of fit may be tested upon statistical equality or homogeneity. A number of tests have been developed and an easily applicable one is Bartlett's χ^2 test [91, 92]. The random variable

$$\chi_c^2 = \frac{\ln \bar{s}^2 \sum_{i=1}^{m} (\mathrm{d.f.})_i - \sum_{i=1}^{m} (\mathrm{d.f.})_i\, s_i^2}{1 + \frac{1}{3(m-1)}\left[\sum_{i=1}^{m} \frac{1}{(\mathrm{d.f.})_i} - \frac{1}{\sum_{i=1}^{m} (\mathrm{d.f.})_i}\right]} \tag{168}$$

is distributed like χ^2 with $m - 1$ degrees of freedom. The symbols have the following meaning:

$(\mathrm{d.f.})_i$: number of degrees of freedom associated with s_i^2, namely $n - p_i$

m: current number of models in competition

s_i^2: estimate of σ^2 plus lack of fit, obtained through model i

$\bar{s}^2$: pooled estimate of σ^2 plus lack of fit, based on all models left in competition:

$$\bar{s}^2 = \sum_{i=1}^{m} (\mathrm{d.f.})_i\, s_i^2 \Big/ \sum_{i=1}^{m} (\mathrm{d.f.})_i \tag{169}$$

If the calculated value of χ_c^2 exceeds $\chi^2(m - 1, 1 - \alpha)$ the m estimates s_i^2 are no longer homogeneous with a probability of $1 - \alpha$ and the lack of fit of certain models is so pronounced that they can be rejected. The numerical value of χ_c^2 alone does not point out which models, however. For this reason the following procedure is adopted: the model giving rise to the largest s_i^2 is rejected and the test is repeated with a new pooled estimate $\bar{s}^2$, based upon the remaining models only; the new χ_c^2 value is compared with the $\chi^2\alpha$-percentage point with one degree of freedom less than in the previous stage, since one degree of freedom is lost each time a model is rejected. This cycle is repeated until the appropriate α-percentage point is not exceeded any longer. The remaining estimates s_i^2 are homogeneous and the corresponding models competitive. The design stage is then entered, followed by the conduction of the experiment and the retesting of the models. This cycle is repeated until one model is left, or until further discrimination appears to be impossible.

The first two tests of this paragraph are true adequacy tests in the statistical sense. The Bartlett test, and the Bayesian tests of both previous paragraphs are not. These simply select the best candidate among a number of models each of which may be inadequate. Additional adequacy verifications are then required, like an analysis of residuals.

Scatter tests appear to be sensitive to deviations from normality. If there is evidence that the experimental error is not normally distributed, Bartlett's test is not recommended in the form 168. Eventually, an alternative, developed by Box [93] and based upon the more robust location statistics may be used.

For a rigorously valid application of Bartlett's test, the s_i^2 estimates should be statistically independent. This is not the case here, since they are obtained from the same sample of experimental data. The repercussion of this dependence upon the outcome of the test cannot be evaluated theoretically. From a number of simulated studies, the technique appears to be reliable and efficient.

Since in the approach described here the model adequacy and design criteria are completely independent of one another, any type of design criterion can be used. An efficient criterion, which is based on model predictions only, is given by

$$D(\mathbf{x}_k) = \sum_{i=1}^{m-1} \sum_{j=i+1}^{m} |\hat{y}^{(i)}(\mathbf{x}_k) - \hat{y}^{(j)}(\mathbf{x}_k)| \tag{170}$$

The double summation is required to take each model consecutively as reference in order to avoid mislocations of optimal discrimination conditions. This is illustrated in Figure 14. If Model 1 would be chosen as the reference for calculating the divergence, the next experiment would be conducted at $x = x_2$. With Model 3 as reference the conditions at $x = x_1$ would appear to be preferable. These are contradictory conditions, however, caused by the fact that not all models have been taken as reference.

Criterion 170 can easily be extended to account for both the model probabilities and the statistical uncertainties associated with the $\hat{y}(\mathbf{x}_k)$ values [94].

b) Case study: dehydrogenation of 1-butene

Dumez and Froment [95] studied the kinetics of the industrially important dehydrogenation of 1-butene into butadiene on a commercial Cr_2O_3/Al_2O_3 catalyst. Various plausible reaction mechanisms and rate determining steps led to 15 rate equations of the Hougen-Watson type which are presented in ref. 95. Experiments were conducted in a differential reactor. The independent variables were the partial pressures of butene, p_B, butadiene, p_D and hydrogen p_{H_2}. Hence, the factor-space is threedimensional. It is limited

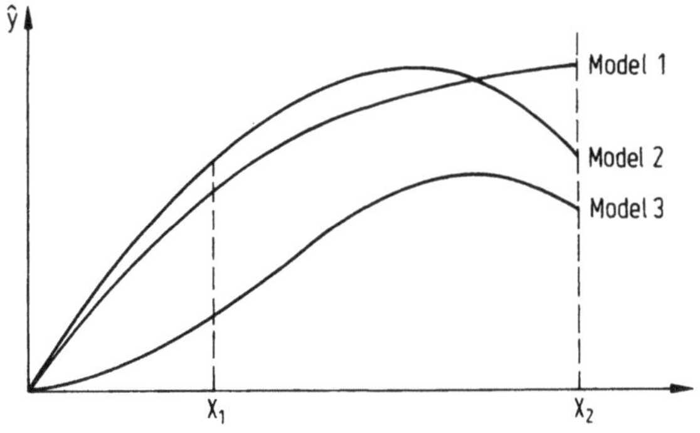

Figure 14. Rival models

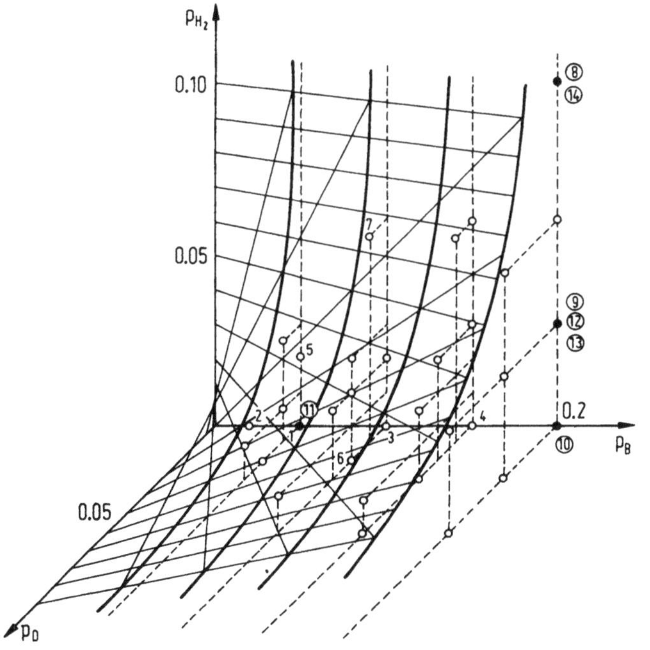

Figure 15. Dehydrogenation of 1-butene. Experimental region and grid. ○: feasible conditions, ●: actually selected conditions 1,2, ..., 7 preliminary experiments; ⑧, ..., ⑭ designed experiments

Table 10. Dehydrogenation of butene. Evolution of sequential Model Discrimination

Location in Grid (Fig. IV-7)		8	9	10	11	12 ≡ 9	13 ≡ 9	14 ≡ 8
Settings of variables (atm)								
p_H	0.1	0.025	0	0				
p_B	0.2	0.2	0.2	0.05				
p_D	0	0	0	0				
χ^2 (m — 1, 0.95)	χ^2_c and eliminated models							
23.68	50.59	68.80	85.31	84.80	98.38	114.14	129.44	139.13
	a3	a3	a3	a3	a3	a3	a3	a3
22.36	43.66	58.78	72.61	69.55	80.84	93.82	106.40	113.60
	b3	b3	b3	b3	b3	b3	b3	b3
21.03	32.50	42.19	50.55	43.37	50.86	58.81	66.39	69.17
	a4	a4	a4	a4	a4	a4	a4	a4
19.68	27.36	35.44	42.47	36.67	43.81	50.87	57.58	59.70
	b4	b4	b4	b4	b4	b4	b4	b4
18.31	20.18	25.82	30.72	28.03	34.75	40.40	45.75	46.92
	c3	c3	c3	c3	c3	c3	c3	c3
16.92	1.91	3.37	4.56	14.04	20.59	23.74	26.66	26.39
					a1	a1	a1	a1
15.51					18.83	21.79	24.53	24.39
					b1	b1	b1	b1
14.07					16.59	19.20	21.78	21.45
					d1	d1	d1	d1
12.59					13.95	16.23	18.32	18.18
					e1	e1	e1	e1
11.07					9.38	10.97	12.42	12.06
							c1	c1
9.49							0.44	1.28

by the hyperboloidic equilibrium surface and is shown in Figure 15 for the experiments at 525 °C. The equilibrium surface is represented by intersections parallel to the three coordinate planes. Feasible experimental combinations are represented by dots. Conditions too close to the equilibrium state were avoided for obvious reasons.

Dumez, Hosten and Froment [96] reported on the performance of a sequential discrimination strategy based on 168 and 170. This study was probably the first in which the model discrimination was entirely and uniquely based upon such a fully systematized procedure. Starting from seven preliminary experiments, the locations of which are shown in Figure 15, the course of the discrimination is summarized in Table 10. The Table mentions the experimental settings selected by the design criterion and the models which were rejected by means of the χ^2-test. After each experiment the model adequacy testing was performed on all the models, including those which were rejected in previous stages. This was done to avoid the possibility that models would be rejected on the basis of some "dishonest" errors. This is especially important in the early stages of the experimental program.

Table 10 shows how models a_3, b_3, a_4, b_4, c_3, a_1, b_1, d_1, e_1 and c_1 were unambiguously eliminated after 14 experiments, seven of which designed. It seems unlikely that the addition of experiments would permit further discrimination between the models a_2, b_2, c_2, d_2 and e_2. Indeed, these models deviate from one another to an extent which is smaller than the experimental error. Models a_2, b_2 and d_2 were eliminated subsequently because they contained at least one parameter which does not differ significantly from zero on the basis of the individual t-test. Deletion of these parameters degenerates the models into either one of the forms c_2 or e_2. Table 11 compares predicted reaction rates for models c_2 and e_2. Whenever possible, observed rates are also provided. In the upper part of the table, partial pressures of the three reaction partners vary simultaneously along the straight line, connecting the origin of the p_B-p_D-p_{H_2} space with the point (0.2, 0.03, 0.03).

Table 11. Dehydrogenation of 1-butene. Confrontation of models c_2 and e_2 with each other and with experimental data

p_B	p_D	p_H	Predicted rates by model		Obsd Rate
			c_2	e_2	
0.04	0.006	0.006	0.00941	0.00859	
0.08	0.012	0.012	0.0106	0.00981	
0.12	0.018	0.018	0.00962	0.00906	
0.16	0.024	0.024	0.00806	0.00773	
0.2	0.03	0.03	0.00644	0.00627	
0.054	0	0	0.0203	0.0204	0.017
0.1	0	0	0.031[illegible]	0.0319	0.0317
0.154	0	0	0.0398	0.0404	0.0436
0.207	0	0	0.0451	0.0458	0.0427
0.202	0	0.036	0.035	0.0327	0.0369
0.19	0	0.11	0.0223	0.0256	0.0237

The lower part of the table is related to that region where the experiments have actually been designed. On the basis of model c_2, the standard deviation of the experimental error is estimated to be 0.0024. This is larger than the difference between the model predictions in any region of the factor space. Models a_2, b_2 and d_2 were not involved in this comparison: since they contain superfluous parameters, they even fit better than their rivals c_2 and e_2.

Finally, model c_2 which fitted slightly better then model e_2 was preferred.

Summarizing: seven preliminary and seven designed experiments, allowed 10 models out of 15 to be discarded. Reliability analyses of the parameter estimates permitted a final selection, without additional experimental effort. This example is clear evidence of the power of a directed sequential design together with a statistical analysis of the data.

5. Sequential Experimental Design For Precise Parameter Estimation

Once the model is selected, it is important to improve the precision with which the parameters are estimated. This chapter deals with the design of an experimental program aiming specifically at reducing the uncertainties associated with the parameter estimates. In other words, from all feasible sequences of experiments, that particular set is selected which defines the parameters with the greatest possible precision. The corner-stone upon which the theory relies is the covariance matrix of the parameter estimates, or some function of it.

A. Single Response Models

It follows from 30 that, when the model is linear, **V(b)** is a function of the error variance σ^2 and of the settings of the independent variables only, but not of the outcomes of the experiments. When the model is nonlinear, the partial derivatives in **X** are evaluated at the converged parameter estimates, which depend on the observations **y**. The influence is only indirect, however, and the criteria to be developed appear to be only weakly sensitive to the actual parameter values. Once the equipment and analytical devices are chosen, σ^2 is generally no longer under control of the experimenter. The set values of the independent variables, on the contrary, can be chosen at will, at least within certain limits.

Assuming that n experiments have been conducted and analyzed already, the effect of k additional experiments on the precision of the parameters, can be assessed by evaluating the *precision matrix* with respect to these n + k runs. The *design matrix* **X** consists of n + k rows, the first n corresponding to the runs already completed, the k last ones to the additional experiments:

$$\mathbf{X}_{n+k} = \begin{bmatrix} \mathbf{X}_n \\ --- \\ \mathbf{X}_k \end{bmatrix} \tag{171}$$

The precision matrix is then written

$$(\mathbf{X}_{n+k}^T\mathbf{X}_{n+k})^{-1} = (\mathbf{X}_n^T\mathbf{X}_n + \mathbf{X}_k^T\mathbf{X}_k)^{-1} \tag{172}$$

This matrix is known before the experiments are actually carried out. An inspection of the elements of this matrix reveals whether or not the k additional experiments are worthwhile.

This principle constitutes the basis of the experimental design for precise parameter estimation: the matrix $\mathbf{X}_k$ is to be constructed in such way that some measure of the reliability of the parameter estimates is optimized. Since the precision matrix consists of $(p^2 + p)/2$ different elements, it rapidly becomes difficult with increasing p to judge if a specific sequence of experiments is significantly better than any other. Overall criteria based on all elements of the covariance matrix are required.

1. Overall Criteria For an Experimental Program For Precise Parameter Estimation

In the absence of specific wishes about the variance of a particular estimate or the covariance between a specific pair of estimates, it seems logical to base the design on the joint confidence region, since this region takes into account the joint variability of all estimates. The boundary surface of this region is defined by all combinations $\boldsymbol{\beta}$ which satisfy

$$(\mathbf{b} - \boldsymbol{\beta})^T\mathbf{X}^T\mathbf{X}(\mathbf{b} - \boldsymbol{\beta}) = \delta \tag{173}$$

(173) can be transformed into

$$\left(\frac{z_1}{\sqrt{\delta/\lambda_1}}\right)^2 + \left(\frac{z_2}{\sqrt{\delta/\lambda_2}}\right)^2 + \ldots + \left(\frac{z_p}{\sqrt{\delta/\lambda_p}}\right)^2 = 1 \tag{174}$$

where λ_i are the eigenvalues of $\mathbf{X}^T\mathbf{X}$. Equation 174 is the standard equation of a hyperellipsoid in the p dimensional parameter space.

a) Minimum Volume criterion

A logical policy for the design of experiments consists of selecting those experimental conditions which minimize the volume of the ellipsoid Θ. From equation 174 it follows that the volume is given by

$$\text{Vol}\,\Theta = \text{Const.} \prod_{k=1}^{p} \sqrt{\delta/\lambda_k} = \text{Const.}\ \delta^{p/2}/\sqrt{\lambda_1\lambda_2\ldots\lambda_p} \sim 1/\sqrt{\det \mathbf{X}^T\mathbf{X}} \tag{175}$$

The volume is minimized by choosing the experimental conditions in such a way that det $\mathbf{X}^T\mathbf{X}$ is maximized. This criterion has been suggested by Box and Lucas [97]. Box and Hunter [98] showed that conditions which maximize det $\mathbf{X}^T\mathbf{X}$ also maximize the posterior parameter distribution. An important property of this criterion is its invariance with respect to changes in scale.

b) Shape criterion

An interesting alternative consists in rendering the confidence ellipsoid as spherical as possible, without worrying explicitly about its volume (Figure 16).

Since the longest axis is mainly responsible for the deviation of the ellipsoid from the sphere, it is this axis that has to be reduced as much as possible. From 174 it follows that the experimental conditions should be chosen as to maximize the smallest eigenvalue of $\mathbf{X}^T\mathbf{X}$.

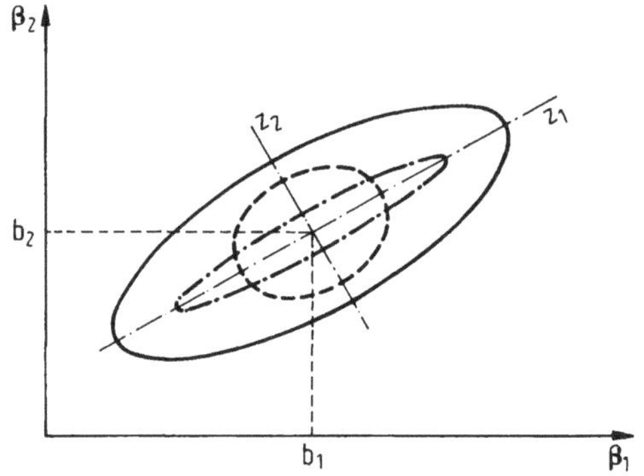

Figure 16. Evolution of the ellipsoidal joint confidence region (——— initial stage; —·—·— minimum volume; ----- shape)

c) Other criteria

Alternative criteria can be found in the literature. One of them consists in selecting those combinations which minimize the trace of $(\mathbf{X}^T\mathbf{X})^{-1}$:

$$\sum_{i=1}^{p} (\mathbf{X}^T\mathbf{X})_{ii}^{-1} \xrightarrow{x} \text{Min.} \tag{176}$$

which in other words minimize the sum of the variances of the parameter estimates. Since the covariances are not involved, this criterion is not related to the joint confidence region.

Another criterion acts upon the precision of the currently least accurate model prediction. This criterion is apparently of an entirely different nature than the ones discussed above, since it involves the precision of the model predictions, instead of the precision of the parameter estimates. Yet, it can be proved that it is identical with the minimum volume criterion.

Pritchard and Bacon [99] minimize the sum of the squares of the binary correlation coefficients between the parameters:

$$\sum_{i=1}^{p} \sum_{j=1}^{p} \varrho_{ij}^2 \xrightarrow{x} \text{Min} \tag{177}$$

The correlation matrix is defined by

$$\boldsymbol{\varrho} = \{\varrho_{ij}\} = \{(\mathbf{X}^T\mathbf{X})_{ij}^{-1}/[(\mathbf{X}^T\mathbf{X})_{ii}^{-1}\,(\mathbf{X}^T\mathbf{X})_{jj}^{-1}]^{1/2}\} \tag{178}$$

2. Design of Experiments

From the theory given above it follows that experiments for precise parameter estimation can be designed in blocks. Some difficulties are associated with such an approach, however. For a nonlinear model, the precision matrix is indirectly dependent on the numerical values of the parameters. It can not be excluded that poor estimates lead to an inefficient experimental program. Even if accurate information concerning the parameters were available, or if the design criteria were only weakly sensitive to the actual values, the design of large blocks of experiments is bound to encounter numerical difficulties. To circumvent these, a true sequential design is recommended. This implies that the parameter estimates and the precision matrix are updated after each experiment, and that the totality of the available data are analysed. The next experiment is then designed on the basis of the updated information. A flow diagram of such a sequential design is given in Figure 17.

3. A Case Study: Enzyme Reaction

The design of experiments based upon the Minimum Volume and the Shape criterion will now be illustrated by means of constructed data on the hydrolysis of saccharose into glucose and fructose [100]

$$S \rightarrow G + F$$

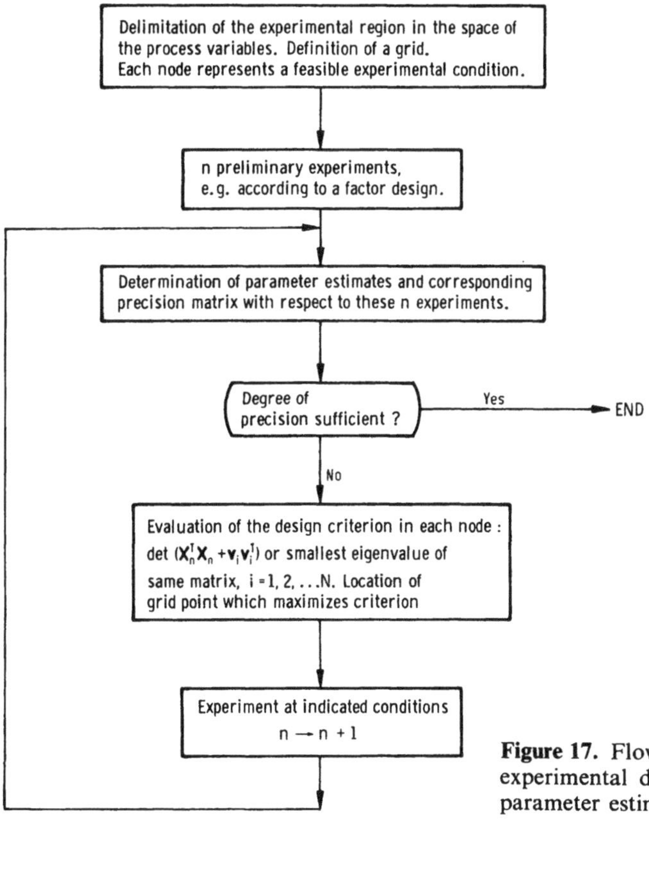

Figure 17. Flow diagram of a sequential experimental design procedure for precise parameter estimation

Michaelis and Menten [101] developed a rate equation for this reaction:

$$r = \frac{kK_{ES}C_EC_S}{1 + K_{ES}C_S + K_GC_G + K_FC_F} \tag{179}$$

Since the enzyme plays the role of the catalyst, it is not surprising that the rate equations for enzymatic reactions are very similar to the Hougen-Watson rate equations. The enzyme concentration C_E is always very small. It is not directly measurable, but is assumed to be constant, so that it can be incorporated into the rate coefficient k, to form one parameter kC_E.

The "experimental" data were generated by means of $kC_E = 3$, $K_{ES} = 60$, $K_G = 10$, $K_F = 20$ and random deviates drawn from $N(\mathbf{0}, \mathbf{I}\,10^{-6})$. The operability region is threedimensional in the variables C_S, C_G and C_F.

The grid was defined as follows:

C_S	0.0214;	0.0822;	0.168;	0.334
C_G	0	(0.05)	0.2	
C_F	0	(0.05)	0.2	

It consists of 100 nodes corresponding to 100 feasible experimental settings. Least squares preliminary parameter estimates were obtained from seven undesigned experiments, covering an important part of the experimental region (Table 12). Table 13 summarizes the results of the estimation applying both criteria, to simulate a program of 23 designed runs. The first two columns show that the parameters are only marginally significant after the preliminary

Table 12. Hydrolysis of saccharose. Course of sequential experimental design for precise parameter estimation

Experimental conditions			Experiment number	
c_S	c_G	c_F		
0.334	0	0	1	
0.334	0.2	0	2	
0.168	0.2	0	3	
0.168	0	0	4 Preliminary	
0.334	0	0.2	5 experiments	
0.168	0	0.2	6	
0.334	0.2	0.2	7	
0.0214	0	0	8, 10, 13, 17, 21, 25, 29	Minimum
0.0822	0.2	0	9, 12, 16, 20, 24, 28	volume
0.0822	0	0.2	11, 14, 18, 22, 26, 30	criterion
0.334	0	0	15, 19, 23, 27	
0.0214	0	0	8, 9, 10, 11, 13, 15, 16, 18, 19, 21, 22, 23, 25, 26, 28, 29	
0.0214	0.05	0	12	Shape
0.334	0	0	14, 17, 20, 24	criterion
0.334	0	0.1	27	
0.334	0	0.05	30	

Table 13. Hydrolysis of saccharose. Results of sequential experimental design for precise parameter estimation

	After preliminary experiments		After 23 designed experiments			
			Minimum Volume		Shape	
	Parameter estimate	Standard deviation	Parameter estimate	Standard deviation	Parameter estimate	Standard deviation
kC_E	4.6079	1.162	2.8863	0.085	2.9665	0.062
K_{ES}	92.2902	24.41	57.4658	1.974	58.9452	1.561
K_G	16.0449	4.222	9.2330	0.540	9.2372	1.065
K_F	33.8887	8.579	18.8841	0.862	21.2352	1.497

experiments. From the s(b) columns, it follows that the Shape criterion yields more precise estimates for kC_E and K_{ES} than the Minimum Volume criterion, whereas the contrary is true for K_G and K_F. This is intuitively clear from a scrutiny of the designed experiments (Table 12): out of a total of 23 runs, the Shape criterion has designed only three experiments whereby C_G and/or C_F is different from zero, whereas the Minimum Volume criterion selects 12 such combinations. Note that kC_E and K_{ES} are the parameters to which the model is most sensitive. Indeed, kC_E acts as a multiplier in the rate equation, while K_{ES} multiplies the concentration of the reactant, C_S. After the preliminary experiments, all estimates seem to be strongly correlated, but kC_E and K_{ES} in particular (Table 14).

After 23 designed runs, all correlation coefficients obtained from the Shape criterion are smaller than those from the alternate criterion. The Shape criterion reduces all correlation coefficients to extremely low values

Table 14. Hydrolysis of saccharose. Binary correlation coefficients between parameter estimates after a) preliminary experiments b) 23 designed experiments by Minimum Volume criterion c) 23 designed experiments by Shape criterion

	kC_E	K_{ES}	K_G	K_F	
kC_E	1.	0.9998	0.977	0.993	a)
K_{ES}	0.9998	1.	0.974	0.991	
K_G	0.977	0.974	1.	0.974	
K_F	0.993	0.991	0.974	1.	
kC_E	1.	0.979	0.565	0.667	b)
K_{ES}	0.979	1.	0.477	0.6	
K_G	0.565	0.477	1.	0.459	
K_F	0.667	0.6	0.459	1.	
kC_E	1.	0.969	0.136	0.115	c)
K_{ES}	0.969	1.	0.056	0.022	
K_G	0.136	0.056	1.	0.010	
K_F	0.115	0.022	0.01	1.	

except for $\varrho(kC_E, K_{ES})$. This could be an indication that the Shape criterion determines K_G and K_F more precisely than is suggested by the individual confidence limits alone.

It is interesting to investigate to what extent the initial elipsoidal joint confidence region has ultimately been modified by the action of the two criteria. The evolution of the volume of the region, in terms of det $\mathbf{X}^T\mathbf{X}$, and the length of the principal axes of the ellipsoids in terms of the eigenvalues of $\mathbf{X}^T\mathbf{X}$ for both types of design are given in Table 15. The ultimate volume of the confidence region obtained with the Minimum Volume criterion is approximately four times smaller than that obtained with the Shape criterion. This has been achieved by some superfluous contraction of axes which were much shorter than the longest at the start already. Owing to its specific action, the Shape criterion arrives at a longest axis which is approximately 15% shorter than that obtained by the Minimum Volume criterion.

Table 15. Hydrolysis of saccharose. Characteristics of approximate joint ellipsoidal confidence region

	After the first designed experiment	After 23 designed experiments	
		Minimum volume	Shape
det $\mathbf{X}^T\mathbf{X}$	1.035×10^{-25}	4.721×10^{-21}	3.399×10^{-22}
λ_1	6.210×10^{-4}	4.333×10^{-3}	4.352×10^{-3}
λ_2	1.818×10^{-7}	3.728×10^{-6}	6.919×10^{-7}
λ_3	1.073×10^{-7}	1.651×10^{-6}	3.509×10^{-7}
λ_4	8.541×10^{-9}	1.770×10^{-7}	3.217×10^{-7}

B. Multiresponse Models

1. Models of the Algebraic Type

From a comparison of the basic formulas of single and multiresponse regression analysis, it is intuitively felt that the basic quantity, for an experimental design for precise parameter estimation is the matrix $\Sigma\Sigma\sigma^{ij}\mathbf{X}_i^T X_j$. The overall criteria become:

Minimum Volume: the volume of the joint confidence region is minimized, by selecting those experimental conditions which maximize det $\Sigma\Sigma\sigma^{ij}\mathbf{X}_i^T\mathbf{X}_j$. This criterion has been derived by Draper and Hunter [102] on different theoretical grounds.

Shape: the region is rendered as spherical as possible by selecting those conditions which maximize the smallest eigenvalue of $\Sigma\Sigma\sigma^{ij}\mathbf{X}_i^T X_j$.

2. Models of the Differential Equation Type

In systems continuously varying in space or time, also misleadingly called in the jargon "with distributed parameters" sampling in several points along the coordinate axis represented by the symbol t in the differential equation (viz. equation 109) is possible. Clearly, the number of sampling points and their distribution will play an important role in the precision with which the parameters can be estimated. With this in mind Seinfeld et al. [103, 104] investigated the optimal location of these sampling points. Hosten and Emig [105] studied the problem in a general perspective that can be stated as follows: determine the experimental conditions which yield the most favorable profiles for precise parameter estimation. In the following it is assumed that the number and location of the sampling points are judiciously chosen, to yield the relevant information contained in the profiles. Optimal location of sampling points could, of course, be incorporated into the approach. In chemical and catalytic reaction engineering, this problem is often of pure academic interest, however, since each combination of the process variables leads to different optimal locations. In many practical situations, like in industrial tubular reactors these cannot be changed from experiment to experiment.

a) Experimental design criteria for precise parameter estimation

When the model is given in the form of a set of coupled first order ordinary differential equations, the experimental design is based on the matrix **H**, defined in (116) and auxiliary formulae.

Minimum volume criterion: the volume of the joint confidence region (122) is minimized when the experimental conditions are chosen to maximize det **H**.

Shape criterion: to render the shape of the region as spherical as possible, the experimental conditions have to maximize the smallest eigenvalue of **H**.

b) A case study: dehydrogenation of ethylbenzene [105]

The production of styrene is carried out in the presence of steam in an adiabatic fixed bed catalytic reactor.

The main reaction

$$\text{Ethylbenzene} \rightleftharpoons \text{Styrene} + \text{Hydrogen}$$

is accompanied by two undesirable parallel reactions:

$$\text{Ethylbenzene} \rightarrow \text{Benzene} + \text{Ethylene}$$

$$\text{Ethylbenzene} + \text{Hydrogen} \rightarrow \text{Toluene} + \text{Methane}$$

The following rate equations proved to be adequate:

$$\begin{aligned} r_1 &= \exp(S_1 - E_1/RT)(p_E - p_S p_{H_2}/K) \\ r_2 &= \exp(S_2 - E_2/RT)\, p_E \\ r_3 &= \exp(S_3 - E_3/RT)\, p_E p_{H_2} \end{aligned} \tag{180}$$

The reactor is modeled by means of continuity equations for ethylbenzene, benzene and toluene, an energy equation and a momentum equation [31]:

$$\frac{dx_1}{dz} = r_1 \Omega \varrho_b / F_{E_0}$$

$$\frac{dx_2}{dz} = r_2 \Omega \varrho_b / F_{E0}$$

$$\frac{dx_3}{dz} = r_3 \Omega \varrho_b / F_{E0} \tag{181}$$

$$\frac{dp_t}{dz} = -\frac{v_0(1-\varepsilon)}{d_p \varepsilon^3}\left[\frac{150\eta(1-\varepsilon)}{d_p} + 1.75 v_0 \varrho_g\right]$$

$$\frac{dT}{dz} = F_{E0} \sum_{i=1}^{3} (-\Delta H_i) \frac{dx_i}{dz} \Big/ \sum_{j=1}^{7} (n_j C_{pj})$$

with initial conditions

$$\begin{aligned} x_1 &= x_{1_0} \qquad p_t = p_{t_0} \\ x_2 &= x_{2_0} \qquad T = T_0 \qquad \text{at} \qquad z = 0 \\ x_3 &= x_{3_0} \end{aligned} \tag{182}$$

The conversion of ethylbenzene, x_1, the conversion into benzene, x_2, the conversion into toluene, x_3, the total pressure and the axial gas temperature are measured in three equidistant points along the axis. "Experimental data" are generated by adding normal random deviates to the model values. These are obtained by integrating the model equations with a set of parameter values considered as true values. The errors are drawn from $N(\mathbf{0}, \boldsymbol{\Sigma})$, where the error covariance matrix $\boldsymbol{\Sigma}$ is a diagonal matrix with elements $\sigma_{11} = 0.0001$, $\sigma_{22} = \sigma_{33} = 0.000025$, $\sigma_{44} = 0.0001$ and $\sigma_{55} = 100$. The parameter values used are those of Sheel and Crowe [106], derived from conversions measured in the effluent of an industrial reactor:

$$\begin{aligned} S_1 &= 0.40709 \qquad S_2 = 13.2392 \qquad S_3 = 0.2961 \\ E_1/R &= 10{,}925 \qquad E_2/R = 25{,}000 \qquad E_3/R = 11{,}000 \end{aligned}$$

An idea of the behavior of the state variables and of the effect of the experimental error is obtained from Figure 18, which shows error free axial profiles of the dependent variables (solid lines) for a particular set of experimental conditions, and "measured" values of two of these responses (○, ●). The variables determining the performance of the industrial reactor are: the inlet temperature T_0, and the molar feed flow rates of ethylbenzene, F_{E0}, and water, F_{W0}. Hence, the operability region is three-dimensional. The experimental grid is defined by

F_{E_0}	0.008	(0.001)	0.011
F_{W_0}	0.08	(0.02)	0.14
T_0	875	(25)	950

and shown in Figure 19. The internal nodes are omitted since it has never been experienced in previous studies of this type that such combinations were selected as optimal conditions. This reduces the number of combinations to be considered from 64 to 56 and decreases the computational effort in the design stage by approximately 12.5%. Obviously, this saving becomes of increasing importance with increasing dimension of the operability region.

Starting from two preliminary experiments a design sequence of 8 experiments is simulated by means of both criteria considered here. The results are summarized in Table 16. The policies derived from the two criteria are

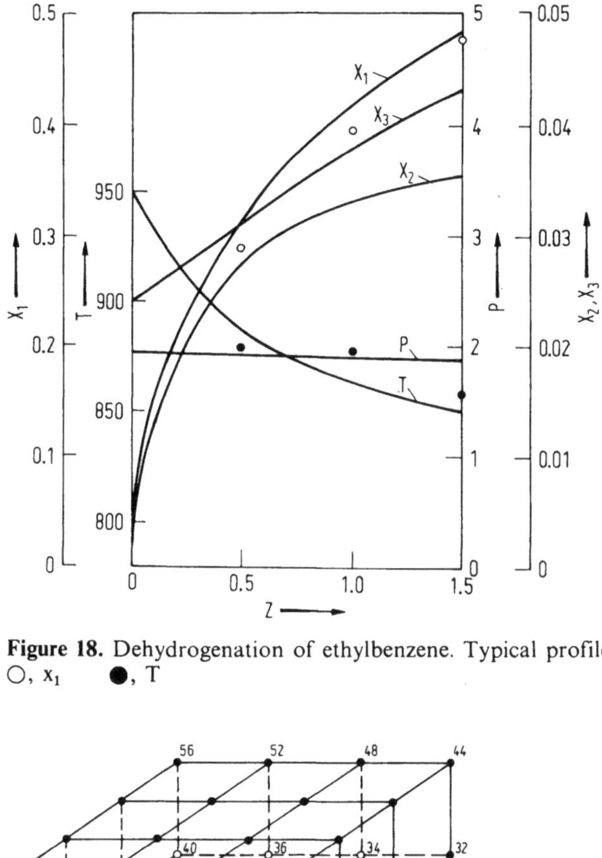

Figure 18. Dehydrogenation of ethylbenzene. Typical profiles. "Observed" values: ○, x_1 ●, T

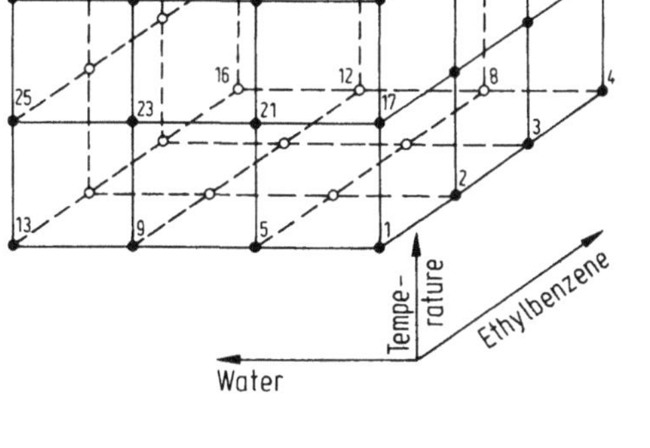

Figure 19. Dehydrogenation of ethylbenzene. Experimental region and grid

Table 16. Dehydrogenation of ethylbenzene. Results of directed sequential and random experimental design with respect to precise parameter estimation

	Preliminary experiments	Minimum Volume		Shape		
		4 designed experiments	8 designed experiments	4 designed experiments	8 designed experiments	10 non designed experiments
S_1	4.2606	1.0968	0.098472	0.95016	0.33199	0.26421
$s(S_1)$	2.55	0.67	0.50	0.63	0.49	0.62
$t(S_1)$	1.67	1.63	0.19	1.5	0.68	0.43
E_1/R	14365.	11529.	10658.8	11391.	10867.	10805
$s(E_1/R)$	2294.	603.	450.	560.	429.	546
$t(E_1/R)$	6.26	19.1	23.7	20.3	25.3	198
S_2	30.621	14.124	13.3995	13.481	12.543	13.416
$s(S_2)$	7.62	3.03	2.29	2.80	2.08	2.43
$t(S_2)$	4.02	4.66	5.86	4.81	6.03	5.52
E_2/R	40825.	25761.	25157.	25159.	24369.	25163
$s(E_2/R)$	7016.	2786.	2105.	2565.	1904.	2211.
$t(E_2/R)$	5.82	9.25	12.	9.81	12.8	11.4
S_3	3.4178	−2.4278	2.1157	−0.68766	8.9853	−5.9965
$s(S_3)$	20.4	4.51	4.02	4.46	4.32	5.43
$t(S_3)$	0.17	−0.54	0.53	−0.15	2.08	−1.1
E_3/R	13899.*	8586.	12713.7	10124.	18741.	5562.3*
$s(E_3/R)$	18182.	3985.	3573.	3916.	3856.	4748
$t(E_3/R)$	0.76	2.15	3.56	2.59	4.86	1.18
Sequence of experiments, in order of conduction		20, 53, 4, 53;	20, 53, 20, 53	20, 4, 53, 4;	4, 53, 20, 20	1, 10, 15, 21, 23a, 32, 35a, 43, 49, 52

* non significant activation energy.
23a F_{E_0}: 0.01, F_{W_0}: 0.12, $T_0 = 900$.
35a F_{E_0}: 0.009, F_{W_0}: 0.12, $T_0 = 925$

slightly different. A comparison of t-values with reference zero in the third and fifth columns of Table 16 shows that the parameters are better determined by means of the Shape criterion than by the Minimum Volume Criterion, but the differences are small. The main drop in uncertainty occurs during the first designed experiments. Further improvement by continued design is to be attributed primarily to the increasing number of experiments. Table 17 summarizes the characteristics of the joint ellipsoidal confidence region before and after completion of 8 designed experiments by means of both design criteria. The volume is expressed in terms of det **H**, the length of the principal axes in terms of the eigenvalues of **H**. It follows that the ultimate volume of Θ, obtained by means of the Shape criterion is about 1.13 times larger than that obtained with the Minimum Volume criterion. Compared with the Minimum Volume Criterion the Shape criterion achieves an additional reduction of the long axes corresponding to λ_4, λ_5 and λ_6 of 6, 10 and 6%.

Evidently, these figures can not be claimed to be general: they strongly depend upon the mathematical structure of the model, the range of variation of the experimental variables and the nature of the parameters. The last two reasons exclude the possibility of reshaping the joint confidence region into a true sphere. The question may now be raised what has been gained by such a directed approach as compared with a classical experimental program whereby the experiments are distributed over the entire experimental space. The last column of Table 16 shows results derived from 10 non designed experiments, the locations of which are given in the last row. The parameters are significantly less precise than those obtained by means of the specific experimental design. It should be added also that the superiority of the design is not overwhelming. This can be explained in terms of the rather high number of measurements. The performance of the specific design methods would be more pronounced if the number of sampling nodes would be smaller. Sequential experimental design is not a real necessity when the system can be observed in many sampling points, provided of course the classical design is judicously chosen. In the present example the sampling points were located in fixed points. It would be possible also to select for

Table 17. Dehydrogenation of ethylbenzene. Characteristics of the approximate joint ellipsoidal confidence region

	After preliminary experiments	After 8 designed experiments	
		Minimum Volume	Shape
det **H**	$4.0675\ 10^{-17}$	$3.1783\ 10^{-11}$	$2.2712\ 10^{-11}$
λ_1	947.94	4365.0	3840.8
λ_2	185.97	838.41	540.37
λ_3	19.434	95.986	80.078
λ_4	$1.9358\ 10^{-7}$	$4.9728\ 10^{-6}$	$5.5665\ 10^{-6}$
λ_5	$2.0276\ 10^{-8}$	$2.2748\ 10^{-7}$	$2.7355\ 10^{-7}$
λ_6	$3.0248\ 10^{-9}$	$7.9984\ 10^{-8}$	$8.9744\ 10^{-8}$

each grid point the optimal set of sampling nodes by maximizing the criterion with respect to the unknown values of t. This would certainly be an optimal procedure, but the succes of the search for the optimum of the objective function is not guaranteed, for local maxima may occur. In addition, the computational effort increases rather sharply so that this strategy is probably not justified today.

C. Miscellaneous Experimental Design Criteria

The above basic criteria have been refined to incorporate special features. Hill et al. [107] combined the discrimination and precise parameter estimation criteria in an attempt to achieve more efficient modeling. In the initial stages, the greatest weight is given to the contribution resulting from the discrimination term. As information is accumulated, emphasis is shifted towards precise parameter estimation. Draper and Hunter [108] extended the determinant criteria for precise parameter estimation to account for prior information on the parameter distribution. Hunter et al. [109] developed a criterion for precise parameter estimation of a subset of model parameters.

6. Concluding Remarks

This text has emphasized rigorous methods of data treatment in kinetic analysis. Use is made of a mathematical background that does not exceed present days standards in the majority of the chemical engineering departments.

The effort required to produce good experimental data is such that there is no justification for not exploiting the full potential of these data by a naive or incomplete data treatment, now that the appropriate mathematical tools have been developed and that modern computers are there to apply them.

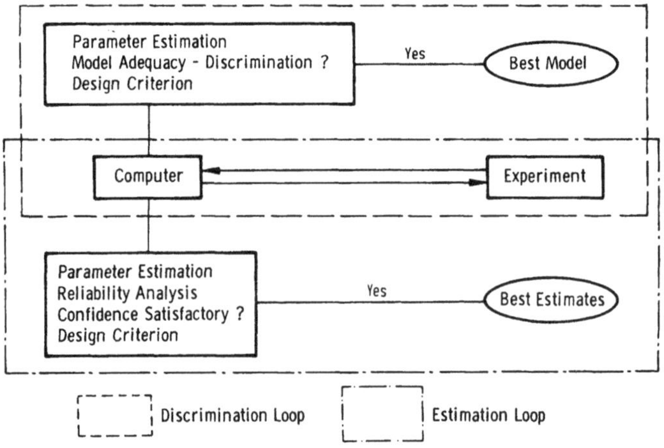

Figure 20. Philosophy of mathematical modelling

The literature on related areas like process control contains an impressive arsenal of powerful techniques of data analysis that should not be ignored by those active in kinetics. Let these words not be misunderstood: they are no plea for the application of sophisticated methods to poor data. There is no mathematical technique that will upgrade deficient data. Higher standards and further experimentation are the only remedies for this situation.

Too often also kinetic data suffer from a poor experimental design. Emphasis was also laid in this text on sequential design techniques. These are powerful tools for locating significant experiments and are particularly useful when the number of experiments has to be limited. Again, sound judgment should come in the first place and contribute at least in early stages to an efficient design. The more complex the problem, however, the more computerized methods of design will have to be involved.

Evidently, at all stages of kinetic analysis a close interaction between the research team and the computer is required. Such an interaction may be of the type represented in Figure 20. In fact it is not limited to the area dealt with in this text: it is the approach that scientific investigation has selected to increase its efficiency and to accelerate progress.

Acknowledgment: L. H. Hosten is grateful to "Nationaal Fonds voor Wetenschappelijk Onderzoek" for an appointment as "Onderzoeksleider".

Notation

$\mathbf{1}$	(m × 1) vector whose elements are all 1.
A, B	constants for the likelihood ratio test, defined in (150).
A, R, S	ethylalcohol, acetaldehyde and hydrogen respectively.
A_i	preexponential factor of i-th parameter.
A_i^*	reparameterized preexponential factor with respect to i-th parameter.
a, b, c, d, e	regression coefficients in rearranged rate equation.
$\mathbf{a}$	m × 1 vector of estimates of parameters in Wilks test.
$\mathbf{b}$	p × 1 vector of parameter estimates.
$\bar{b}_i$	constant reference value for i-th parameter.
C	degree of confidence, defined equation in 164.
C_1, C_2	intrinsic parameters defined in equations 127, 128.
C_{p_j}	molar specific heat of j-th component.
C_X	concentration of species X.
$\mathbf{C}$	vector of concentrations.
C_l	concentration of unoccupied active sites.
C_t	total concentration of active sites.
C_{Xl}	concentration of adsorbed complex Xl.
$(d.f.)_i$	number of degrees of freedom with respect to the i-th model.
d_p	diameter of catalyst particle.
$D(\mathbf{x}_k)$	divergence between rival models at settings $\mathbf{x}_k$.
E_i	activation energy/adsorption enthalpy with respect to i-th parameter.
$\mathbf{e}$	n × 1 vector of residuals $\mathbf{y}-\hat{\mathbf{y}}$.
$\mathbf{e}_j$	vector $\mathbf{e}$ with respect to model j.
F_{A_0}	feed molar flow rate of species A.

$F(p, n-p, 1-\alpha)$ α-percentage point of the F-distribution with p and n—p degrees of freedom.
ΔH reaction respectively adsorption enthalpy.
$\mathbf{I}$ unit matrix.
$\mathbf{J}_i$ $n \times p$ Jacobian matrix of partial derivatives of dependent variable towards the parameters with respect to the (i + 1)-th iteration.
$\mathbf{J}_*$ $n \times p$ Jacobian matrix $\mathbf{J}_i$ at converged parameter estimates.
K overall equilibrium constant.
$\mathbf{K}$ matrix of first order rate coefficients.
K_i equilibrium constant of i-th step in the sequence of reactions.
K_X adsorption equilibrium constant of component X.
k global forward rate coefficient for a single reaction.
k_i forward rate coefficient for the i-th step in a sequence of reactions.
$L(\boldsymbol{\beta}/\mathbf{y})$ likelihood function of the parameters.
l active catalytic site.
m number of rival models.
N number of nodes in the experimental grid.
n number of observations.
n_e number of replicated experiments.
n_j number of moles of j-th component.
p number of parameters.
p_i probability density of the next observation, y_{n+1}, under model i, defined by equation 161.
p_j number of parameters in model j.
p_t total pressure.
p_X partial pressure of component X.
p_{X_0} inlet partial pressure of component X.
r_A overall reaction rate for species A.
r_{A_0} reaction rate of species A in initial conditions, i.e. when $x_A = 0$.
r_i reaction rate of i-th step in a sequence of reactions.
$\mathbf{r}_i$ $n \times 1$ vector of residuals $\mathbf{y} - \mathbf{f}(\mathbf{x}, \mathbf{b}_i)$ with respect to the (i + 1)-th iteration.
$\mathbf{r}_{j_s}$ $n \times 1$ vector of current residuals $\mathbf{y}_j - \mathbf{f}_j(\mathbf{x}, \mathbf{b}_s)$ with respect to the (s + 1)-th iteration.
R gas constant.
s number of state variables.
s^2 estimate of error variance, defined by equation 31.
s_e^2 estimate of pure error variance from replicated experiments.
s_i^2 estimate of error variance plus eventual lack of fit under model i.
$\bar{s}^2$ pooled estimate of error variance plus eventual lack of fit, defined by equation 169.
$s(b_i) = [\hat{V}(\mathbf{b})_{ii}]^{1/2}$ standard deviation on parameter estimate b_i.
ΔS reaction respectively adsorption entropy.
$\mathbf{S}$ $v \times v$ matrix of sums of squares and cross products of deviations between observed and calculated response values, defined by equation 93.
$|\mathbf{S}|$ determinant of S
$|\mathbf{S}|^{min}$ minimum value of $|\mathbf{S}|$ with respect to the parameters.
t independent variable like time, reactor length.
$t(n - p, 1 - \alpha/2)$ $\alpha/2$-percentage point of the t-distribution with n — p degrees of freedom.
T absolute temperature.
$\bar{T}$ mean absolute temperature.
v number of dependent variables.
v_o linear velocity at inlet conditions.
$\mathbf{v}_i$ $p \times 1$ vector of partial derivatives of dependent variable with respect to the parameters, taken at the i-th node of the experimental grid and converged parameter estimates.
$\mathbf{v}_k^{(j)}$ $p \times 1$ vector of partial derivatives of the j-th model with respect to the parameters, taken at settings $\mathbf{x}_k$ and current least squares parameter estimates.

$\mathbf{V}$ $n \times n$ weighting matrix in weighted univariate regression.
$\mathbf{V(b)}$ variance-covariance matrix of parameter estimates $\mathbf{b}$.
$\hat{\mathbf{V}}(\mathbf{b})$ estimate of $\mathbf{V(b)}$.
W weight of catalyst.
x, u, v, z fractional conversion of X, to T, PT and PA respectively.
x_A, x fractional conversion (of species A).
x_{kij} setting of the j-th independent variable in the i-th response equation for the k-th experiment when the model is linear, respectively $\left.\frac{\partial f_i(\mathbf{x}_k, \boldsymbol{\beta})}{\partial \boldsymbol{\beta}_j}\right|_{\boldsymbol{\beta}=\mathbf{b}}$ when the model is nonlinear.
$\mathbf{x}$ vector of independent variables, respectively $(s \times 1)$ vector of state variables.
$\mathbf{x}_k$ specific settings for the independent variables.
X, T, PT, PA o-xylene, o-tolualdehyde, phthalide, phthalic anhydride respectively.
Xl adsorbed complex of component X.
$\mathbf{X}$ $n \times p$ matrix of settings of independent variables.
$\mathbf{X}_i$ $n \times p$ matrix $\mathbf{X}$ with respect to the i-th model.
$\mathbf{X}_j$ $n \times p$ matrix $\mathbf{X}$ with respect to the j-th response.
$\mathbf{X}_n$ design matrix $\mathbf{X}$ after n experiments.
$\mathbf{X}_{j_s}$ matrix $\mathbf{X}_j$ with respect to the $(s + 1)$-th iteration.
y dependent variable.
y_i i-th observed value of the dependent variable y.
y_j j-th dependent variable.
$\mathbf{y}$ $n \times 1$ vector of observed values of the dependent variable y.
$\hat{y}^{(j)}(\mathbf{x}_k)$ predicted value of response variable under model j at settings $\mathbf{x}_k$.
$\mathbf{y}_j$ $(n \times 1)$ vector of observed values of the j-th response variable.
y_{ki} observed value of the i-th response variable during the k-th experiment.
z variable defined in 132.
$z_1, z_2, \ldots, z_p$ new coordinate system after translation and rotation of the original $\boldsymbol{\beta}$ coordinate system.

Greek Symbols

$\boldsymbol{\alpha}$ $m \times 1$ vector of non-intrinsic parameters in Wilks test.
$\boldsymbol{\beta}$ $p \times 1$ vector of parameters.
γ molar ratio water/ethanol in the feed.
ε void fraction of catalyst bed.
$\boldsymbol{\varepsilon}$ $n \times 1$ vector of unobservable experimental errors.
$\boldsymbol{\varepsilon}_j$ $n \times 1$ vector of unobservable errors with respect to the j-th response.
$\varepsilon^{(i)}$ $v \times 1$ vector of unobservable errors with respect to the v response variables and the i-th experiment.
ε_{kj} unobservable error associated with the k-th observation of the j-th response.
η dynamic viscosity.
η_i true, unknown, value of the dependent variable for the i-th experiment.
$\boldsymbol{\eta}_j$ $(n \times 1)$ vector of true, unknown, values of the j-th response variable.
η_{kj} true, unknown, value of the j-th response variable corresponding to the k-th observation.
λ non-intrinsic parameter in Williams and Kloot test, respectively Lagrangian multiplier in Wilks test.
λ_i parameter in Marquardt procedure with respect to the $(i + 1)$-th iteration, respectively i-th eigenvalue of $\mathbf{X}^T\mathbf{X}$.
μ molar ratio acetaldehyde/ethanol in the feed.
$\Pi_{i,n}$ probability of model i after n experiments.
$\boldsymbol{\varrho}$ $p \times p$ symmetric correlation matrix of parameter estimates.
ϱ_b bulk density of catalyst.
σ^2 error variance, respectively proportionality factor according to equation 35.

σ_{ij} covariance between the experimental errors associated with measurements of the i-th and j-th response variables.
σ^2_{ki} variance of $\hat{y}^{(i)}(x_k)$.
$\Sigma = \{\sigma_{ij}\}$ $(v \times v)$ covariance matrix of experimental errors in the multivariate regression situation.
$\chi^2(a, 1 - \alpha)$ α-percentage point of the χ^2 distribution with a degrees of freedom.
Ω reactor section.

References

1. Bard, Y., Lapidus, L.: Catal. Rev. **2**, 67 (1968)
2. Kittrell, J. R.: Advan. Chem. Eng. **8**, 97 (1970)
3. Mezaki, R., Happel, J.: Catal. Rev. **3**, 241 (1969)
4. Reilly, P. M., Blau, G. E.: Can. J. Chem. Eng. **52**, 289 (1974)
5. Froment, G. F.: A. I. Ch. E. J. **21**, 1041 (1975)
6. Froment, G. F.: Proc. 4th Internat. Symp. Chem. Reaction Engng., Heidelberg 1976, Frankfurt: Dechema 1976
7. Boudart, M.: A. I. Ch. E. J. **18**, 465 (1972)
8. Boudart, M.: Kinetics of Chemical Processes, Englewood-Cliffs-N.J.: Prentice Hall 1968
9. Bischoff, K. B., Froment, G. F.: Ind. Eng. Chem., Fundam. **1**, 195 (1962)
10. Bradshaw, R. W., Davidson, B.: Chem. Eng. Sci. **24**, 1519 (1969)
11. Froment, G. F., Bischoff, K. B.: Chemical Reactor Analysis and Design, New York: John Wiley 1979
12. Franckaerts, J., Froment, G. F.: Chem. Eng. Sci. **19**, 807 (1964)
13. Yang, K. H., Hougen, O. A.: Chem. Eng. Progr. **46**, 146 (1950)
14. Mezaki, R., Kittrell, J. R.: A. I. Ch. E. J. **13**, 176 (1967)
15. Draper, N. R., Smith, H.: Applied Regression Analysis, New York: John Wiley 1966
16. Graybill, F.: Introduction to linear Statistical Models, Part I. New York: McGraw-Hill 1961
17. Himmelblau, D. M.: Process Analysis by Statistical Methods, New York: John Wiley 1970
18. Bowker, A. H., Lieberman, G. J.: Engineering Statistics, Englewood Cliffs, N.J.: Prentice Hall 1972
19. Walpole, R. E., Myers, R. H.: Probability and Statistics for Engineers and Scientists, New York: MacMillan 1972
20. Lindgren, B. W., McElrath, G. W.: Introduction to Probability and Statistics, New York: MacMillan 1969
21. Volk, W.: Applied Statistics for Engineers, New York: McGraw-Hill 1969
22. Cooper, B. E.: Statistics for Experimentalists, Oxford: Pergamon Press 1969
23. Green, J. R., Margerison, D.: Statistical Treatment of experimental data, New York: Elsevier North Holland 1977
24. Andersen, L. S.: Statistics in Chemical Engineering. In: CE Refresher, Chem. Eng., Oct. 29, 1962 to Sept. 2, 1963
25. Ullmann's Encyclopädie der Technischen Chemie: Weinheim, Verlag Chemie 1978
26. Bard, Y.: Nonlinear parameter estimation, New York: Academic Press 1974
27. Levenberg K.: Q. Appl. Math. **2**, 164 (1944)
28. Marquardt, D. W.: Soc. Ind. Appl. Math. J. **11**, 431 (1963)
29. Wilde, D. J., Beightler, C. S.: Foundations of optimization, Englewood-Cliffs, N.J.: Prentice Hall 1967
30. Beveridge, G. S., Schechter, R. S.: Optimization, Theory and Practice, New York: McGraw-Hill 1970
31. Hoffmann, U., Hofmann, H.: Einführung in die Optimierung, Weinheim: Verlag Chemie, 1971

32. Beale, E. M. L.: Rep. N1/ML-64, Admiralty Res. Lab. Teddington, England 1960
33. Guttman, I., Meeter, D. A.: Technometrics **7**, 625 (1965)
34. Hattikudur, U. R., Thodos, G.: Proc. 3rd Internat. Symp. Chem. Reaction Engng. Evanston, Washington D.C.: Amer. Chem. Soc. 1974
35. Hosten, L. H., Froment, G. F.: Ind. Eng. Chem., Process Des. Develop. **10**, 280 (1971)
36. Fox, L.: Numerical solution of ordinary and partial differential equations, Addison-Wesley, Mass. 1962
37. Froment, G. F., Mezaki, R.: Chem. Eng. Sci. **25**, 293 (1970)
38. Peterson, T. I., Lapidus, L.: ibid. **21**, 655 (1966)
39. Shacham, M., Kehat, E.: ibid. **27**, 2099 (1972)
40. Box, G. E. P.: Ann. N.Y. Acad. Sci. **86**, 792 (1960)
41. Box, G. E. P., Draper, N. R.: Biometrika **52**, 355 (1965)
42. Box, G. E. P., Hunter, W. G.: Proc. 2nd Internat. Symp. Chem. Reaction Engng. Amsterdam, Amsterdam: Elsevier 1972
43. Box, G. E. P. et al.: Technometrics **15**, 33 (1973)
44. Box, G. E. P., Cox, D. R.: J. Roy. Stat. Soc. **B 26**, 211 (1964)
45. Mezaki, R., Butt, J. B.: Ind. Eng. Chem., Fundam. **7**, 120 (1968)
46. Hosten, L. H., Froment, G. F.: Periodica Polytechnica, (Chemical Engineering — Budapest) **19**, 123 (1975)
47. Seinfeld, J. H., Lapidus, L.: Mathematical Methods in Chemical Engineering, Vol. 3: Process Modeling Estimation and Identification, Englewood Cliffs, N. J.: Prentice Hall 1974
48. Himmelblau, D. M., Jones, C. R., Bischoff, K. B.: Ind. Eng. Chem., Fundam. **6**, 539 (1967)
49. Cresswell, D. L., Quon, D.: Can. J. Chem. Eng. **53**, 334 (1975)
50. Van den Bosch, B., Hellinckx, L.: A. I. Ch. E. J. **20**, 250 (1974)
51. Cresswell, D. L., Paterson, W. R.: Proc. 4th Internat. Symp. Chem. Reaction Engng. Heidelberg 1976, Frankfurt, Dechema 1976
52. Hosten, L. H.: Computers and Chemical Engineering, **3**, 117 (1979)
53. Rosenbrock, H. H., Storey, C.: Computational Techniques for Chemical Engineers, New York, Pergamon Press 1966
54. Heineken, F. G., Tsuchiya, H. M., Aris, R.: Math. Biosci. **1**, 115 (1967)
55. Emig, G., Hosten, L. H.: Chem. Eng. Sci. **29**, 475 (1974)
56. Wei, J., Kuo, J. C.: Ind. Eng. Chem., Fundam. **8**, 114 (1969)
57. Bailey, J. E.: Chem. Eng. J. **3**, 52 (1972)
58. Hutchinson, P., Luss, D.: ibid. **1**, 129 (1970)
59. Luss, D., Hutchinson, P.: ibid. **2**, 172 (1971)
60. Golikeri, S. U., Luss, D.: A. I. Ch. E. J. **18**, 277 (1972)
61. Golikeri, S. U., Luss, D.: Chem. Eng. Sci. **29**, 845 (1974)
62. Luss, D., Golikeri, S. U.: A. I. Ch. E. J. **21**, 865 (1975)
63. Jacob, S. M. et al.: A. I. Ch. E. J. **22**, 701 (1976)
64. Liu, Y. A., Lapidus, L.: A. I. Ch. E. J. **19**, 467 (1973)
65. Luenberger, D. G.: IEEE Trans. Mil. Electron, Vol. MIL-8, **74**, April (1964)
66. Ozawa, Y.: Ind. Eng. Chem., Fundam. **12**, 191 (1973)
67. Lee, H.: A. I. Ch. E. J. **23**, 116 (1977)
68. Lee, H.: ibid. **24**, 116 (1978)
69. Hunter, W. G., Mezaki, R.: ibid. **10**, 315 (1964)
70. Kittrell, J. R., Hunter, W. G., Mezaki, R.: ibid. **5**, 1014 (1966)
71. Kittrell, J. R., Mezaki, R.: ibid. **13**, 389 (1967)
72. Hosten, L. H.: Preprints Colloque sur les modèles mathématiques Pau (France) 1970
73. Williams, E. J., Kloot, N. H.: Aust. J. Appl. Sci. **4**, 1 (1953)
74. Mezaki, R., Kittrell, J. R.: Can. J. Chem. Eng. **44**, 285 (1966)
75. Wilks, S. S.: Ann. Math. Stat. **17**, 257 (1946)
76. Barnard, G. A., Jenkins, G. M., Winston, C. B.: J. Roy. Stat. Soc. **A 125**, 321 (1962)
77. Reilly, P. M.: Can. J. Chem. Eng. **48**, 168 (1970)
78. Rao, M. S. et al.: ibid. **49**, 354 (1971)
79. Wong, C. W., Scharer, J. M., Reilly, P. M.: ibid. **52**, 645 (1974)

80. Wald, A.: Sequential Analysis, New York: John Wiley 1947
81. Chou, H. C.: Ind. Eng. Chem. **50**, 799 (1958)
82. Boudart, M., Mears, D. E., Vannice, M. A.: Ind. Chim. Belge **32**, 281 (1967)
83. Hunter, W. G., Reiner, A. M.: Technometrics **7**, 307 (1965)
84. Box, G. E. P., Hill, W. J.: ibid. **9**, 57 (1967)
85. Hunter, W. G., Hill, W. J.: Univ. of Wisconsin, Dept. Statist. Tech. Rept. No. 65 (1966)
86. Hougen, O. A., Watson, K. M.: Chemical Process Principles, Part III, New York: John Wiley 1947
87. Blakemore, J. W., Hoerl, A. E.: Chem. Eng. Progr. Symp. Ser. **59**, 14 (1963)
88. Hicks, C. R.: Fundamental Concepts in the Design of Experiments, New York: Holt, Rinehart and Winston 1973
89. Hsiang, T. C., Reilly, P. M.: Can. J. Chem. Eng. **49**, 865 (1971)
90. Hosten, L. H., Froment, G. F.: Proc. 4th Internat. Symp. Chem. Reaction Engng. Heidelberg 1976. Frankfurt: Dechema 1976
91. Bartlett, M. S.: Proc. Roy. Soc. A **160**, 268 (1937)
92. Hartley, H. O.: Biometrika **31**, 249 (1940)
93. Box, G. E. P.: ibid. **40**, 318 (1953)
94. Hosten, L. H.: Mathematische Modelbouw in de Chemische Reaktietechniek. Aggregatiethesis voor het Hoger Onderwijs, Gent (1978)
95. Dumez, F., Froment, G. F.: Ind. Eng. Chem., Process. Des. Develop. **15**, 291 (1976)
96. Dumez, F., Hosten, L. H., Froment, G. F.: Ind. Eng. Chem., Fundam. **16**, 298 (1977)
97. Box, G. E. P., Lucas, H. L.: Biometrika **46**, 77 (1959)
98. Box, G. E. P., Hunter, W. G.: Proc. IBM Sci. Computing Symp. Statistics, p. 113 (1965)
99. Pritchard, D. J., Bacon, D. W.: Chem. Eng. Sci. **33**, 1539 (1978)
100. Hosten, L. H.: Chem. Eng. Sci. **29**, 2247 (1974)
101. Michaelis, L., Menten, M. L.: Biochem. Z. **49**, 333 (1913)
102. Draper, N. R., Hunter, W. G.: Biometrika **53**, 525 (1966)
103. Seinfeld, J. H., Chen, W. H.: Chem. Eng. Sci. **26**, 753 (1971)
104. Ravimohan, A. L., Chen, W. H., Seinfeld, J. H.: Can. J. Chem. Eng. **48**, 420 (1970)
105. Hosten, L. H., Emig, G.: Chem. Eng. Sci. **30**, 1357 (1975)
106. Sheel, J. G., Crowe, C. M.: Can. J. Chem. Eng. **47**, 183 (1969)
107. Hill, W. J., Hunter, W. G., Wichern, D. W.: Technometrics **10**, 145 (1968)
108. Draper, N. R., Hunter, W. G.: Biometrika **54**, 147 (1967)
109. Hunter, W. G., Hill, W. J., Henson, T. L.: Can. J. Chem. Eng. **47**, 76 (1969)

Chapter 4

Texture of Catalysts

A. J. Lecloux

Solvay & Cie., Central Research Laboratory,
Catalysis Department
Rue de Ransbeek, 310, B-1120 Brussels, Belgium

Contents

1. Introduction

The description of a catalyst is first of all given by its chemical composition: nature of the active compounds, type of support, presence of promoters or doping agents. Each component has generally a well known influence on the catalytic properties of the system, on its activity, selectivity, stability and so on. But the study of all these effects are too often limited to the pure chemical aspect of the catalyst ability to increase the rate of a reaction or to modify the reaction steps in such a way that a particular product is preferentially formed.

This tendency has been reinforced during the last decade by the development of coordination chemistry and its application to homogeneous catalysis. As this type of chemistry has been successfully applied to the understanding of a certain number of heterogeneously catalysed reactions, numerous studies have been made on the nature of the active sites present on the catalyst surface. But the number of these sites, their geometry and their chemical efficiency will depend on *the main physical features* which go to make up the catalyst surface:

— its *structure*, i.e. the nature and the distribution of atoms or ions in the material part of the catalyst and in particular on its surface
— its *texture*, i.e. the detailed geometry of the void space in the catalyst particles, from the intergranular voids in the agglomerates down to the pore size distribution at the finest level of resolution, from the shape of the particles and their external surface area down to the pore shapes and the extent of the accessible internal surface.

As the structure and the texture of a catalyst are strongly dependent on the way it has been prepared and subsequently treated, it is most desirable to know how these physical properties can be parametrised in terms of measurable quantities.

The development of methods for the measurement of structure — or texture — related parameters will thus provide a tool for controlling the preparation of catalysts.

But the knowledge of the physical properties of a heterogeneous catalyst is also a key-factor for its practical utilization. For example, the number, the shape and the size of the pores greatly affect the extent of the accessible internal area of the catalyst. They also govern the heat and mass transfer phenomena which generally occur in practical conditions and sometimes can modify the selectivity of the reaction and markedly influence the thermal stability of the catalyst.

The aim of this paper is not to present an exhaustive review of the relevant litterature published since the well known paper of Brunauer, Emmett and Teller [1] describing the method for determining the specific surface area of solids, but to propose some useful guidelines and methods for a systematic investigation and a coherent description of the catalysts texture. Such a description requires the specification of a very large number of parameters and implies the use of "models" involving assumptions and simplifications. Even if, in the present state of our knowledge, it is impossible to obtain an exact, complete and detailed description of a porous solid, oversimplifications must be avoided. Very often, indeed, the range of validity of crude approximations is strictly limited and the comparison of results obtained on such a basis for different solids is generally difficult or meaningless. The use of an over-sophisticated model must also be avoided because it gives the illusion of a deep understanding of the problem and generally needs too many measurements for the usual practice.

The general approach proposed here for determining the porous texture of solids is based on a few techniques. Their results are cross analyzed in such a way that a self consistent picture of the porous texture of solids is obtained.

Mercury porosimetry and physical adsorption isotherms analysis are the two principal means to get a reasonnably good description of the texture of a solid, provided the information coming from particle-size and -shape analysis and from structural and crystallographical studies of the solid are taken into account. Only a self consistent combination of all these results can give a coherent pattern of the solid texture.

The general method presented and illustrated here is based on this fundamental principle.

2. Parameters Describing the Catalyst Texture

It is impossible to give an exact and detailed description of a porous solid. Consequently different average properties are generally used [2]:

— *the specific surface area* is the measure of the accessible surface area per unit mass of solid; this surface is the sum of the internal surface area associated with pores and of the external surface area developped by the outer boundary of the particles

— *the specific porosity* refers to the accessible pore void space in the particles per unit mass of solid; it eventually includes the intergranular void space in agglomerates like pellets or extrudates

— *the pore shape* is generally difficult to describe with a rather good precision. It is the reason why the pores are usually classified into groups corresponding to different types of shapes; cylindrical, ink-bottle and slit-shaped pores are the main classes. But in practice, pores of different kinds may be present in a solid and interconnections may exist between them, so that in some cases the assigment of a definite pore shape may in a certain way be meaningless. However, as in the case of zeolite, the pore shape can also be completely determined by the crystal structure of the solid

— *the pore size distribution* is the distribution of the pore volume (or pore area) versus the pore size. It is an important factor for the kinetic behaviour of a porous catalyst, but the computation of such a distribution always involves somes assumptions that we must always bear in mind in the everyday practice. In general, due to the experimental methods used (adsorption of fluid, mercury porosimetry), the pore size distribution does not give the exact volume (or area) of the pores having a given pore size but the volume (or area) of the pores accessible through the pores of a given size. The pore size distribution thus depends on the shape of the pores present in the solid

— *the mean pore size* either corresponds to the mean value of the pore size distribution or is obtained by dividing the specific porosity by the specific surface area, the result being multiplied by a factor depending on the pore shape. It is therefore expedient to classify pores according to their mean sizes. They are called macropores, micropores or mesopores if their widths respectively exceed 30 to 50 nm, do not exceed 2 nm or are of intermediate size [2, 3]

— *the particle size distribution* is the distribution of the mass of the particles as a function of their size. Generally the particles are supposed to be "spheroidal" but in some cases a shape-factor must be taken into account, in particular for sheet — or fiber — like materials. The particle size distribution is kinetically important when the catalyst is used as a powder and when external diffusion phenomena occur, as for example in industrial fluid-bed or slurry reactors [4, 5, 6]. A mean particle size is very often used: it generally corresponds to the mean value of the particle size distribution

— *the shapes and the sizes of agglomerates of particles* like pellets, granules, extrudates and so on are generally rather well defined and consequently can be easily characterized by simple geometrical parameters. These physical quantities are kinetically important when external diffusion phenomena occur as for example in industrial fixed-bed or trickle-bed reactors [4, 5, 6].

3. Analysis of Physical Adsorption Isotherms to Determine the Porous Texture of Solids

A. Experimental Remarks

Although an extensive account of experimental techniques will appear in a later chapter, it is worth emphasizing here some practical points most important for a correct determination of the porous texture of solids.

1. Choice of the Adsorptive

Only physical adsorption is considered here i.e. adsorption for which the forces involved between the solid, called *adsorbent*, and the fluid, called *adsorbate*, are intermolecular forces of the van der Waals type. These forces are of the same kind as those responsible for deviation from ideality of the real gases and the condensation of vapours; they do not significantly change the electronic orbital patterns of the species involved. In this case the adsorption phenomenon is strongly dependent on the physical properties of the solid and in particular on its porous texture. But if the adsorbent-adsorbate interactions involve valence forces, strong hydrogen bonds or charge transfer complexes, the adsorption phenomenon is no longer representative of the sole physical properties of the solid and consequently can no longer be used to determine the texture of the adsorbent.

This fundamental remark implies that the adsorbate-adsorbent interactions must be kept at a level such that the physical properties of the solid are shown through adsorption but are not masked by its chemical behaviour. Consequently,

(i) the adsorptive must be as chemically inert as possible
(ii) vapors must be prefered to liquids or gases as adsorptive
(iii) adsorption temperature and pressure will usually be low.

Furthermore, if a high level of resolution in the pore structure is needed, the adsorptive molecules (or atoms) must be small enough so as to enter the whole pore space.

In these conditions, the most commonly used adsorptives are nitrogen, argon or krypton.

As pure nitrogen is readily available and less expensive than argon and krypton, nearly all the textural studies of solids and catalysts are carried out on the basis of nitrogen adsorption isotherms.

However a number of other gases and vapours such as O_2, CO, CO_2, CH_4, CCl_4, H_2O have also been adsorbed on solids of different types [7, 11]. Their adsorption isotherms are usually compared to those of nitrogen but seldom for textural analysis.

It is interesting to note here that water vapour has been claimed to be the best adsorbate in numerous studies on clay minerals and concretes [12–14]. In these particular cases, the aim is not to determine the texture of the solid itself. The main purpose is to characterize the hydrophilic properties of the adsorbent, its structure or texture being very often strongly modified during the adsorption [13–14]. For pure textural analyses, water vapour must be avoided as an adsorptive.

2. *Pretreatment of the Adsorbent*

In all cases, the adsorbent surface will be cleared out of any adsorbed molecules — generally by outgassing — before the physical adsorption isotherm can be determined. The outgassing process of the solid sample can be performed at room temperature, or at a higher temperature to promote the desorption of strongly adsorbed species. It is important to point out that the outgassing temperature must be kept below the temperature of the last thermal treatment to which the solid has been submitted, otherwise the sample texture could be modified. Whatever the temperature, the solid must be submitted to a vacuum of about 1.3×10^{-2} Nm^{-2} ($\sim 10^{-4}$ Torr) during at least one hour. In general, the lower the temperature, the longer the outgassing process.

The pretreatment conditions must remain sufficiently mild to avoid any modification of the textural characteristics of the sample but severe enough to obtain a good surface clean-up.

3. *Adsorption Isotherm Determination*

When the adsorptive is chosen and the adsorbent correctly outgassed, the adsorption isotherm is the function which relates at constant temperature the amount of substance adsorbed at equilibrium to adsorptive pressure in the gas phase [2]. This adsorbed amount can be given by a number of moles or by the corresponding volume at STP conditions. As the temperature of adsorption is usually lower than the critical temperature of the adsorptive, the adsorbed amount is generally given as a function of the relative pressure, i.e. the pressure in the gas phase, p, divided by the saturation pressure, p_s, at the same temperature. In these conditions the adsorption isotherm can be represented for a given gas-solid pair by:

$$v = f(p/p_s)_T \qquad (1)$$

where v is the amount adsorbed, p/p_s the relative pressure and T the temperature.

According to whether the v values are determined for increasing the pressure (from $p/p_s \simeq 0$ up to $p/p_s = 1$) or decreasing the pressure (from $p/p_s = 1$ down to $p/p_s \simeq 0$), the adsorption branch or the desorption branch of the isotherm is obtained. In general, the two branches of the isotherm do not coincide with each other: there is an hysteresis loop, characteristic of the "mean shape" of the pores present in the solid.

To obtain an accurate determination of the adsorption-desorption isotherms two different methods can be used:

— *the gravimetric method* where the weight variations of the solid sample are determined for increasing or decreasing the vapour pressure. This method is very accurate but it calls for a very sensitive and sophisticated microbalance

— *the volumetric method* where the volume of adsorbed species is determined as a function of the pressure. This method is in principle less accurate than the gravimetric method but it is extensively used because only a rather simple glass apparatus is needed

Whatever the experimental method used, the reliability of the results will depend on the knowledge of the actual working temperature [14]. If nitrogen (or argon) is the adsorptive, the adsorption isotherms are determined at liquid nitrogen (or argon) temperature, by using a dewar containing the corresponding liquid. To keep the temperature constant throughout the experiment, the liquid bath must be kept in the boiling state, for example by bubbling the corresponding gas through the liquid. However, even in these conditions, a slight variation of the boiling temperature of the liquid nitrogen (or argon) bath can be observed due to the slow but regular dissolution of the air components in the liquid bath and, in a lesser extent, to modifications of the atmospheric pressure. *These small temperature variations can strongly affect the value of the saturation pressure* p_s and, consequently the whole relative pressure scale of the isotherm. In practice, the saturation pressure must be measured at the temperature of the liquid bath containing the sample cell. A simple way of doing this is to use a gas manometer filled with the adsorptive used for the adsorption experiment and to keep it in the liquid bath containing the sample cell.

The temperature of the solid sample is supposed to be the same as the liquid bath temperature. However, an extensive study of Nicolaon [15] showed that important temperature gradients may exist between the sample and the liquid bath when the gravimetric method is used. This is due to the fact that in this method there is no direct contact between the liquid bath and the basket of the microbalance containing the solid and, consequently an appreciable time may be required to reach the equilibrium.

B. Classifications of Isotherms According to Their Shape or Their Hysteresis Loop

In 1940, Brunauer et al. [17] proposed to classify the physical adsorption isotherms in five different groups numbered from *I* to *V*. This well known classification relates the shape of the isotherms to both the mean pore size of the adsorbent and the intensity of adsorbate-adsorbent interactions [10,

18—20]. The five types of adsorption isotherm are schematically represented in Figure 1. It is generally admitted that:

— *isotherms of type I* are characteristic of microporous adsorbents
— *isotherms of types II and III* are observed for macroporous adsorbent. Type *II* isotherms frequently occur. On the contrary type *III* isotherms are rather unusual and correspond to very weak adsorbate-adsorbent interactions
— *isotherms of types IV and V* are obtained for mesoporous adsorbents. Type *IV* isotherms are widespread whereas type *V* isotherms (which correspond to very weak adsorbate-adsorbent interactions) are seldom encountered.

In everyday practice, the shape of the isotherm usually gives a rather good idea of the mean size of the pores present in the solid. Sometimes, however, an isotherm cannot be related to one definite group of this classification because different types of pores are present in the adsorbent. These "mixed" isotherms are also indicative of the general porosity of the solid.

An other easy way to get information on the porous texture of the adsorbent is to look at the shape of the hysteresis loop. The presence of an hysteresis loop is always observed with isotherms of types *IV* and *V* and sometimes with isotherms of types *II* and *III*. A classification of the hysteresis loops into five main groups has been proposed by de Boer [21]. Each group is related to the general shape of the pores present in the solid, as illustrated in Figure 2.

— *Type A hysteresis* is associated with "cylinder shaped" pores of rather constant cross section
— *Type B hysteresis* points to the presence of slit-shaped pores
— *Types C and D hysteresis* derive from type *A* and *B* respectively and they indicate the presence of cone-shaped pores (type *C*) and pores formed by non parallel plates (type *D*). These two types of hysteresis loops are rather unusual
— *Type E hysteresis* very frequently occurs. It corresponds to "ink-bottle" pores, spheroidal cavities or voids between close-packed spherical-like particles.

This well known classification is widely used [9, 10, 20, 22, 23] even if in practice the hysteresis loop shape may appear rather different from the ideal curves presented in Figure 2. The main reason for these differences

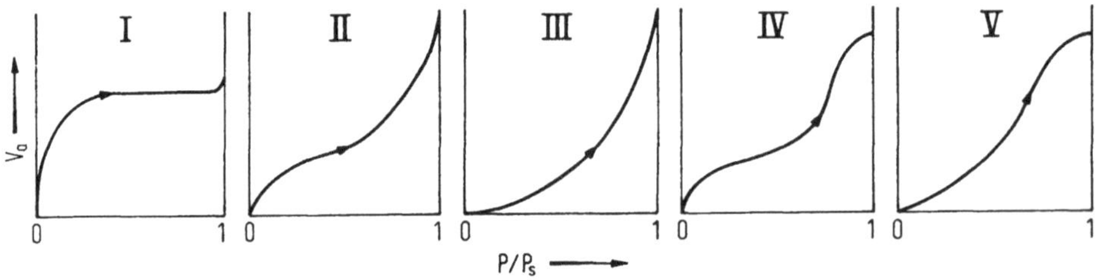

Figure 1. The five types of adsorption isotherms according to Brunauer et al. [17]

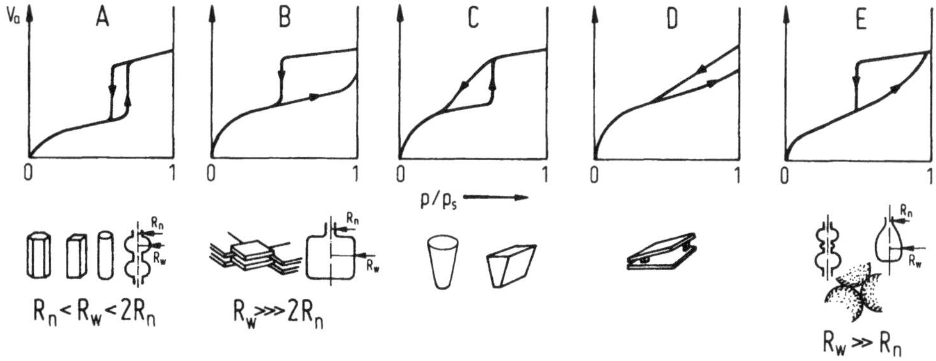

Figure 2. The five types of hysteresis loops and the associated pore shapes, as proposed by de Boer [21]

is the existence in real solids of a more or less broad pore size distribution, whereas the models used are based on ideal solids where all the pores have the same size. In general the broader the pore size distribution, the wider the hysteresis loop. Furthermore, "mixed" hysteresis loops may often be observed if different pores shapes are present in the same solid.

The absence of hysteresis may indicate either that the solid is microporous (type *I* isotherm) or that the pores are nearly perfect cylinders closed at one end, or that the solid is non porous.

A general inspection of the isotherm shape coupled with the analysis of the shape and the width of the hysteresis loop can indeed give the main outlines of the textural characteristics of a solid. This first picture must be kept in mind as a framework for a further and more detailled analysis.

C. The BET Method

Since the publication, in 1938, by Brunauer, Emmett and Teller [1], of their famous model for physical adsorption of vapours on solids, the BET-theory and its use as a tool for determining the specific surface area of solids have formed the subject of many papers, reviews and books [9, 18—20, 24]. The limitations and the range of applicability of the BET-model have been critically discussed by several authors. In this respect, the book of Gregg and Sing [10] and the review paper of Nicolaon [15] are particularly worth reading.

1. The Hypothesis of the BET Theory

The BET theory is based on various hypothesis the knowledge of which is important for a good understanding of the model limitation and its correct use in the textural analysis of solids. The main assumptions of this theory can be summarized as follows:

— the adsorption is supposed to be localized on well defined sites; all the sites have the same energy (homogeneous surface) and each of them can only accomodate one adsorbate molecule
— a multilayer adsorption is supposed to occur even at very low pressure.

the adsorbed molecules in the first layer acting as adsorption sites for the molecules of the second layer and so on; there is no lateral interaction between adsorbed molecules

— an adsorption-desorption equilibrium is supposed to be effective between molecules reaching and leaving the solid surface; the desorption is an activated process, the activation energy being E_1 for the molecules adsorbed in the first layer and E_L for the molecules adsorbed in the other layers. The E_1 value is taken as the differential heat of adsorption of adsorbate molecules on the solid surface. The E_L value corresponds to the heat of liquefaction of the adsorbate at the adsorption temperature. Except for the first layer, the adsorbate molecules are thus supposed to be as in the liquid state.

2. The BET Equation

The physical and mathematical treatment of these hypothesis leads to the following equation of an isotherm

$$v_a = \frac{v_m c x}{1 - x} \cdot \frac{1 - (n + 1) x^n + n x^{n+1}}{1 - x + c x (1 - x^n)} \tag{2}$$

where v_a is the adsorbed volume of the adsorbate per unit mass of adsorbent;
v_m is the volume of adsorbate just sufficient to cover (with a complete monolayer of molecules in close packed array) the surface developped by the unit mass of adsorbent;
x is the relative pressure p/p_s;
n is the total number of layers which, partly or fully filled, can be adsorbed;
c is a constant varying with the adsorbent-adsorbate interactions. It is related to the differential heat of adsorption E_1 and to the heat of liquefaction E_L by the following relation

$$c = \exp\left(\frac{E_1 - E_L}{RT}\right); \tag{3}$$

where R is the ideal gas constant and
T the absolute temperature.

The relation (2), called the "n-layer BET equation", is seldom used as such except in the border-line case where the number of adsorbed layers is very large. In these conditions, as the x values are always smaller than one, the relation (2) becomes

$$v_a = \frac{v_m c x}{(1 - x)(1 - x + c x)}. \tag{4}$$

With this relation, called the "BET equation", it is possible to describe the five different types of adsorption isotherms encountered in practice and represented in Figure 1, simply by using the appropriate values of c and v_m [1, 10, 18–20]. However the theoretical and experimental isotherms coincide only over a limited pressure range. This limitation comes from the assumptions on which the BET-theory is based. This theory indeed takes into account neither the heterogeneity of the solid surface, which mainly affects the low pressure part of the isotherm, nor the lateral interactions between adsorbed molecules which become more and more important as the pressure increases. Consequently, the BET equation is normally able to correctly describe a physical adsorption isotherm only in the relative pressure range between 0.05 and 0.35.

These limits are indicative [10, 14, 23] for they can vary slightly from one adsorbent-adsorbate pair to an other.

3. Specific Surface Area of Solids According to the BET-Model

The BET-equation (4) is more conveniently written in the following form

$$\frac{x}{v_a(1-x)} = \frac{1}{v_m c} + \frac{c-1}{v_m c} x \qquad (5)$$

It shows that the graph of

$$y = \frac{x}{v_a(1-x)}$$

versus x, for x varying from 0.05 to 0.35 should give a straight line

$$y = \alpha x + \beta \qquad (6)$$

if the BET model is valid. From this so-called BET-plot it is easy to obtain, by graphical or numerical methods, the values of α and β from which the values of the two parameters of the BET-equation, c and v_m, can be derived:

$$c = \frac{\alpha}{\beta} + 1 \qquad (7)$$

$$v_m = \frac{1}{\alpha + \beta} \qquad (8)$$

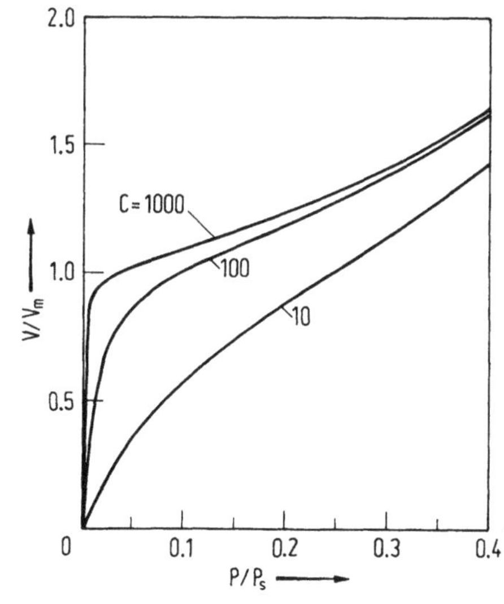

Figure 3. Influence of the c-value on the shape of adsorption isotherms in the BET model

The c-*value* is characteristic of the intensity of the adsorbate-adsorbent interactions. As the c parameter is related to the differential heat of adsorption (see equation 3), the stronger the adsorbate-adsorbent interactions, the higher the c-value, whatever the type of porosity present in the solid. As shown in Figure 3, the c-value greatly affect the shape of the isotherm in the low pressure region ($p/p_s < 0.2$), whatever the v_m value [9, 10, 15]. The importance of the c-value in relation with the adsorbate-adsorbent interactions will be discussed later on.

The v_m-*value* is directly proportionnal to the specific surface area of the adsorbent, because v_m is, by definition, the amount of adsorbate just sufficient to cover with a complete monolayer the whole surface developped by the unit mass of adsorbent. The specific surface area of the solid is thus equal to the area occupied by one adsorbate molecule multiplied by the number of adsorbate molecules contained in v_m, so that

$$s = a_m \cdot \frac{v_m \cdot N_A}{V_m} \tag{9}$$

where s is the specific surface area
v_m is the monolayer capacity of the unit mass of solid
N_A is the Avogadro constant
V_m is the molar volume of the adsorbate
a_m is the part of surface occupied by one molecule of adsorbate in a closed layer.

In the case of nitrogen adsorption at liquid nitrogen temperature (77 K) the most widely admitted [10, 22–24] and used value of the area occupied by a nitrogen molecule is

$$a_m(N_2) = 16.2 \times 10^{-20}\ m^2 \tag{10}$$

In these conditions, if N_A and V_m are taken respectively equal to 6.025×10^{23} mole^{-1} and $22.414 \times 10^{-3}\ m^3 \cdot mole^{-1}$, the specific surface area of the solid (expressed in m^2/kg of solid) is given by

$$s_{BET} = 4.37 \times 10^6\ v_m \tag{11}$$

v_m being the monolayer capacity expressed in m^3 STP of N_2/kg of solid.

The subscript BET indicates that the specific surface area has been obtained throught the BET model.

If other adsorbates than nitrogen are used, it is rarely possible to assign a fixed value to the molecular surface area a_m of a given adsorptive. In general a range of value is given for each adsorptive [9, 10] but, from their comprehensive review of the literature, McClellan and Harnsberger [25] "recommend" the following values of a_m expressed in $10^{-20}\ m^2$:

Ar	at 77 K:	13.8
Kr	at 77 K:	20.2
nC_4H_{10}	at 273 K:	44.4
C_6H_6	at 293 K:	43.0

However great caution should be exercice while using these values because the molecular surface area of an adsorbate depends somewhat on the naturc of the adsorbent. The a_m values proposed here have been adjusted in such a way that the specific surface area obtained by using the BET model and one of these adsorbates are in good agreement with the s_{BET} values calculated from nitrogen adsorption. In all cases, the nitrogen molecular area of 16.2×10^{-20} m^2 is considered as the reference value.

In practice, the use of nitrogen as absorbate in textural studies is thus generally more reliable and technically simpler than the utilization of other adsorbates.

4. Range of Applicability and Limitations of the BET-Model

Although the BET theory is based on a very simple model and can only describe a rather small part of adsorption isotherms, it works quite satisfactorily for determining the specific surface area of solids. In fact, for non porous, macroporous and mesoporous solids, s_{BET}, obtained from nitrogen adsorption isotherm must be considered in the present state of our knowledge as the best measure of the specific surface area. However the s_{BET} value can be determined without any limitations and with a good accuracy only from isotherms of types *II* and *IV* (cf. Figure 1) which fortunately are the most common isotherms.

In principle, the BET equation can also describe the other types of adsorption isotherm but the specific surface area determination is subjected to drastic limitations.

On the one hand, with isotherms of type *III* and *V* typical of very weak adsorbent-adsorbate interactions, it is impossible to determine with a sufficient accuracy the s_{BET} value for two main reasons:

- the value of the c constant in the BET equation is nearly equal to one and in this condition, it has been shown [10, 15] that for mathematical reasons the v_m value is very sensitive to the slightest variation of c
- the adsorbed amounts are generally very small so that the determination of the interesting part of the isotherm may not be very accurate [24].

These two effects have such a cumulative unfavorable influence on the accuracy of s_{BET} determination, that it is not worth calculating a specific surface area from isotherms of type *III* or *V*. The only possibility to get round this difficulty is to change the nature of the adsorbate in order to increase its interaction with the adsorbent and thus to obtain an isotherm of type *II* or *IV*.

On the other hand, with isotherms of type *I*, the calculation of s_{BET} can be made with a very good mathematical accuracy but the physical meaning of the value obtained in this way is a much debated question [10, 26—32]. Type *I* isotherms are characteristic of microporous solids in which the pore sizes are of the same order of magnitude as the size of adsorbate molecules. A free multilayer adsorption as assumed in the BET theory cannot take place without being disturbed in these narrow pores, for the pore walls put a limitation to the number of adsorbed layers. How much this limitation affects the v_m value calculated from the BET-equation is an important subject of controversy, for on the given answer depends the range of applicability of the BET-model. According to Dubinin [29] the isotherms of type *I* correspond neither to a multilayer adsorption nor to the progressive completion of a monolayer like in the Langmuir mechanism, but to the filling up of the micropore volume with adsorbate in a liquid like condition. The volume of micropores in thus considered by Dubinin [30] as the one and only quantity which can be determined from type *I* adsorption isotherms. Nevertheless, after a long debate through the literature, Brunauer and Dubinin [33] came to an agreement on the ranges of applicability of their models: if nitrogen is used as the adsorbate, the BET-model may be applied to microporous solid down to a micropore width of 1 nm whereas the Dubinin model has to be used if "supermicropores" are present i.e. if the pore width is less than 1 nm.

However, in the every day practice, the BET specific surface area is commonly used whatever the type of microporous solids under investigation but its conventional character must be kept in mind and the corresponding s_{BET} values only considered in comparative experiments.

D. The Dubinin-Radushkevich (DR) theory

Starting from his point of view that the micropore volume is the only significant quantity which can be derived from type *I* adsorption isotherm, Dubinin [34, 35] has proposed together with Radushkevich a theory for the calculation of the micropore volume from the low pressure part of the isotherm. This approach is based on the Polanyi thermodynamical theory of adsorption [36] and leads to the following expression of the isotherm

$$\log_{10} v_a = \log_{10} v_o - D\,(\log_{10} x)^2 \qquad (12)$$

where v_a is the adsorbed volume of adsorbate per unit mass of adsorbent (m^3 STP/kg)

v_o is the largest volume of adsorbate which can be condensed in the micropores of an unit mass of adsorbent (m^3 STP/kg)

x is the relative pressure p/p_s

D is a factor varying with the temperature and the adsorbate-adsorbent pair.

A plot of $\log_{10} v_a$ against $(\log_{10} x)^2$ should therefore, be a straight line. The intercept, equal to $\log_{10} v_o$, would lead immediately to the micropore volume V_o

$$V_o = v_o \cdot (\varrho_g/\varrho_l) \qquad (13)$$

ϱ_g and ϱ_l being the density of the adsorbate in the gas and liquid phase respectively. In the case of nitrogen at 77 K the ratio ϱ_g/ϱ_l is equal to 0.0015.

For a large number of adsorption systems [37, 38] the DR plot yields indeed a perfect straight line.

Frequently, however, the plot is found to deviate from linearity, especially in the high pressure part of the isotherm. In this case, the micropore volume is obtained by extrapolation of the linear part of the DR plot corresponding to the low pressure region of the isotherm.

Nevertheless the micropore volume determination by the DR plot must be limited to nearly pure microporous solids [39, 40] (type *I* isotherm). If larger pores are also present in the adsorbent (isotherms of type *II* or *IV*) the DR method is no longer applicable, for it is not possible to estimate the relative contributions to the adsorbed volume made by micropore filling and surface coverage. The DR plot has thus a very restricted area of application.

E. Standard Isotherms as a Tool in Textural Studies

1. Concept of Standard Isotherm

For a large number of non porous solids, the nitrogen adsorption isotherms of which did not give any indication of capillary condensation, several authors like Shull [41], Cranston and Inkley [42], Pierce [43], Lippens and de Boer [44], showed that the ratio between the adsorbed volume, v_a, and the volume of the unimolecular layer, v_m, if plotted versus the relative pressure p/p_s could be represented by a single curve called "*standard isotherm*". Some discrepancies appear between the different published versions of the "standard isotherms". These discrepancies may partly occur from the fact that the various solids used for adsorption experiments are not necessarily free of pores, for it is not easy to establish the absence of porosity in a solid. A second reason for the discrepancies is due to the fact that the strength of interaction of a given adsorbate with the adsorbent varies somewhat from one adsorbent to another and that the standard isotherms do not account for these variations.

If a sufficient precision is desired, therefore, one must abandon the search of a truly universal standard isotherm valid for a given adsorptive with all adsorbents.

At this stage, two different approaches have been taken in the literature. On the one hand Sing [26, 45–49] and his co-workers put forward the necessity for building up a series of standard isotherms, each of them corresponding to several samples of the same chemical nature. A standard isotherm for silica [47] and an other one for alumina [48] have been proposed by these authors.

On the other hand, Lecloux [50–52] and his group proposed a new set of standard isotherms including the reference data of Sing and demonstrated that each of them corresponds to a definite strength of interactions between adsorbent and adsorbate. The fact of classifying the standard isotherms according to the intensity of adsorbate-adsorbent interactions has been proved to be a very interesting and general one [52]. The same set of standard isotherms may indeed be successfully applied whatever the nature of the adsorbent-adsorbate pair.

2. The t-Method of Lippens and de Boer [44]

Lippens and de Boer [22, 44, 53, 54] showed that, if the multimolecular layer of adsorbed nitrogen could be formed freely on the whole surface, the statistical thickness, t (in nm), of the adsorbed layer could be calculated according to the equation

$$t = 0.354\,(v_a/v_m) = f_1(p/p_s) \qquad (14)$$

where v_a and v_m are respectively the volume of nitrogen adsorbed at a relative pressure p/p_s and the volume of nitrogen required to cover the whole surface of the solid substance unimolecularly, calculated by the BET-equation [1]. According to de Boer et al. [53–54], the thickness, t, of the adsorbed layer of nitrogen, as given by equation (14), is nearly independent of the nature of the adsorbent if the latter is non-porous. Consequently, in the absence of capillary condensation, the variation of the statistical thickness of the adsorbed layer of nitrogen with the relative pressure can be filled by a single curve, so called "*the universal t-curve*".

The basic principle of the t-method is to compare the nitrogen adsorption isotherm obtained for a given porous solid with a particular reduced form of a standard isotherm (t-curve) corresponding to a non porous solid of the same type. To make this comparison, the nitrogen volume, v_a, adsorbed at a relative pressure p/p_s on a given sample is plotted as a function of the thickness, t, of the adsorbed layer, as given by the "universal t-curve" for the same relative pressure. Through this v_a-t-plot, valuable information can be obtained about the porous structure of the adsorbent.

As long as the multilayer of adsorbate is formed unhindered on the solid surface, the v_a-t-plot is a straight line passing through the origin, the slope of which is a measure of the surface area of the adsorbent, according to the equation (for nitrogen at 77 K)

$$s_t = 1.547 \times 10^6 \, dv_a/dt \tag{15}$$

where t is expressed in nm, v_a in m^3 STP/kg and s_t in m^2/kg. The subscript t indicates that the specific surface area has been obtained by the t-method.

At higher relative pressures (higher t values), deviations from the straight line may occur

— *an upward deviation* indicates the presence of a capillary condensation in the pores of the adsorbent; this is characteristic of cylinder-shaped pores, "ink-bottle" pores or spheroidal cavities

— *a downward deviation* is observed when micropores or slit-shaped pores are present in the solid.

The v_a-t-plot is an interesting tool in textural studies for it gives a measure of the specific surface area of the adsorbent as well as valuable information on the shape of pores. Moreover it is worthwhile to point out that the manner in which the v_a-t-plot deviates from the straight line is also characteristic of the pore-size distribution in the solid. For example, the higher the pressure at which deviation occurs, the larger the size of the pores.

As showed in section (3. E. 5. b), the shape of v_a-t-plot may be use in certain cases to obtain the pore-size distribution.

3. *The* α_s*-Method of Sing* [45]

The same principle of comparing an experimental isotherm to a standard one is applied in a slightly different way by Sing [45]. This author modifies the t-method by replacing the thickness, t, of the adsorbed layer by the ratio (v_a/v_x), so called α_s, where v_x is the amount adsorbed by a non porous reference solid at the selected relative pressure $(p/p_s)_x$. By doing this, he defines a new type of reduced standard isotherm by plotting α_s versus p/p_s, according to the equation

$$\alpha_s = (v_a/v_x) = f_2(p/p_s) \tag{16}$$

In principle α_s could be set equal to one at any convenient point on the standard isotherm. According to Sing [9, 26, 53] it is ususally convenient in practice to set $\alpha_s = 1$ for $(p/p_s) = 0.4$ since monolayer coverage and micropore filling occur at $p/p_s < 0.4$ whereas capillary condensation takes place at $p/p_s > 0.4$.

If the v_a-α_s-plot is contructed as the amount adsorbed on the test sample versus the α_s-values for the standard, another means of comparing the mathematical form of isotherms is obtained. The shapes of the v_a-α_s-plots are essentially the same as those of the v_a-t-plots and the deviations from linearity are explained in the same way.

There are however two main differences between the t- and the α_s-method:

1. The first one, as already mentioned, lies in the choice of the standard isotherm. According to de Boer, the "t-curve" is nearly independent of the nature of the adsorbent and consequently the standard isotherm is obtained by averaging several reduced isotherms corresponding to various non porous solids. On the contrary, Sing chooses the standard isotherm adapted to the nature of the test sample. He indeed determined standard α_s-curves for nonporous hydroxylated silica [47] and for non porous alumina [48].
2. The second difference between α_s- and t-methods, as claimed by Sing [46], lies in the fact that the α_s-method is independent of the BET-method, while the t-method obviously fall into the field of the BET-method by using, in the definition of t, the value, v_m, of the BET monolayer capacity. See equation (14). This question has been much debated in the literature [55–57], but recently, Lecloux [52] showed that the t- and α_s-methods are strictly relevant to the BET-theory and that these two methods are two equivalent ways of analyzing adsorption isotherms.

4. The n-Method of Lecloux [52]

The basic principle of the t- and α_s-methods, i.e. the comparison between test and standard isotherms, was somewhat extended by Lecloux who introduces a general criterion in the selection of standard isotherms. He pointed out that the thickness of the adsorbed layer on a non porous solid not only depends on the relative pressure, but also on the magnitude of the adsorbent-adsorbate interactions [51—52]. Consequently, the standard isotherm must be chosen according to these interactions and not merely on the basis of the chemical nature of the test sample.

As one of the easiest way of expressing the intensity of adsorbate-adsorbent interactions is to use the c_{BET} constant value [58, 60] a new set of five standard isotherms was proposed, each of them corresponding to a definite range of the c_{BET} value between 20 and 3000. These reference isotherms are used in the following reduced form:

$$n(c_{BET}) = v_a/v_m = f(p/p_s, c_{BET}) \tag{17}$$

where v_m is calculated by the BET equation. These standard data are illustrated in Figure 4 and the corresponding numerical values are given in Table 1.

The comparison between test and standard isotherms is made by constructing the v_a-n(c_{BET})-plot where v_a is the amount adsorbed on the test sample and n(c_{BET}) is the standard value chosen according to the c_{BET} constant characteristic of the test isotherm.

The shapes of the v_a-n(c_{BET})-plots are essentially the same as those obtained with the v_a-t- or the v_a-α_s-plots and the deviations from linearity are explained in the same way.

But the main advantages of the n-method can be summarized as follows [52]. The simple reduced form defined by the equation (17) is the only siutable form to compare isotherms obtained with different adsorbates. Moreover, the whole set of standard isotherms obtained from nitrogen adsorption has been demonstrated to be suitable for other adsorbates [52]. *The reference data, presented in Figure 4 and Table 1, may thus be used whatever the nature of the adsorbent and of the adsorbate.*

The n-method puts forward that the only parameter for choosing the right standard isotherm is the adsorbate-adsorbent interaction intensity espressed by the c_{BET} constant value. If this choice is correctly made, the straight line corresponding to the low relative pressures in a v_a-n(c_{BET})-plot always pass through the origin. But if the right reference data are not used, this straight line cuts the v_a-axis for a positive or a negative value of the adsorbed volume according to whether the c_{BET} constant of the test isotherm has a

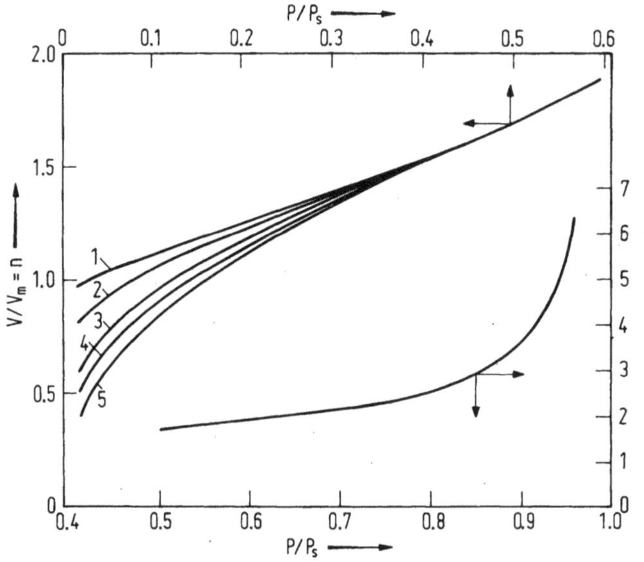

Figure 4. Reference isotherms in the n reduced form. Each curve corresponds to a definite range of the c_{BET}-value: n_1 for $c_{BET} \geqq 300$, n_2 for $300 \geqq c_{BET} \geqq 100$, n_3 for $100 \geqq c_{BET} \geqq 40$, n_4 for $40 \geqq c_{BET} \geqq 30$ and n_5 for $30 \geqq c_{BET} \geqq 20$

Table 1. Numerical values of the reference isotherms in the n reduced form

Range of validity	$c_{BET} \geqq 300$	$300 \geqq c_{BET} \geqq 100$	$100 \geqq c_{BET} \geqq 40$	$40 \geqq c_{BET} \geqq 30$	$30 \geqq c_{BET} \geqq 20$	Range of validity
p/p_s	n_1	n_2	n_3	n_4	n_5	p/p_s
0.02	0.972	0.805	0.593	0.503	0.401	0.02
0.03	0.992	0.853	0.669	0.575	0.489	0.03
0.04	1.017	0.893	0.718	0.635	0.545	0.04
0.05	1.034	0.920	0.763	0.686	0.602	0.05
0.06	1.048	0.946	0.802	0.730	0.647	0.06
0.07	1.062	0.966	0.839	0.768	0.692	0.07
0.08	1.076	0.992	0.870	0.804	0.729	0.08
0.09	1.090	1.011	0.901	0.838	0.768	0.09
0.10	1.102	1.037	0.929	0.868	0.808	0.10
0.12	1.127	1.068	0.983	0.924	0.872	0.12
0.14	1.155	1.105	1.028	0.976	0.932	0.14
0.16	1.184	1.136	1.071	1.022	0.983	0.16
0.18	1.209	1.167	1.113	1.066	1.034	0.18
0.20	1.234	1.203	1.153	1.110	1.082	0.20
0.22	1.260	1.234	1.192	1.154	1.127	0.22
0.24	1.285	1.263	1.232	1.195	1.175	0.24
0.26	1.311	1.297	1.266	1.235	1.218	0.26
0.28	1.339	1.325	1.305	1.275	1.257	0.28
0.30	1.367	1.356	1.339	1.320	1.299	0.30
0.32	1.395	1.384	1.373	1.370	1.342	0.32
0.34	1.427	1.415	1.407	1.407	1.384	0.34
0.36	1.452	1.452	1.444	1.444	1.424	0.36
0.40	1.511	1.511	1.508	1.508	1.497	0.40
0.42			1.536			0.42
0.44			1.565			0.44
0.46			1.596			0.46
0.48			1.630			0.48
0.50			1.667			0.50
0.55			1.757			0.55
0.60			1.864			0.60
0.65			1.983			0.65
0.70			2.127			0.70
0.75			2.280			0.75
0.80			2.528			0.80
0.85			2.853			0.85
0.90			3.588			0.90
0.95			5.621			0.95

higher or a lower value than the c_{BET} value of the standard isotherm used. An illustration of such a phenomenon is presented in Figure 5.

The n-method is a generalisation of the t- and the α_s-method. The t- and the α_s-values are simply proportional to the corresponding n-values. In the case of *nitrogen adsorption* at 77 K the proportionality factors are given by

$$t = 0.354\, n(c_{BET}) \tag{18}$$

$$t = 0.538\, \alpha_s \tag{19}$$

$$\alpha_s = 0.658\, n(c_{BET}) \tag{20}$$

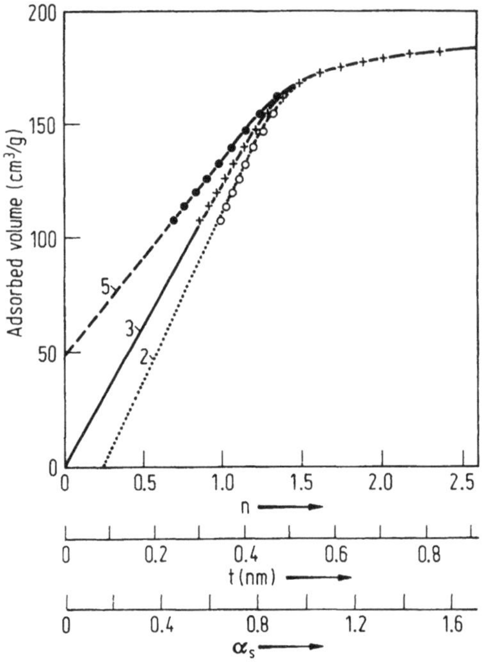

Figure 5. Nitrogen adsorption isotherm on an amorphous silica-alumina characterized by a BET specific surface area of 539 m^2/g^{-1} and a c_{BET} constant value of 68. Comparison of the v_a — t, v_a — α_S and v_a — n plots and importance of the correct choice of the reference isotherm. The numbers on the curves indicate the reference isotherms used (same numbers as in Figure 4)

These values come from the definition of t (equation 14) and from a systematic comparison of the t- and α_s-methods made by Lecloux [52]. A consequence of these proportionalities is that the only difference between the v_a-t, the v_a-α_s and the v_a-$n(c_{BET})$ plot is a scale factor on the x-axis if the right standard isotherm is chosen. This is illustrated in Figure 5 where these different scales are represented.

In the n-method, the set of standard isotherms has been determined once for all and covers a very large range of adsorbent-adsorbate interactions. In fact, the use of the n-method suggest the existence of a continous variation of the standard isotherms as a function of the c_{BET} values. This continuum has been proved [52] to be fitted with a good accuracy (within 5%) up to a relative pressure of 0.65, by the following equation

$$n(c_{BET}) = \frac{v_a}{v_m} = \frac{kx}{1-kx} \cdot \frac{[1-(m+1)(kx)^m + m(kx)^{m+1}]}{\left[\frac{1}{c_{BET}} + \frac{c_{BET}-1}{c_{BET}} kx - (kx)^{m+1}\right]} \tag{21}$$

where x is the relative pressure p/p_s. The values of k and m were calculated so as to be the same for all the curves and equal to 0.95 and 4.18 respectively. In these conditions, the only characteristic parameter is the c_{BET} constant.

5. *The n-Method and the Assessment of Porosity*

a) Specific surface area measurement

If the right reference isotherm is used to analyze an unknown sample, the v_a-n(c_{BET})-plot is, in the low pressure region, a straight line passing through the origin. The slope of this line is a measure of the specific surface area according to the equations (15) and (18):

$$s_n = 0.465 \times 10^6 \, dv_a/dn(c_{BET}) \qquad (22)$$

the numerical value of the proportionality factor being given for nitrogen at 77 K and the n-subscript indicating that the n-method has been used to calculate the specific surface area.

The only condition to obtain an excellent agreement between the s_n and the s_{BET} value of the surface is the correct choice of the standard isotherm [58, 60]. If the right reference data are not used, the only way to determine the specific surface area is from the slope of the tangent to the curve v_a-n(c_{BET}) through the origin. The surface area value obtained in this way is lower or higher than the BET-value if the chosen reference isotherm is characterized by a c_{BET}-value respectively higher or lower than the c_{BET} value of the isotherm under test. This incorrect way of calculating the surface is too often practised in the literature when the t- or the α_s-method is used with incorrect standard data.

b) Assessment of microporosity — Method of Brunauer [58, 61]

It has already been noted that the way in which a v_a-t, a v_a-α_s or a v_a-n plot deviates from linearity is indicative of the pore size distribution of the sample. But experimentally, it appears (see Figure 5 for example) that the shape of these plots can be strongly affected by the choice of the reference data. Consequently, if such a plot is used to characterize the porosity of a sample, it is essential to choose the right standard isotherm according to the c_{BET} constant value before performing any analysis of the curve-shape. A similar remark has been made by Brunauer et al. [58] in their paper describing a method of analyzing the microporosity of a sample on the basis of the v_a-t-plot shape.

The basic principle of *Brunauer's method* is to consider that equation (15) can be applied to every point of the v_a-t curve, if this curve shows a downward deviation from linearity. In other words, if micropores or slitshaped pores are present in the solid, at any point of the isotherm, the surface area accessible to further adsorption decreases as the thickness of the adsorbed layer increases and the slope of the tangent to the v_a-t curve at a given t-value is proportional to the accessible surface area after a layer of thickness t has been adsorbed on the solid. As no capillary condensation occurs in slit-shaped pores or in micropores, the filling of a pore takes place by the coalescence of the adsorbed layers on its opposite walls. At this moment, the thickness of the adsorbed layer is equal to half a pore width.

$$t = d/2 \qquad (23)$$

To evaluate the pore size distribution [61–64] on this basis, the v_a-t plot is divided into k steps, each corresponding to an increase Δt in the thickness of the adsorbed layer. At the beginning and the end of each step a tangent to the v_a-t-curve is drawn and the respective slopes are converted into surface area s_k and s_{k-1} by use of equation (15). The difference $\Delta s_k = s_k - s_{k-1}$ is equal to the surface of those pores having a width lying between $(2\,t_k - \Delta t)$ and $(2\,t_k + \Delta t)$. The curve $\Delta s_k/2\,t_k$ against $d_k = 2\,t_k$ directly express the distribution of pore surface as a function of pore width. To calculate the pore volume distribution it is necessary to define a relationship between the volume, v, and the surface, s, of a pore i.e. to choose a pore model. In the case of slit-shaped pores of width d, the ratio v/s is equal to d/2, whereas for cylindrical pores of radius r, the ratio v/s is equal to r/2.

Although the soundness of Brunauer's approach has been questioned, in particular by Gregg and Sing [9, 65] and Dubinin [29], this analysis of the v_a-t plot is at the present time the only mean of calculating the pore size distribution of micropores from one isotherm. Even if the theoretical bases of the method are within the limits of the BET theory [64], it is the only practical way of comparing various microporous solids providing that the reference isotherm used to construct the v_a-t plot is chosen, in each case, on the basis of the c_{BET} value [52].

It is worth noting that Brunauer's analysis may be extended to the v_a-n plots and thus performed whatever the adsorbent-adsorbate pair. The distribution of pore surface as a function of pore width, can be calculated if the statistical thickness of a monolayer of adsorbate is known.

6. Limitations in the Use of Standard Isotherms

The use of standard isotherms to characterize the porosity of solids through the t-, the α_s- or the n-method is submitted to the same limitations as the BET model [29, 66], for all these methods have been demonstrated to be directly or indirectly relevant to the BET-theory [52, 64]. But, in spite of its theoretical simplicity the BET-model has been experimentally proven [52, 64–70] to have a very wide range of applicability. In the same way, the practical usefulness of the standard isotherms and the methods based upon largely exceeds their theoretical validity, for at the present time there is no other method which can give, in such a simple way, the same qualitative and quantitative information on the porous texture of solids. It is necessary however to keep in mind that the porous solids have been "seen" through a rather simple model and that the reality may be more complex than the pictures obtained in this way. An interesting but sophisticated new approach has been recently developed by Gregg [71, 72] and his school. This author proposes to progressively fill in the micropores with preadsorbed nonane before performing the nitrogen adsorption.

F. Thermodynamics of Adsorption and the Porous Texture of Solids in the Mesopore-Size Region

1. *Introductory Remarks*

A porous texture can be considered as a collection of elements in pore space, each characterized by shape, size and relative allocation with respect to other elements in pore space. In practice however, the information on pore texture will seldom be as detailed as that. Only information concerning size and number of elements in pore space will generally be available. The shape of the pores for actual porous texture is often so difficult to define that, in common practice a predominant pore shape is assumed rather than established independently. But this assumed pore shape has a critical influence on the pore size distribution calculated from nitrogen adsorption data.

In the same way the pores are supposed to be independent, i.e. in direct communication with the external environment but without any interconnection with each other. If this condition is not obeyed, the procedures generally used for obtaining pore size distribution from sorption data will be no more valid.

Even in the simplest case of independent pores of simple well-defined shape, there are difficulties in analyzing adsorption-desorption isotherms in terms of porous structure.

Pore-size analysis is traditionally based upon the application of Kelvin's relation between vapour pressure of a capillary condensed phase and pore size. But Kelvin's relation was developed for large size capillaries and basically ignores physical adsorption of the adsorbate on the pore walls. This phenomenon is of importance for pores in the mesosize range and necessitates therefore additional correction terms to Kelvin's relation. This fact was already pointed out by Deryagin [73] in 1940 on the basis of thermodynamic reasoning and emphasized by Broekhoff and de Boer [74—75] who developped a complete thermodynamical analysis of physical adsorption and capillary condensation phenomenon. Unfortunately, this approach has too largely been ignored by subsequent authors in the field.

2. *Thermodynamics of Vapour Sorption in Capillaries*

In mesopores, sorption of a vapour around its normal liquefaction temperature generally shows hysteresis: less vapour is taken up at a given relative pressure after an increase in pressure from a low value than after a decrease in pressure from saturation. Although hysteresis is a source of information on the size and even shape of the adsorbent pores, it is also a fundamental difficulty for the interpretation of sorption measurement. The existence of a hysteresis loop in the isotherm indeed points to the fact that either the adsorption branch or the desorption branch or both are essentially irreversible in nature. This irreversibility is of importance for the accuracy of pore sizes determination from vapour sorption measurements, at it is shown hereafter.

By including physical adsorption into the thermodynamic treatment of capillary condensation and evaporation and putting forward the Gibbs

Maximum Work principle, Broekhoff [74, 76] demonstrated the following general relations between n_a, the number of moles of vapour adsorbed in the porous system at a given pressure, p, and s, the surface area developped by the pores:

$$s \geqq \frac{RT \int_{p_i}^{p_s} n_a \, d(\ln p)}{\sigma_{sv} - \sigma_{sl}} \quad \text{during adsorption} \tag{24}$$

$$s \leqq \frac{RT \int_{p_i}^{p_s} n_a \, d(\ln p)}{\sigma_{sv} - \sigma_{sl}} \quad \text{during desorption} \tag{25}$$

In these relations

R is the gas constant

T is the absolute temperature

p_s is the saturation pressure

p_i is the highest value of the pressure which can be applied without the pores being filled up with liquid

σ stands for the interfacial free energy per unit surface area, the subscripts sv and sl respectively indicating the solid-vapour and the solid-liquid interfaces. The σ values vary with the amount adsorbed at the interface and thus with the pressure, p.

As s is the true mesopore surface area, it appears from equations (24) and (25) that the analysis of the adsorption branch of the isotherm leads to underestimates of the pore surface area whereas the use of the desorption branch gives overestimates of it. Only in the exceptional case where there is no hysteresis loop, the two estimates coincide with the true mesopore surface area.

The practical application of equations (24) and (25) requires a quantitative knowledge of $(\sigma_{sv} - \sigma_{sl})$. Fortunately, this quantity can be calculated by applying the same line of reasoning to the analysis of standard isotherms [75]. In this case, the adsorption is measured on non porous solids, no pore filling takes place even at high relative pressures and the $(\sigma_{sv} - \sigma_{sl})$ value is given by:

$$\sigma_{sv} - \sigma_{sl} = \sigma_{lv} + RT \int_{p_i}^{p_s} \frac{n_a}{s} \, d(\ln p) \tag{26}$$

the integration being made along a standard isotherm. As the statistical thickness of the adsorbed layer is defined by

$$t = \frac{n_a}{s} \cdot V_m \tag{27}$$

where V_m is the molar volume of liquid adsorbate, the relation (26) may be written as follows: ·

$$\sigma_{sv} - \sigma_{sl} = \sigma_{lv} + \frac{RT}{V_m} \int_{p_i}^{p_s} t \, d(\ln p) \tag{28}$$

In practice, the standard isotherms are used in the following form:

$$\ln (p/p_s) = F(t, c_{BET}) \tag{29}$$

where $F(t, c_{BET})$ is an analytical function of t, the c_{BET} value being a parameter. In this way, the integration of (28) is facilitated. Various empirical expressions for $F(t, c_{BET})$ have been given in the literature [41–44, 50, 52, 57]. One of the simplest form proposed for a fixed c_{BET} value is:

$$\ln (p/p_s) = -a/t^k \tag{30}$$

Substituting (30) into (28) leads to:

$$\sigma_{sv} - \sigma_{sl} = \sigma_{lv} + \frac{RT}{V_m} \cdot \frac{ak}{(k-1)\, t_i^{(k-1)}} \tag{31}$$

where t_i is calculated from equation (30) at the pressure p_i.

As the value of k is higher than one, the relation (31), shows that for large values of t_i (i.e. at high pressures p_i) the $(\sigma_{sv} - \sigma_{sl})$ value tends to the value σ_{lv} of the liquid surface tension of the adsorbate. In this borderline case, the thermodynamically correct espressions (24) and (25) tend to

$$s = \frac{RT \int_{p_i}^{p_s} n_a \, d(\ln p)}{\sigma_{lv}} \tag{32}$$

which is the well-known Kiselev [78]-Brunauer [79] form of the famous Kelvin-equation.

It must be pointed out at this stage that the Kelvin equation, in spite of its large use for pore size distribution calculations, is only applicable if:

- sorption in the mesopore system is completely reversible, which is exceptional
- and $(\sigma_{sv} - \sigma_{sl})$ equals σ_{lv}, condition which is only satisfied if physical adsorption in the multilayer region can be neglected (see equation 26 and 28).

From the thermodynamics of vapour sorption in capillaries, it appears that Kelvin equation cannot be used as such to calculate the pore size distribution in the mesopore range, but it must be "corrected" to account for the multilayer adsorption which takes place even on non porous solids.

3. Calculation of Pore Size Distributions — The Generalized Broekhoff-de Boer Method, the "BdB Method" [74]

The thermodynamical approach presented in the preceding section gives the formal framework for estimating the pore size distribution and shows how the key quantity $(\sigma_{sv} - \sigma_{sl})$ can be calculated from standard isotherms.

In fact, in Broeckhoff and de Boer's approach the thermodynamic potential, μ_A, of the adsorbed layer on a *non porous solid* is assumed to depend on the thickness, t, of the adsorbed layer according to the equation

$$\mu_A = \mu_L - RT \cdot F(t, c_{BET}) \tag{33}$$

where μ_L is the thermodynamic potential of the liquid phase and $F(t, c_{BET})$ is the analytical expression of the "standard isotherm" as defined by relation (29). From (33), it appears that the thermodynamic potential of the adsorbed layer varies with the intensity of adsorbate-adsorbent interactions, as expressed by the c_{BET}-constant value.

Moreover, for a *porous solid*, the thickness of the adsorbed layer also depends on the curvature radius of the surface i.e. on the shape and size of the pore in which the adsorption takes place.

Although Broekhoff and de Boer published separate papers on adsorption in- and desorption from capillaries of various shapes, it is possible to summarize their results [80] by introducing a *shape factor*, f, which has to be taken equal to 1 for slit shaped pores, to 2 for tubular type pores and to 3 for "ink-bottle" type pores or spheroidal cavities. In these conditions de Broekhoff-de Boer model leads to the following results:

1. the thickness t of the adsorbed layer in a pore of radius r_p is given, at a relative pressure p/p_s, by:

$$-\ln (p/p_s) + F(t, c_{BET}) = \frac{(f-1)\,\sigma_{lv} V_m}{RT\,(r_p - t)} \tag{34}$$

where σ_{lv} and V_m are respectively the surface tension and the molar volume of the liquid adsorbate at the absolute temperature T. It is worth noting that equation (34) tends to equation (29) if r_p tends to infinity (non porous solids or slit shaped pores).

2. *during the adsorption*, the capillary condensation in a pore of radius r_p occurs for a critical thickness t_{cr} of the adsorbed layer, the value of which is solution of:

$$\frac{dF(t, c_{BET})}{dt} = \frac{(f-1)\,\sigma_{lv} V_m}{RT\,(r_p - t_{cr})^2} \tag{35}$$

the relative pressure (p/p_s) at which the capillary condensation takes place in a pore of radius r_p is obtained by introducing in equation (34) the value of t_{cr} given by equation (35).

3. *during the desorption*, the evaporation of the condensed vapour from a pore of radius r_p occurs at a relative pressure p/p_s solution of

$$r_p - t_e = \frac{f\sigma_{lv} V_m}{RT \ln (p_s/p)} + \frac{\int_{t_e}^{r_p} f.(r_p - t)^{f-1}\, F(t, c_{BET})\, dt}{(r_p - t_e)^{(f-1)} \ln (p/p_s)} \tag{36}$$

where t_e is the value of t given by the relation (34). Equation (36) is called by Broekhoff and de Boer "*the modified Kelvin equation*".

To calculate the distribution of pore area and pore volume as a function of pore size, the isotherm is divided into a suitable number of intervals, each of them corresponding to a small relative pressure range defined by its mean value. The calculation is worked downwards from saturation and for the kth interval, characterized by a pressure p_k, the specific surface area, s_k and the volume V_k of the pores of radius r_{pk} are given by:

$$\begin{cases} s_k = \dfrac{f r_{pk}^{(f-1)}}{(r_{pk} - t_{k,k})^f} \left[\varrho \cdot \Delta v_k - \sum_{i=1}^{k-1} \dfrac{s_i}{f \cdot r_{pi}^{(f-1)}} \left[(r_{pi} - t_{i,k})^f - (r_{pi} - t_{i,k-1})^f \right] \right] & (37) \\ V_k = s_k \cdot r_{pk}/f & (38) \end{cases}$$

where Δv_k is the variation of the adsorbed or desorbed volume on the kth interval (in m^3 STP gas/kg of adsorbent)

ϱ is the ratio of the molar volumes of the adsorbtive in the liquid and in the gas phase

f is the shape factor as defined at the beginning of this section

r_{pk} and $t_{k,k}$ are the solutions of equations (34) and (35) if the adsorption branch of the isotherm is considered, and of equations (34) and (36) if the desorption branch is used

$t_{i,k}$ for $i = 1, ..., (k - 1)$, are the solutions of equation (34) for each pair p_k, r_{pi}

The relations (37) and (38) have been obtained according to a line of reasoning similar to the one developed by Barrett, Joyner and Halenda [81] in their simple method based on a cylindrical pore model and the uncorrected Kelvin-equation. The relations proposed by these authors are only a particular limiting case of equations (37) and (38). The latter are of general interest, for they are applicable to both adsorption and desorption isotherms, whatever the pore shape and the adsorbent-adsorbate interactions.

By plotting as a function of r_{pk} the values of s_k and V_k calculated by equations (37) and (38), the distributions of pore area and pore volume with respect to pore size are obtained.

The *cumulative* specific surface area or pore volume are defined by the sum on *k* of all the s_k or V_k values.

$$s_{cum} = \sum_k s_k \tag{39}$$

$$V_{cum} = \sum_k V_k \tag{40}$$

4. Validity of the Generalized "BdB Method"

First of all, the method developed by Broekhoff-de Boer [74] and generalized by Lecloux [80] is strictly limited to pore size distributions calculation in the **mesopore range** (pore size between 2 and 30 nm).

For narrower pores, the capillary condensation phenomenon does not exist and the BdB model is purposeless. For wider pores, the capillary condensation phenomenon occurs at pressures very near the saturation i.e. at relative pressures close to unity. For these p/p_s values, the equations (34) to (36) of the BdB model have such a mathematical structure that the pore radius value is very sensitive to a small variation of the pressure. Consequently, a slight inaccuracy in the experimental determination of the pressure near the saturation point leads to an unacceptable lack of precision in the definition of the pore size.

The main limitation of the applicability of the BdB method comes from the irreversible nature of vapour sorption in mesopores. In fact, theoretical [82, 83] and experimental [84, 85] evidences are available to demonstrate that in general both adsorption and desorption branches are irreversible and thus that application of equilibrium thermodynamics to either of these branches will lead to over-or underestimates of pore sizes if no pore model is adopted (see equations (24) and (25)).

But this uncertainty can be removed by choosing specific simple pore models. Equations (34) to (36) have been developed for three types of pores [74, 80]:

— *for slit shaped pores* ($f = 1$), only the desorption branch can be used to calculate pore size distribution from equations 34 and 36
— *for cylindrical pores* ($f = 2$), the pore size distribution can be calculated both from adsorption branch (with equations 34 and 35) and desorption branch (with equations 34 and 36) of the hysteresis loop. Ideally, i.e., if the actual porous system is very well described by the cylindrical pore model, both distribution curves should coincide and in both cases the total cumulative pore surface area s_{cum} should be nearly equal to s_{BET} (unless micropores are present, in which case s_{cum} remains smaller than s_{BET}). In practice, some discrepancies may occur because of the inadequacy of the cylindrical pore model for the solid considered. For example, the fact that s_{cum} appears substantially larger than s_{BET} for the desorption branch but not for the adsorption branch is indicative of bottle-neck effects in the pore and of irreversibility of the desorption branch.
— *for "ink-bottle" pores or spheroidal cavities* ($f = 3$), only the adsorption branch can be used to calculate pore size distribution from equations 34 and 35. According to Broekhoff [76], the same applies to pores of irregular shape for, in this case one can seldom expect evaporation to be a reversible process.

An other aspect of the irreversibility of vapour sorption arises from the fact that in actual porous systems, pores of different sizes are mutually interconnected, so that emptying of an element of the pores network can only occur if there is at least one connecting, emptied pathway to the external environment. Evaporation and desorption from the interior of a porous particle can only take place if the outermost shells of the pores network have been emptied [86]. In these conditions, large deviations from reversibility for evaporation from any interior pore may be considered as a rule. There is however reason to believe that for pore filling during adsorption, network effects are much less important and a relation of the type of equation (35) leads to a quite satisfactory estimate of the pore size distribution.

From the foregoing discussion, the necessity of proceeding with caution in any attempt to calculate pore size distribution from vapour sorption data clearly appears. However, the systematic use of BdB method with a good working model for the pore shape and the right choice of the standard isotherm generally leads to a qualified and selfconsistent picture of the porous texture of the analyzed sample. In fact, the generalized "BdB method" is, at the present time, the only method which can be successfully applied to both adsorption and desorption branches of the isotherm for various pore shapes and whatever the adsorbate adsorbent interaction.

One of the main interest of this method is also the fact that, in all cases, *the cumulative specific surface area,* s_{cum} *is much more consistent with the* s_{BET} *value than the* s_{cum} *value obtained by any other method.* Moreover, by using synthetic samples of well defined porous structure, Scholten [87, 88] demonstrated that the BdB model gives pore size distributions much closer to the reality than those given by other models. Finally, by introducing in the BdB model various reference t-curves, directly derived from the $n(c_{BET})$ standard data given in Table 1, it is possible to calculate pore size distribution whatever the adsorbate-adsorbent pair. In these conditions the pore size distribution of a given solid can be calculated from sorption isotherms of differents vapours and the results compared to each other. It appears that a very good agreement can be obtained if, in the BdB model, the reference isotherm $F(t, c_{BET})$ is chosen according to the adsorbate-adsorbent interactions [89].

G. Practical Analysis of Adsorption Isotherms

The various theories and models presented in the previous sections are applicable whatever the nature of the vapour physically adsorbed on the surface of a given solid. The practical applications of these theories however are almost limited to the analysis of adsorption-desorption isotherm of nitrogen at 77 K. This adsorptive is widely used because of its chemical inertness at low temperature, its availability at a high purity level, and also because the small size of the nitrogen molecule allows itself to penetrate in very narrow pores.

1. *Experimental Determination of Sorption Isotherms*

Whether the adsorption-desorption isotherms are determined by a gravimetric or a volumetric method, it is worth pointing out the following remarks:

- the *adsorbent pretreatment* (handling in an eventually controlled atmosphere, thermal treatment, outgassing, ...) must be adapted to each particular case and never modify the porous texture of the solid
- the *actual saturation pressure* of the adsorptive at the temperature of measurement must be known with very good precision. Its value is indeed of importance not only for the accurate determination of the isotherm but also for its correct analysis
- the larger the *number of experimental points* used to define the sorption isotherm, the more efficient its analysis will be and, consequently the

more detailed the description of the textural properties of the adsorbent. For applying the BET or the Dubinin model at least 6 to 10 experimental points well distributed between $p/p_s = 0$ and $p/p_s = 0.35$ are needed. The use of the t-, the α_S- or the n-method needs about 20 experimental points in the relative pressure range between 0 and 0.9. Finally, the BdB method for calculating the pore size dustribution also requires at least 20 experimental points to define each branch of the isotherm.

These three remarks must be taken into account to appreciate in what extend an experimental adsorption-desorption isotherm is worth being used to characterize the porous texture of the adsorbent. It is indeed meaningless to apply a sophisticated method of analysis to a not very carefully determined isotherm. The validity of the porous texture description derived from sorption isotherm first of all depends on the quality of the experimental determination.

2. Numerical Treatment of Sorption Isotherms

To turn the analysis of sorption isotherms to the best account it is necessary to systematically apply the various models and methods described in the previous sections. This type of methodical investigation is only possible if the mathematical treatment of experimental data is performed with the aid of a computer program. This program must be sufficiently sophisticated to give rapidity and flexibility of execution by using, for example, a set of subroutines which can be separately or successively called by an index system.

With a general program of this type [80] it would be interesting to calculate:

— from the measured values, all the data needed to plot the adsorption-desorption isotherm, the BET and the Dubinin transforms and the generalized t-, α_S- or n-plot

— the pore size distributions from the adsorption branch of the isotherm according to the generalized BdB model (equations 34, 35, 37 and 38) for 3 shapes of capillaries: open cylindrical pores, cylindrical ink-bottle type pores and spheroidal cavities

— the pores size distributions from the desorption branch of the isotherm according to the generalized BdB model (equations 34 and 36 to 38) for cylindrical and slit-shaped pores.

It is generally very convenient to present the pore size distributions in two normalized forms:

— an *integral form* which gives the percentage of the surface area (or volume) developed by the pores having a radius higher (or lower) than the abscissa value

— a *differential form* which gives, in percent/nm, the part of the surface area (or volume) corresponding to the pores of a given radius.

The normalized forms are advantageous to compare the porous texture of various samples having specific surface areas.

3. General Scheme of Analysis of Sorption Isotherms

In the laboratory practice, it is very useful to always apply the same systematic procedure to analyze the sorption isotherms [76–80]. It is the only way to obtain the most realistic picture of the porous texture of the solid under test. It is very convenient to proceed as follows.

a) General inspection of the isotherm

First of all, the adsorption-desorption isotherm must be defined by an adequate number of experimental points. The measured isotherm must be smooth and shows no unexpected irregularities and no permanent hysteresis at relative pressure below $p/p_s = 0.4$. Otherwise, experimental errors must be suspected and the determination repeated.

After that, the type of the isotherm is determined by comparing its shape to the five model isotherms of Brunauer's classification (Figure 1). From that comparison, the adsorbent can be inferred as a microporous, mesoporous or macroporous solid.

Moreover an analysis of the shape and the width of the hysteresis loop according to the classification of de Boer (cf. Figure 2) can give as stated before interesting information about the shape of the pores present in the solid and, to some extent, about the broadness of the pore size distribution.

b) Application of the BET model

First, s_{BET}, the BET value of the specific surface area is determined from the BET-plot, ideally over the pressure range $0.05 \leqq p/p_s \leqq 0.30$. Restricted linearity over this pressure range is often indicative of the presence of micropores. In this case it has already be mentioned that the BET model must be used with caution.

From the same BET-plot, the value of the c_{BET} constant which is a measure of the intensity of adsorbate-adsorbent interaction is easily calculated. As already pointed out, the knowledge of the c_{BET} value is essential for the right use of the n-method as well as for the application of the generalized BdB model.

c) Application of the n-method

The v_a-n-plot of the adsorption branch of the isotherm is constructed over the pressure range $0.05 < p/p_s < 0.9$, the standard isotherm being chosen on the basis of the observed c_{BET} value. If the right reference data are used, the v_a-n-plot appears in the low pressure region as a straight line passing through the origin. From the slope of this line, the value s_n, of the specific surface area can be calculated with the aid of relation (22). At higher pressure, the shape of the v_a-n-plot is indicative of the type of pores present in the solid:

- a straight line corresponds to a pure macroporous solid
- a upward deviation can be associated with cylindrical or spheroidal mesopores
- a downward deviation indicates the presence of micropores or slit shaped pores.

In fortunate cases, a second straight line is found at higher relative pressures. This may be considered as representing adsorption in wide pores, the surface area of which, s_w, can be calculated from the slope of this second straight line:

$$s_w = 0.465 \times 10^{\prime}\ dv_a/dn \qquad (41)$$

Moreover, as stated before, the downward deviation of the v_a-n-plot can be used to evaluate the micropore size distribution by the Brunauer method (cf. section 3. E. 5. b) or at least, the micropore volume by the Dubinin plot (cf. section 3. D).

d) Assessment of the mesopore volume

An accurate estimate of the mesopore volume V_{mp} is essential to yield a correct calculation of the pore size distribution.

If the sorption isotherm shows a clearly defined plateau in the vicinity of $p/p_s = 1$, then complete mesopore filling may be assumed and the total mesopore volume V_{mp} is determined by the adsorbate uptake at the plateau. In a number of cases, however, the adsorption isotherm shows asymptotical behaviour around $p/p_s = 1$. In this case the upper closing point of the hysteresis loop may be used for assessing the mesopore volume V_{mp}, but interference with the presence of macropores cannot be excluded. If no clearly developed hysteresis loop is found, then the macropore system is predominant and no mesopore volume can be calculated. A common practice, however, tends, in this case, to define V_{mp} as the volume of adsorbate taken up at a relative pressure very close to the saturation, c.g. $p/p_s = 0.98$. This practice is potentially dangerous and should be exercised with care for it could lead to false conclusions on the mesopore distribution.

e) Pore size distribution calculation

The BdB method based on equations (34) to (38) will systematically be applied to adsorption and desorption branches of isotherms, for different shapes of capillaries, as described in the section concerning the numerical treatment of isotherms. The standard isotherm from which the thickness of the adsorbed layer is calculated will be chosen according to the c_{BET} value characteristic of the isotherm under test.

In this way, different curves giving the distribution of pore volume and pore area with respect to pore size are obtained. The problem arises now of choosing the curve which will lead to the most realistic picture of the pore structure of the analyzed sample. To make this choice, all the information acquired during the preceding steps of this analysis must be taken into account. On the one hand, the pore shape model must be consistent with the type of hysteresis and the shape of the n-plot, and on the other hand the cumulative specific surface area, s_{cum}, should be nearly equal both to the s_{BET} and the s_n values.

f) Final discussion

The various methods of analyzing sorption isotherm and the results they give must lead to a coherent picture of the porous texture of the sample under test. Finally a balanced judgement should be made between all the information coming from all these methods, bearing in mind the approximate nature of each of them. It is wise to support this judgement by using independent information coming for instance from structural studies or electron microscopy [50, 87].

The different steps in the analysis of sorption isotherms and the way they are interconnected is summarized in Figure 6. Even if a reasonably self-consistent picture of the porous texture of a solid can be obtained in this way, it is important to keep in mind that such a picture is based upon a series of simplifications of reality. Pictures of different solids obtained through the same approach can however be used to make quantified comparison, which is the most interesting point for practical purposes.

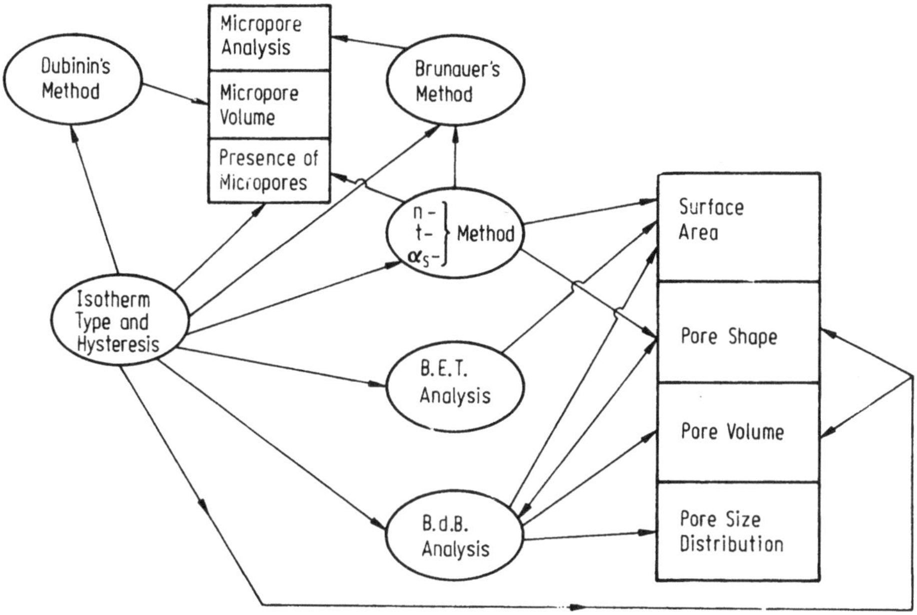

Figure 6. The different steps of the analysis of sorption isotherms and the way they are interconnected

4. Use of the Mercury Penetration Technique to Determine the Porous Texture of Solids

A. Basic Principle of Mercury Penetration

The theoretical bases of the determination of porous texture of solids from mercury penetration are similar to those used for the analysis of sorption isotherm [83]. In the latter case, the adsorbate is supposed to wet perfectly the solid surface so that the condensation phenomenon in small capillaries occurs at a pressure lower than the saturation value. In the former case, the liquid opposes the wetting of the solid and the entrance into the pore spaces, so that an external pressure must applied to overcome the resistance due to the surface tension. As mercury exhibits non wetting properties with a large number of materials, it appears as a most suitable liquid for penetration technique.

In both cases, however, the same thermodynamical approach may be applied.

When mercury is forced into a porous solid, the interfacial free energy change dF_1 for an infinitesimal reversible increase of mercury-solid interface under isothermal conditions is given by

$$dF_1 = \sigma_{sl} \cdot ds_{sl} + \sigma_{sv} \cdot ds_{sv} + \sigma_{lv} \cdot ds_{lv} \tag{42}$$

where σ stands for the interfacial free energy per unit surface area, the subscripts sl, sv and lv respectively indicating the solid-liquid, the solid-vapour and the liquid-vapour interfaces

ds represents an infinitesimal variation of an interface area, the subscripts indicating the type of interface.

The energy dF_2 needed to force an infinitesimal volume dV of mercury into the porous solid in the case of isothermal reversible penetration is given by

$$dF_2 = p \cdot dV \tag{43}$$

where p is the pressure at which the intrusion occurs.

At equilibrium, $dF_1 = dF_2$ so that,

$$p \cdot dV = \sigma_{sl} \cdot ds_{sl} + \sigma_{sv} \cdot ds_{sv} + \sigma_{lv} \cdot ds_{lv} \tag{44}$$

Considering that during intrusion ds_{sl} and ds_{sv} are much greater than ds_{lv} (corresponding to the change of the surface area of the mercury menisci), the last term of equation (44) may be neglected and with the same approximation $ds_{sl} = -ds_{sv}$. In these conditions, equation (44) becomes

$$pdV = ds_{sl} (\sigma_{sl} - \sigma_{sv}) \tag{45}$$

By introducing in (45) the ratio between the volume and the surface of a pore

$$dV/ds_{sl} = r_p/f \tag{46}$$

it comes

$$p = \frac{f(\sigma_{sl} - \sigma_{sv})}{r_p} \tag{47}$$

where r_p is the pore "radius" and f, a shape factor. By integration of equation (45) it follows that

$$S_{s,l} = \frac{\int_0^{p_{max}} p\,dV}{\sigma_{sl} - \sigma_{sv}} \tag{48}$$

The $s_{s,l}$ value is a measure of the specific surface area of the solid, if at the pressure p_{max} the solid surface is completely covered with mercury.

If the volume of mercury which has entered the porous texture of the solid is measured as a function of the externally applied pressure, with the aid of equations (47) and (48), the volume and area of pores accessible through pores of "radius" r_p can easily be calculated.

B. The Use of Mercury Penetration Curves

Even if the principle of determining pore size distributions from equations (47) and (48) appears quite simple, there are two main limitations in the correct use of these equations. It is indeed difficult both to determine the right $(\sigma_{sl} - \sigma_{sv})$ value and to know the pressure p_{max} at which the whole solid surface is covered by mercury.

At this stage, it is worth pointing out the similarity between mercury penetration and vapour adsorption. In fact, the problem of the $(\sigma_{sl} - \sigma_{sv})$ value has already been encountered in the thermodynamical study of vapour sorption and there is an obvious analogy between equation (48) for mercury penetration and equation (24) for vapour adsorption. It thus appears that the problems raised by the practical application of mercury penetration could be solved by developing a type of reasoning similar to the one followed in the analysis of sorption isotherms.

Two different developments may be considered. The simplest and most commonly used approach is based on the assumption of a complete reversibility of mercury penetration and thus corresponds to the analysis of vapour sorption by the classical Kelvin equation. The second approach must be sufficiently sophisticated to take into account the existence of irreversibility in mercury penetration. But, at the present time, no thermodynamical treatment analogous to the BdB sorption theory has ever been applied to mercury intrusion-extrusion curves. Indications for future work in this direction is tentatively given in a later section.

1. The Classical Analysis

Since Washburn [90] in 1921 suggested the use of mercury penetration to determine pore size distribution of porous solids, equation (47) has been used in the following form

$$p = (-f \cdot \sigma_{lv} \cdot \cos\theta)/r_p \tag{49}$$

where θ is the contact angle between the solid and the mercury and σ_{lv} the surface tension of mercury.

This relation implies that

$$\sigma_{sl} - \sigma_{sv} = -\sigma_{lv} \cdot \cos\theta \tag{50}$$

which is only true if mercury penetration is a reversible process [76].

The practical application of equation (49) requires the triple choice of a pore shape factor, a contact angle value and a mercury surface tension value.

a) Choice of a shape factor

Defining a shape factor is assuming a pore shape model. In the most common practice, pores of cylindrical shape are supposed to be present in the solid. This assumption leads to a shape factor of 2. This value is also suitable for slit-shaped pores. The r_p value calculated from (49) then corresponds to the "cylinder" radius or the slit width respectively.

Many porous materials, like powders, pellets or extrudates frequently used as catalysts, are however better approximated by packed spheres than an assemblage of tube — or sheet — like particles. Penetration and retraction of mercury into and from various packings of non porous uniform spheres have been studies by several authors [91–94]. Mayer and Stowe [93] correlate the radius of the packed spheres with the pressure required for initial mercury penetration in the packing. The relation between this so called "breakthrough pressure", p_b, and the sphere radius r_s is of the same type as equation (49), the shape factor f varying from one type of packing to another. Kruyer [91] analyzed the early stage of mercury retraction curve from assemblies of uniform spheres. From his analysis, he defined for various types of spheres packing a "critical retraction pressure" at which the retracting mercury will leave free the whole pore-space. The "critical pressure" value depends on the spheres radius and the results obtained are in good agreement with those of Mayer and Stowe [93]. Finally, through the analysis of the mercury penetration curve proposed by Frevel and Kressley [92], it is possible to get a rough idea of the largest and smallest particles present in a solid (generally a powder), if these particles are supposed to be spheroidal. According to this analysis, the radii r_M and r_m of the largest and smallest particles are given by

$$r_M = (-f_M \cdot \sigma_{lv} \cdot \cos\theta)/p_M \tag{51}$$

$$r_m = (-f_m \cdot \sigma_{lv} \cdot \cos\theta)/p_m \tag{52}$$

where $f_M = \dfrac{3\sqrt{2}+4}{2}$ and $f_m = \dfrac{5\sqrt{2}+9}{2}$

p_M is the pressure value corresponding to the beginning of the mercury penetration in the solid

p_m is the pressure value corresponding to the appearance of a plateau in the mercury penetration in the solid. Of course, p_m is higher than p_M.

In some particular cases, it is thus possible to obtain information about the particle size distribution from the mercury intrusion-extrusion curves.

b) Choice of a contact angle

Reported measurements of contact angle values between mercury and a large variety of materials range from about 120° to about 150°. Most investigators [95–100] usually assume either $\theta = 130°$ or $\theta = 140°$, depending on their preference.

To compare porous materials of the same type, it does not really matter which number is chosen, unless an exact measure of the pore opening is needed. In this case, the right value of the contact angle must be determined. This determination is generally not easy to carry out, especially with powdered or porous materials. But an elegant procedure was proposed by Rootare [95]. By combining equation (48) and (50) and solving them for cos θ, he obtained

$$\cos\theta = \frac{-\int_0^{p_{max}} p \cdot dV}{\sigma_{lv} \cdot s_{s,l}} \tag{53}$$

If the surface area $s_{s,l}$ is measured in an independent way (by the BET method for instance), the θ value may be calculated from the simple integration of the mercury penetration curve and the σ_{lv} value. But equation (53) is only appicable if at the pressure p_{max} the whole solid surface is covered with mercury, which implies the use of macroporous samples. However, as stated by Rootare [95], “once the contact angle has been calculated for a given material, this value can then be used in subsequent pore-size distribution determinations for samples of the same kind of material but of different porosity or particle size“.

Even if this approach can give more reliable results than the use of a mean standard value of θ, it must be kept in mind that important deviations from the contact angle value may occur [95, 97, 101] from sample to sample and thus give rise to large differences in the apparent pore size calculated from equation (49).

It is also worth pointing out at this stage that the use of equations (49) to 53 presumes that the contact angle value is pressure independent.

c) Choice of the surface tension value of mercury

There are rather small variations in the measured values of the surface tension of pure mercury reported in the literature [95, 102]. One of the best values appears to be $\sigma_{lv} = 0.485\ Nm^{-1}$ at 298 K. However as the effect of impurities on the surface tension value is not negligible, it is important in practice to use *very pure mercury* (tridistilled or at least bidistilled mercury). Like the contact angle, the surface tension value is supposed to be pressure independent.

d) Pore size distribution calculation

The values of f, θ and σ_{lv} being chosen, it is easy to obtain the pore size distribution from the mercury penetration curve. In fact the volume of pores having an opening “radius” greater than r_p is equal to the measured volume of mercury which has entered the porous solid at the pressure p given by equation (49). At this stage, it is obvious to calculate the distribution of pore volume with respect to pore size and, through the pore shape model, to obtain the surface area distribution.

The total area of the surface of the solid covered by mercury is given by

$$s_{s,l} = \frac{\int_0^{p_{max}} p \cdot dV}{-\sigma_{lv} \cos\theta} \tag{54}$$

which is a simple combination of equation (48) and (50). Even if the p_{max} value is very high, equation 54 generally leads to an erroneous estimation of the specific surface area of the solid [97, 101, 103], for the reversible pore filling assumed in its derivation is seldom encountered in practice.

When the solid under test is a powder, it is not always easy to differenciate the actual pore volume inside the particles from the void space between them, in particular if the solid is characterized by a broad particle size distribution.

2. *Limitations of Mercury Porosimetry*

a) Range of applicability

To define the pore size range accessible to the mercury penetration technique, the most commonly used value of f, σ_{lv} and θ are introduced in equation (49). It follows, if r_p is expressed in nm and p in Nm^{-2}

$$r_p \simeq \frac{0.75 \times 10^9}{p} \qquad (55)$$

It means that mercury can enter the pores having an opening radius between about 75 μm and about 7.5 nm if the absolute pressure is increased from 10^4 Nm^{-2} to 10^8 Nm^{-2}.

On the one hand the pore size range accessible to mercury penetration corresponds very well to the study of macroporous solids. The mercury porosimetry thus appears from this viewpoint, as an interesting technique complementary to the analysis of vapour sorption isotherms.

On the other hand, the pressure range between 10^4 and 10^8 $N \cdot m^{-2}$ is hereafter demonstrated to be particularly well adapted to a practical and reliable use of mercury porosimetry.

b) Limitations due to the pressure

Mercury porosimetry is based on the determination of the mercury volume which enters the solid as the pressure increases. This volume is generally obtained by measuring the variation of the mercury level in a calibrated capillary tube [104–106] (called the "dilatometer"). This tube is situated at the top of the cell containing the sample so that the latter is always subjected to the pressure corresponding to the mercury column in the dilatometer. In these conditions it is difficult to start the mercury penetration curve at an absolute pressure lower than about 10^4 Nm^{-2} and consequently to examine pores having an opening radius larger than about 75 μm.

At the other side of the pressure range, there are various reasons of limiting the applied pressure to about 10^8 Nm^{-2} even if in these conditions, the pores having an opening radius shorter than 7.5 nm are not accessible. These reasons which have been extensively developed in the literature [95, 97, 107, 108] can be summarized as follows:

— the higher the pressure, the more difficult to reach pressure equilibrium within the porous texture, in particular if some "ink bottle" pores are present

— the higher the pressure, the larger the possible variation of the contact angle between mercury and the solid. The contact angle value is not only sensitive to the pressure itself but also to the roughness and the curvature of the solid surface, the latter being more important in the narrover pores
— the higher the pressure, the higher the risk of modifying or breaking down the porous texture of the sample, particularly if the solid is crumbly or flimsy. The presence of large cavities with bottleneck type constricted openings makes the solid very sensitive to crushing or squeezing
— the higher the pressure, the more sophisticated the apparatus must be.

In practice, for minimising the difficulties and keeping a good efficiency, *a convenient compromise seems to limit the applied pressure to about* 10^8 Nm^{-2}.

c) Possible sources of errors

Although some of them have already been discussed in the preceding sections, it is however interesting to list here the different factors contributing to errors in mercury porosimetry.

The largest source of error in pore size distribution calculation comes from the assumption of some constant, average value for the contact angle between the mercury and the solid surface [95–98, 101]. A difference of 20° between the chosen σ value and the actual one can lead to a 50% error on the calculated pore size value.

The choice of the pore shape model has a big influence on the pore size distribution obtained from mercury penetration-retraction curves. It is illustrative of the problem involved in interpretation of mercury porosimetry data.

Another possible factor of error is the degree of purity of mercury, for it has an important effect on the surface tension of mercury. As this quantity directly affects the calculated pore size value, the use of very pure mercury is generally prefered.

The volumes measured during the mercury penetration-retraction experiments must be corrected to take into account not only the compressibility of mercury but also the compressibility of the solid sample [107–110]. The compressibility of mercury has a well defined value ($\chi_{Hg} = 3.4 \times 10^{-11}$ m^2/N^{-1}) whereas the compressibility of the sample material is generally not very well known. For most solid materials, compressibility values approximately range from 10^{-10} to 10^{-11} m^2/N^{-1}. However, for some porous materials, the presence of open or closed pores may enhance the apparent compressibility in an undefinite way.

To get a right determination of the mercury intrusion-extrusion curve, the vessel containing a definite amount of porous sample must be carefully evacuated [108] down to a pressure of about 10^2 m^2/N^{-1} before introducing the mercury. This is necessary to avoid the presence of any residual entrapped air which can somewhat affect the measurement of volumes.

Sometimes considerable time is required for the mercury to flow into the pores at a given pressure [111]. This kinetic effect, which often results in shifting of the distribution curves, may be reduced by carefully vibrating the vessel containing the solid sample.

Finally, the presence of a volume hysteresis [112, 113] in the penetration-retraction curve is highly indicative of irreversibility. In this case, the inadequacy of the theoretical analysis is an important source of error.

3. Trends for a New Type of Analysis

The largest sources of errors in the classical analysis of mercury penetration curves come from both the hypothesis of reversibility of pore filling and the use of an average constant value for the contact angle σ. In fact, this classical approach is based on an oversimplified model corresponding to the simple Kelvin equation in the case of sorption isotherms analysis. But, a thermodynamical treatment analogous to the one applied by Broekhoff and de Boer to sorption isotherms has never been developed for mercury penetration curves.

It is suggested here that the efficiency of mercury porosimetry could be significantly increased by improving the thermodynamical description of mercury intrusion and extrusion in and from a porous system. Like the BdB model, this description should take into account:

— the Gibbs Maximum work principle which would lead to overestimate the pores surface during intrusion and to underestimate it during extrusion

— some "standard reduced penetration curves" obtained on "non porous" samples of different natures; from these standard data, it would be possible to account for the variations with pressure of surface tension and contact angle

— the irreversibility of pores filling and emptying. This problem could be solved for particular pore shapes on the basis of the analogy, already pointed out by Kruyer [91], between mercury retraction from and capillary condensation of vapors in pores.

Such an analysis of mercury porosimetry could lead to a description of macroporous solids as comprehensive as the picture of mesoporous solids obtained from sorption isotherms with the BdB model.

5. Particle Size Distribution of Catalyst Powders

When a catalytic material is used in the form of pellets, granules or extrudates, the shapes and the sizes of these agglomerates are generally well defined so that they can be characterized by simple geometrical parameters. In many cases however, powders are used as catalysts and it is of practical importance to determine their particle size distributions. Several methods for particle size measurement are available but it must be pointed out that no analysis can be more accurate than the *sampling procedure* preceding it. It is necessary that the selected samples for measurement should be representative of the bulk in grain size distribution. To collect such samples, two general rules of sampling should be adhered to whenever possible [114]:

1. The sample should be taken when the powder is in motion
2. The whole powder stream should be taken during many short increments of time in preference to part of the stream being taken during the whole time.

Observance of these rules coupled with an understanding of the segregation phenomena wich may occur during the handling of the powder will lead to a quite satisfactory sampling procedure. A number of practical devices have been described and are commercially [115, 116] available for sample splitting down to the gram level. The sample amount required for analysis depends on the size measurement technique being used and may vary from a few milligrams for microscopic techniques to about 50 grams for sieving.

A. Particle Size, Shape and Distribution

1. Equivalent Diameters

The size of a spherical homogeneous particle is uniquely defined by its diameter. For irregular particles, the assigned size usually depends upon the method of measurement. In general, *equivalent spherical diameters* are determined by measuring a size dependent property of the particle and relating it to a linear dimension. Several equivalent diameters are of practical importance:

— *the volume diameter*, d_v, is the diameter of a sphere having the same volume as the particle
— *the surface diameter*, d_s, is the diameter of a sphere having the same surface as the particle
— *the surface volume diameter*, d_{sv}, is the diameter of a sphere having the same external surface to volume ratio as the particle, $d_{sv} = d_v^3/d_s^2$
— *the free-falling diameter*, d_f, is the diameter of a sphere having the same density and the same free-falling velocity as the particle in a fluid of the same density and viscosity
— *the stokes' diameter*, d_{st}, is the free falling diameter of a particle in the laminar flow region (Reynolds' number Re < 0.2):

$$d_{st}^2 \simeq d_v^3/d_s$$

— *the projected area diameter*, d_p, is the diameter of a circle having the same area as the projected area of the particle in random orientation. For convex particles $d_p = d_s$
— *the sieve diameter*, d_A, is the width of the minimum square aperture through which the particle will pass.

2. Shape Factors

The variation between these diameters increases as the particles diverge more from the spherical shape. The shape is thus an important factor in the correlation of size analyses made by various procedures and it is necessary to measure and define the shape quantitatively. In order to do this, different shape factors have been introduced which generally account for both surface and volume properties of the particles.

Wadell [117] defines *sphericity* as the ratio between the surface area of the sphere having the same volume as the particle and the surface area of particle.

$$\psi = d_v^2/d_s^2 \leqq 1 \tag{56}$$

The reciprocal of this ratio is known as the *rugosity coefficient* of the particles.

Although the sphericity factor ψ is generally sufficient to characterize the "shape" of catalyst particle, it is worth pointing out here the comprehensive work made by Heywood [118, 119] on particle shape analysis with the aid of surface- and volume-shape coefficients.

3. Size Distributions and Average Diameters

A particle size distribution generally appears in the form of a cumulative percentage frequency curve as a function of particle size. The sizes may be expressed by dividing the systems of particles into small intervals of size, δ_x, centered on diameters values $x_1, x_2, ..., x_n$, all the particles being assumed to have the same shape. If the percentage of the total number of particles in these groupings is $\delta N_1, \delta N_2, ..., \delta N_n$ respectively, various "frequency functions" may be defined:

$$\Phi(x_j) = \sum_{i=1}^{j} \delta N_i \qquad \text{for a number distribution}$$

$$\Phi(x_j) = \sum_{i=1}^{j} x_i \delta N_i \qquad \text{for a size distribution}$$

$$\Phi(x_j) = \sum_{i=1}^{j} x_i^2 \delta N_i \qquad \text{for an area distribution}$$

$$\Phi(x_j) = \sum_{i=1}^{j} x_i^3 \delta N_i \qquad \text{for a volume or weight distribution}$$

The *arithmetic mean size*, $\bar{x}_A$, is given by

$$\bar{x}_A = \frac{\sum_i x_i \, d\Phi(x_i)}{\sum_i d\Phi(x_i)} \tag{57}$$

This mean value is most significant when the distribution is normal.

For a weight distribution, which is the most commonly used in catalytic works, the arithmetic mean size is thus given by:

$$\bar{x}_{AW} = \frac{\sum_i x_i^4 \delta N_i}{\sum_i x_i^3 \delta N_i} \tag{58}$$

The corresponding *geometric mean size* $\bar{x}_g$ is given by:

$$\log \bar{x}_g = \frac{\sum_i d\Phi(x_i) \log x_i}{\sum_i d\Phi(x_i)} \tag{59}$$

This type of mean size is of particular value with log-normal distributions.

Equations (57) and (59) hold whatever the method of measurement used and the equivalent spherical diameter chosen. They lead, depending on the frequency function $\Phi(x)$, to various *average diameters* characterizing the particle size distribution.

It is worth noting that the choice of the size intervals δx is of fundamental importance to obtain a good description of the particle size distribution. The basic requirement is that the resolution defined as the size interval divided by the mean interval size should be kept fairly constant. It is the reason why a geometric progression of size intervals is generally used.

4. Methods of Presenting Size Analysis Data

Several ways of presenting the size distribution data are commonly used in practice. First of all, is the *tabular method* giving, in columns, the particle size ranges, the corresponding mean interval size and the related frequency number.

The simplest pictorial form of presenting size distribution data is the *histogram*. In this method, a rectangle is constructed over each size interval, the height or, more usefully, the area of which is proportional to the number of particles in that interval. If a sufficient number of particles have been counted, a smooth curve may be drawn through the histogram to give a *frequenvy distribution*. It is usual to have more than twelve intervals in this case.

It is often more convenient to plot the information as a *cumulative distribution*, the abscissa being particle size and the ordinate, the percentage smaller than or larger than the size. Moreover, it is a common practice to plot cumulative size distribution in such a way that a straight line results. In principle this can only be done if the distribution fits a standard law such as the *normal law* or the *log-normal law*. But the use of normal or log-normal probability plots is of interest to detect deviations from these laws and in particular the presence of a bimodal distribution [120] which very often correspond to a powder mixture.

B. Some Methods of Particle Size Analysis

1. Sieving

Sieving is probably the easiest and certainly the most popular method of size analysis, but it is restricted to powders where most particles have sizes larger than 75 μm. The method is not generally used for finer powders because of the high cost of producing sieves with uniform small apertures.

A variety of sieve aperture ranges are currently used but the International Organisation for Standardisation has attempted to standardize [121] these ranges for international co-ordination. This standard applies to the sieving of material from about 3 mm to about 0.05 mm in size. The sieving operation may be carried out wet or dry, mechanically or by hand, for a fixed period of time or until less than 0.2% of the sample passes through the mesh in any 5 min seeving period (the rate test).

As the number of particles on the sieve (load), the physical properties of the particles, the method of shaking the sieve and the geometry of the sieving surface may have a definite influence on the size distribution obtained, the *standardization of the procedure* [122] *is very important* to reach a good reproducibility of sieving operation.

The recommended weight of powder for sieving varies from 25 to 100 g according to its density. The weight particle size distribution is directly obtained in a cumulative form. The equivalent diameters correspond to the width of the minimum square aperture of each sieve.

2. *Gravitational and Centrifugal Sedimentation Techniques*

a) The problem of powder dispersion

In many size analyses methods, and in particular in sedimentation techniques, it is necessary to incorporate a powder into a liquid medium such that the particles are evenly dispersed. As the first stage of the dispersion process is the replacement of the solid-air interface by a solid-liquid interface [123], dispersing or surface-active agents are added to increase the affinity between the liquid and the solid so that a good wetting of the surface is obtained. For easily wetted materials, penetration of liquid into the voids between particles my provide sufficient force to break down the aggregates of particles. Often, however, some energy is required and this is usually achieved by spatulation, stirring or ultrasonics. The stability of a wetted dispersed system depends on the forces between the particles and therefore on the nature of the dispersing medium. A list of dispersing agents and dispersing solutions is given by Allen [124] for powders of various materials. The right choice of the dispersing medium is essential to avoid flocculation during the particle size analysis. It is also necessary to adjust the density and viscosity of the medium to suit the density and particle size of the powder so that the largest particle in the suspension can settle under laminar flow conditions (Reynolds' number less than 0.2, i.e. Stokes' region). This condition applies both for graviational and centrifugal sedimentation techniques.

b) The different approaches

The particle size distribution of a powder may be determined by examining a sedimenting suspension of this powder in a well chosen liquid medium. The powder may be introduced as a thin layer on top of a clear liquid, this is the two layer technique; or it may be uniformly dispersed in the liquid, this is the homogeneous suspension technique.

Depending on the technique used for size analysis, two different types of methods can be considered:

— in the *incremental methods*, changes with time in the concentration or density of the suspension at known depths are determined and from these the size distribution is calculated. Incremental methods may be divided into fixed time and fixed depth methods, the latter being more popular

— in the *cumulative methods*, the rate at which the powder is settling out of suspension is determined and, from this, the size distribution is calculated.

Ususally it is preferable to use incremental methods since analyses can be carried out more rapidly than by cumulative methods. The big advantage, however, of cumulative techniques is that the amount of powder required is small (about 0.5 g). This is particularly useful when only a small quantity of powder is available.

All these methods apply both to gravitational and centrifugal techniques and are based in the Stokes' law of settling velocities assuming spherical particles moving slowly (Reynold's number less than 0.2) in an homogeneous infinite fluid. The settling velocity, u_{st}, for a sphere of diameter D and density ϱ_s falling in a fluid of density ϱ_f and viscosity η is given, for gravitational settling by:

$$u_{st} = \frac{(\varrho_s - \varrho_f)\, gD^2}{18\eta} \qquad (60)$$

where g is the acceleration due to gravity.

For a particle settling in a centrifugal field, the settling velocity, u_c, is given by

$$u_c = \frac{(\varrho_s - \varrho_f)\, D^2}{18\eta} \cdot \omega^2 x \qquad (61)$$

where ω is the speed of rotation of centrifuges in radians per second and x is the distance from the axis of rotation to the particle.

From (60) and (61) it appears that in centrifugal methods, the settling velocity of a particle depends on its position in the suspension as well as its size, whereas for gravity sedimentation the settling velocity is dependent solely on size. It is the reason why the calculations of size distributions from centrifugal data are more difficult than calculations from gravitational data.

c) Practical methods of size analysis

The most widely used among the gravitational methods of particle size analysis are shortly described here after.

— *The pipette method*

In this incremental method, the concentration changes occuring as a function of time within a settling homogeneous suspension of the powder are followed by drawing off definite volumes of the suspension by means of a pipette. The concentration of solids in the samples may be determined simply by drying and weighing. In general, samples are withdrawn at a fixed level of the sedimentation vessel for time scales in a 2:1 progression to give a $\sqrt{2}$ progression in particle size D as calculated from Stokes' equation. Various types of pipette have been proposed in the literature [125–127).

— *The photosedimentation technique*
In a photosedimentometer, the concentration changes occuring as a function of time within a settling homogeneous suspension of the powder are followed by the variation of intensity of a narrow parallel light beam projected through the suspension at a known depth. As the attenuation of the light beam is related to the projected surface area of the particles in the beam, the particle-size distribution may be determined from this relationship and the Stoke's law. Various types of photosedimentometers [128] have been developed some of them being commercially available. A natural extension to the use of light is to use X-rays, but in this case the X-ray density is proportional to the weight of powder in the beam [129, 130].

— *The sedimentation balances*
In the balance methode, which is the most popular method for carrying out a cumulative sedimentation size analysis, the weight of powder settling out on to the balance pan is recorded versus time. If P is the weight percent of the suspension which has settled out at time, t, the weight percentage W of particles having a "diameter" greater than D_t (as calculated by Stoles' law) is given by

$$W = P - \frac{dP}{d\ln t} \tag{62}$$

Several methods have been suggested for the determination of W using this equation [131–133]. Various models of sedimentation balances have been described [134–138].

The most commonly used *centrifugal sedimentation techniques* involve incremental methods of analysis, in which the concentration changes occuring within the settling suspension are followed, either by withdrawing samples of suspension by means of a pipette, or by determining the intensity variations of a light beam projected through the suspension. In both cases, equation (61) is used to calculate the particle diameters. Several instruments based either on the two-layer [139–144] technique or on the homogeneous [145–148] suspension technique are described in the literature. X-ray density has also been used to determine the weight concentration of powder without [149] disturbing the suspension in the centrifuge.

3. *Microscopy*

Microscopy is often used as an absolute method of particle size analysis since it is the only method where the individual particles are observed and measured and which enables examination of the shape of particles. But the most difficult problem in microscopy is the preparation of a representative sample of the powder. Many methods have been suggested, but the final result often depends more on the skill of the user than on the procedure itself.

When microscopy is used for determining the particle size distribution, a large number of fields of view are generally required for accurate results. As at least one thousand particles must be measured, the use of an automatic or semi-automatic counting and sizing device is recommended. These devices use to determine the projected area diameter, d_p, of each particle and to present the size distribution data in an histogram form [150–155].

In catalysis, however, the main interest of microscopy lies in the possibility of studying the surface topography and morphology of particles. Transmission electron microscopy (TEM) and scanning electron microscopy.(SEM) are indeed outstanding tools in textural analysis of solids. The information given by these techniques is generally very useful for the critical discussion of the data coming from adsorption isotherms and mercury penetration curves analysis.

6. Practical Applications

Hereafter, two practical examples are given to illustrate how mercury porosimetry and physical adsorption isotherms analysis can be combined with particle-size and -shape measurements in such a way that a self consistent pattern of the porous texture of a catalyst is obtained.

A. Textural Study of a Pelletized Alumina

The results presented in this section were obtained on a commercially available (from HALCON) dehydration catalyst.

This catalyst is sold in the form of *cylindrical pellets*. By taking the measurements of about fifty pellets and calculating the mean values and the standard deviations, the diameter, d, and the height, h, of the cylinders were determined:

$$d = 4.612 \pm 0.050 \text{ mm}$$
$$h = 4.405 \pm 0.320 \text{ mm}$$

These pellets have thus a particularly regular shape: only 1% variation on the diameter and about 7.3% on the height.

Chemical analysis and X-ray diffraction pattern indicate that this catalyst is a very pure γ-alumina.

As the normal working temperature of this catalyst is between 573 and 673 K, the textural study was carried out after a standard-pretreatment of the samples, i.e. 16 h at 573 K under a dry air stream.

The nitrogen adsorption-desorption isotherm obtained on the sample outgassed at 573 K is shown in Figure 7. It is clearly a type *IV* isotherm with a type *E* hysteresis loop, indicating the presence of a mesoporosity created by a packing of spheroidal particles or by "ink-bottle" pores.

The application of the BET model leads to a BET specific surface area value of 264 m^2/g^{-1} and a c_{BET} constant of 82.

The v_a-n-plot, given in Figure 8, is constructed with the n_3 reference isotherm. There is no indication of the presence of a microporosity in the sample

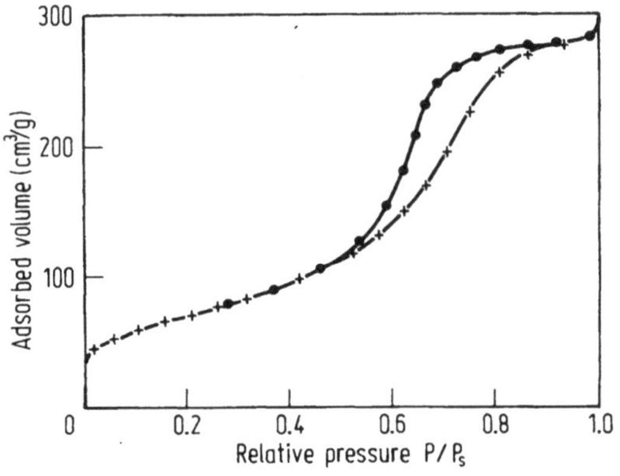

Figure 7. Nitrogen adsorption isotherm obtained on a pelletized γ-alumina, pretreated 16 h at 573 K under a dry air stream

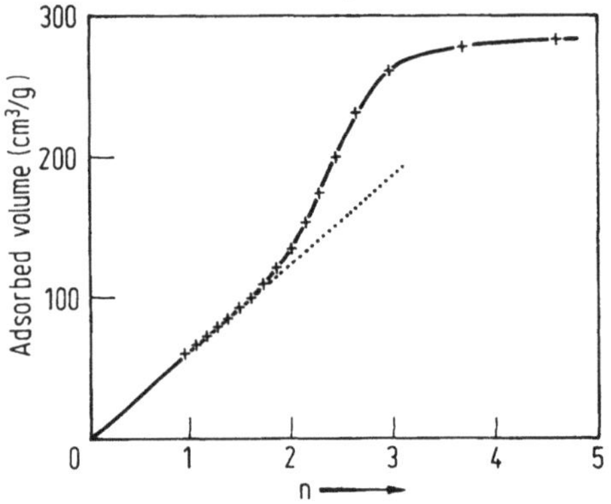

Figure 8. The v_a-n-plot corresponding to the isotherm in Figure 7.

but the plot clearly shows the existence of an important capillary condensation phenomenon. The slope of the straight line passing through the origin gives a specific surface area of 269 m^2/g^{-1}, in perfect agreement with the BET value.

The BdB method, applied to the adsorption branch of the isotherm in the case of "ink bottle" pores, gives the distributions of surface area versus pore "radius" presented in Figure 9 in a differential and an integral form. Both forms are normalized and the cumulative surface area corresponding to the hundred percent level is 256 m^2/g^{-1}, a value nearly equal to those obtained by the BET and the n-plot methods. The differential distribution shows two maxima at 4 and 4.8 nm, the "radius" of pores varying from 1.8 to 10 nm. All these results are reproducible.

Curve 1 in Figure 10 gives the distribution of pore volume versus pore size as obtained by the mercury penetration technique. Pores having an opening "radius" between 0.1 and 1 μm are observed. This macroporosity cannot be described by nitrogen sorption isotherm analysis, but it must be pointed out that in this particular case, the mercury penetration technique is unable to indicate the presence of a mesoporosity in the solid, even if the applied pressure is increased up to $2 \cdot 10^8$ Nm^{-2}. This phenomenon is

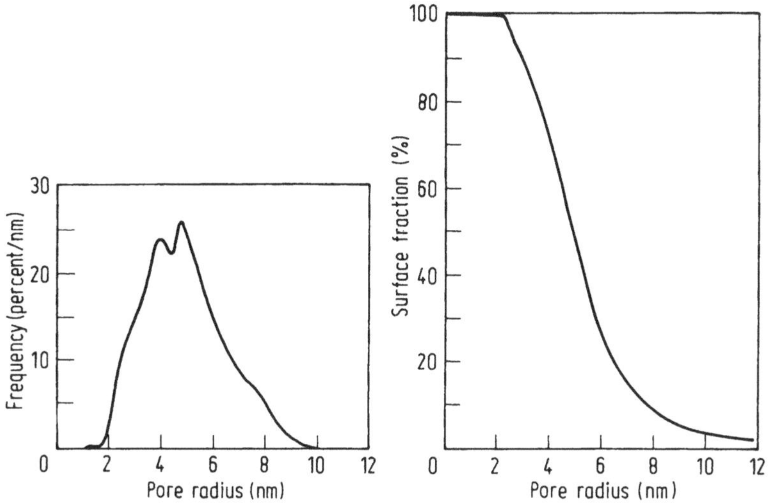

Figure 9. Normalized differential and integral distributions of surface area, obtained by the BdB method on the adsorption branch of the isotherm described in Figure 7

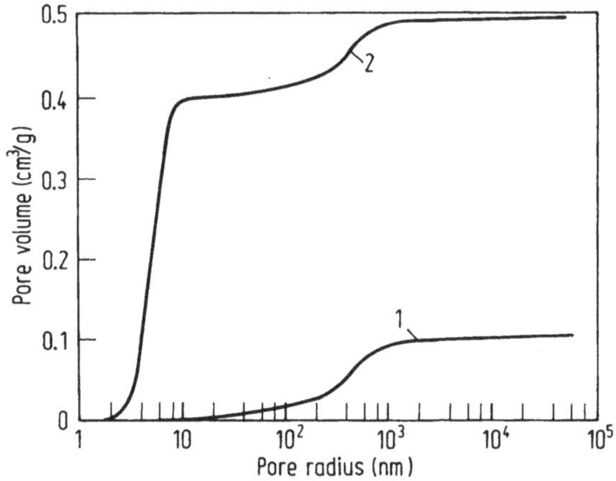

Figure 10. Pore volume distribution of a pelletized γ-alumina pretreated 16 h at 573 K under a dry air stream. Curve 1: mercury penetration only. Curve 2: mercury penetration combined with BdB analysis of the adsorption branch of the isotherm described in Figure 7

probably due to the difficulty of reaching a pressure equilibrium with "ink bottle" pores and also to the poor wetting of this kind of solid by mercury.

As shown by curve 2 in Figure 10, it is possible to get a complete picture of the pore volume distribution of the sample by combining mercury porosimetry and BdB analysis of nitrogen adsorption isotherm. This curve shows that about 4/5 of the pore volume is developed by mesopores and only 1/5 by macropores. The total pore volume obtained in this way is 0.49 cm^3/g^{-1}.

To complete the description of the textural properties of this catalyst an electron microscopy analysis has been carried out. Some of the micro-

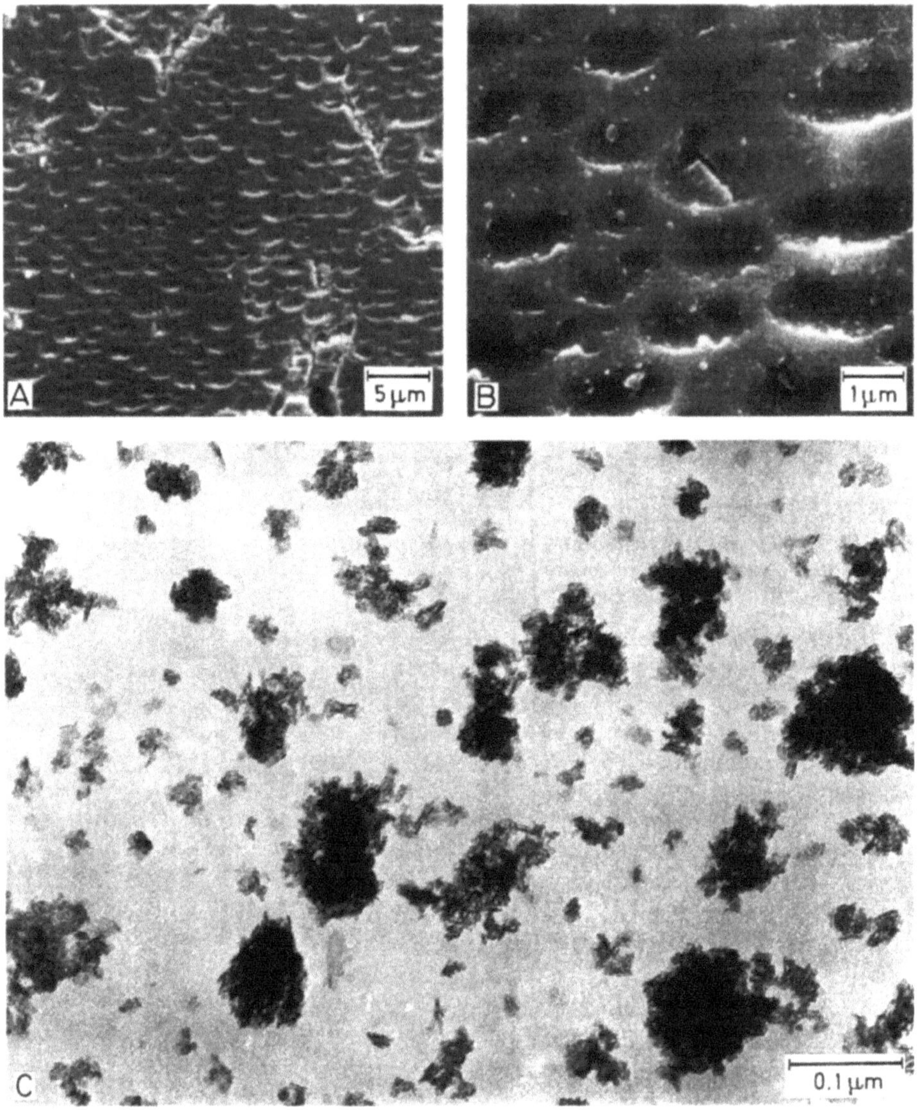

Figure 11. Electron micrographs obtained on a pelletized γ-alumina pretreated 16 h at 573 K under a dry air stream. A and B are SEM pictures of a cut made through a pellet. C is a TEM picture of a pellet crushed manually in a mortar. Magnifications appear on the micrographs

graphs are presented in Figure 11. The micrographs A and B, obtained by scanning electron microscopy (SEM), show a cut made through a pellet by a diamond knife. The macroporosity indicated by the mercury penetration technique is clearly illustrated on these pictures and the pore size ranges are in good agreement with each other.

The micrograph C in Figure 11 is obtained by transmission electron microscopy (TEM) on a pellet crushed manually in a mortar. This high magnification picture shows several agglomerates of very small particles. The voids between these particles can account for the mesoporosity of the sample. The pore shape deduced from the nitrogen sorption isotherm analysis is well confirmed by this microscopic examination.

B. Textural Comparison of Two Hydrogenation Catalysts

In this section an extensive comparison on the textural characteristics of two commercially available hydrogenation catalysts is proposed.

Catalyst number 1 is the 2 wt % Pd/SiO_2-Al_2O_3 MG 5 from *Montedison*, while *catalyst number 2* is the 2 wt % Pd/*Doucil* from *Laporte*. Both are free flowing powders having nearly the same chemical composition. In the textural analysis presented here, no attention is paid to the dispersion of the metal on the support. This problem is widely discuss in an other chapter of this treatise.

In Figures 12 and 13 are shown the nitrogen adsorption-desorption isotherms obtained on catalysts 1 and 2 respectively, both samples having been outgassed at room temperature. These isotherms are of type *IV* with a type A hysteresis loop indicating the presence of a mesoporosity formed by "cylindrical" pores. At this stage, the main difference between the two samples is the total amount of adsorbed nitrogen at the upper closing point of the hysteresis loop. The mesopore volume is larger in catalyst 1 than in catalyst 2.

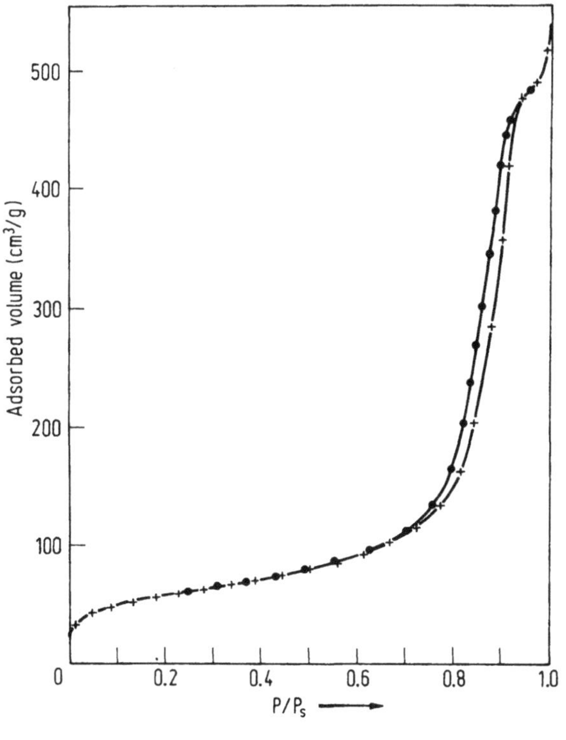

Figure 12. Nitrogen adsorption isotherm obtained on the 2 wt% $Pd/SiO_2 \cdot Al_2O_3$ MG 5 from MONTEDISON (catalyst 1)

The application of the BET model to these isotherms leads to the following figures:

catalyst 1: $s_{BET} = 208\ m^2\ g^{-1}$, $c_{BET} = 154$
catalyst 2: $s_{BET} = 179\ m^2\ g^{-1}$, $c_{BET} = 153$

The v_a-n-plots, given in Figure 14, are constructed with the n_2 reference isotherm. The shape of these plots (a downward-followed by an upward deviation from the straight line) indicate the presence in the samples of both a micro- and a mesoporosity. The slopes of the straight lines passing through the origin give the following values for the specific surface area: 202 $m^2\ g^{-1}$ for catalyst 1 and 173 $m^2\ g^{-1}$ for catalyst 2. These figures agree quite satisfactorily with the corresponding BET values.

As the v_a-n-plots show two types of porosity in the samples, two different methods of analysis must be used: the Brunauer method for the micropores and the BdB method for the mesopores.

The distributions of mesopore surface area of the two catalysts are shown in Figure 15 in the normalized differential form. The BdB analysis has been applied to the desorption branch of the isotherm, assuming cylindrical pores. Each distribution shows two maxima at 8 and 12 nm for catalyst 1 and at 5.8 and 7.2 nm for catalyst 2. The cumulative surface area of mesopores is 182 $m^2\ g^{-1}$ for catalyst 1 and 157 $m^2\ g^{-1}$ for catalyst 2.

The Brunauer analysis of the v_a-n-plots indicates that the micropore distributions are very narrow; the maxima of the differential distribution is observed at a "radius" of 0.93 nm for catalyst 1 and of 0.88 nm for cata-

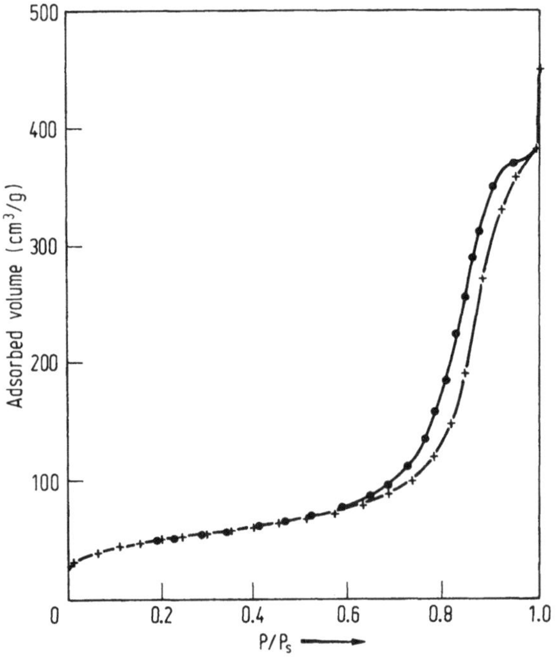

Figure 13. Nitrogen adsorption isotherm obtained on the 2 wt% Pd/DOUCIL from LAPORTE (catalyst 2)

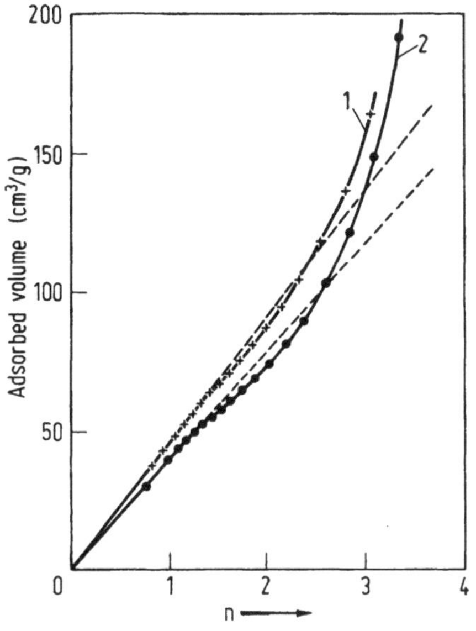

Figure 14. The v_a-n-plots corresponding to the isotherms in Figure 12 (curve 1) and Figure 13 (curve 2)

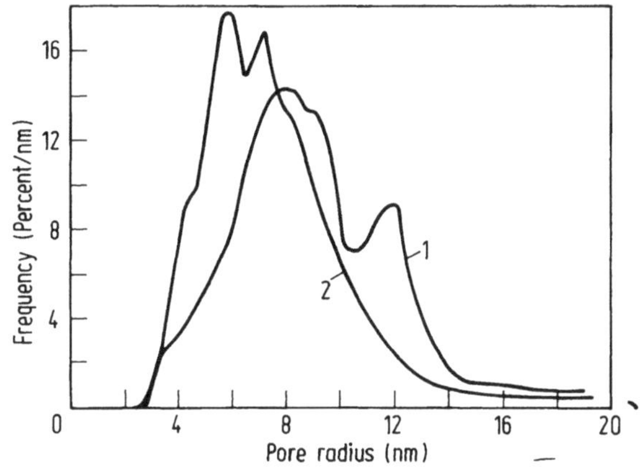

Figure 15. Normalized differential distributions of mesopore surface area obtained by the BdB method on the desorption branches of the isotherms described in Figure 12 (curve 1) and Figure 13 (curve 2). Pores are supposed to be cylindrical

lyst 2. The specific surface area corresponding to these micropores are 19 $m^2 g^{-1}$ and 25 $m^2 g^{-1}$ respectively.

By summing the surface areas developed by the meso- and the microporosity, the following total cumulative values are obtained: 201 $m^2 g^{-1}$ for catalyst 1 and 182 $m^2 g^{-1}$ for catalyst 2. These figures are in a very good agreement with the corresponding s_{BET} and s_n values.

The pore volume distributions calculated from mercury penetration experiments are presented in Figure 16. The two distributions have the same general shape: two sharp variations separated by a plateau. The first variation corresponds to the mesoporosity while the second one is due to the mercury penetration in the voids between the packed powder particles.

As for the mesopore volume distributions, a very good agreement is observed between the results of the BdB analysis of nitrogen sorption isotherms and those of the mercury porosimetry, the latter being represented in Figure 16. The main difference between the two catalysts is the mesopore volume: 0.85 $cm^3\ g^{-1}$ for catalyst 1 and 0.55 $cm^3\ g^{-1}$ for catalyst 2.

The part of mercury penetration curves corresponding to the filling of intergranular voids can be used to get an idea of the particle size distribution limits. The application of equations (51) and (52) indicate that the diameters of the largest and the smallest particles are about 0.3 mm and about 0.06 mm respectively for both catalysts, if the particles are supposed to be spheroidal.

These figures are in rather good agreement with the weight particle size distributions obtained by mechanical dry sieving of the powders. The cor-

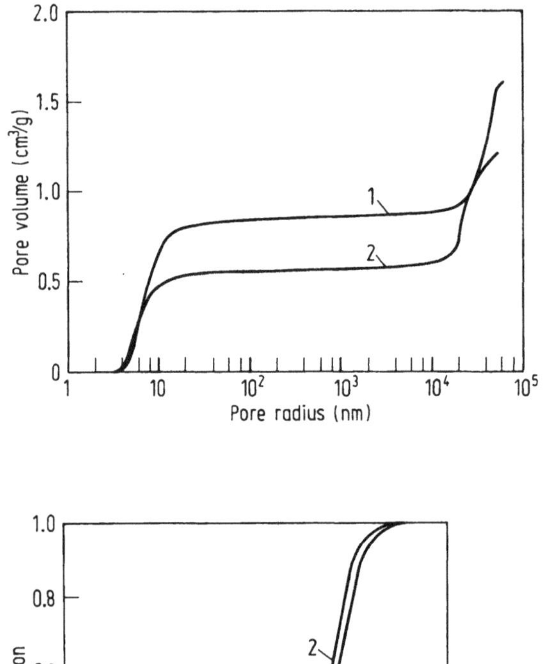

Figure 16. Pore volume distributions obtained by mercury porosimetry on catalysts 1 and 2 as defined in Figures 12 and 13

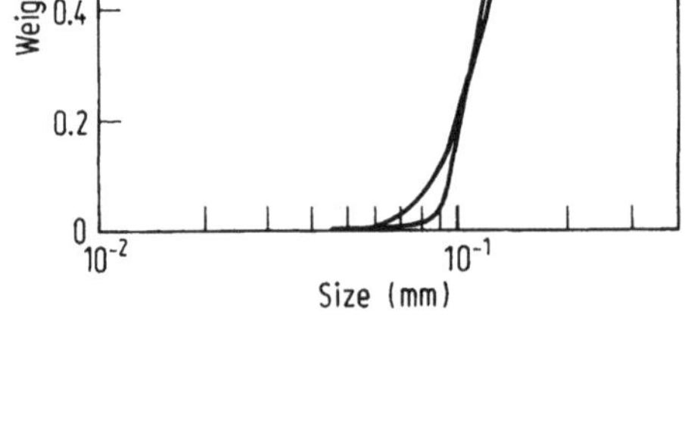

Figure 17. Particle size distributions obtained by mechanical dry sieving on catalysts 1 and 2, as defined in Figures 12 and 13

responding cumulative distributions are given in Figure 17 where the weight fraction of particles smaller than a given size is ploted as a function of the logarithm of this size. In Figure 18 the same distributions have been drawn on a log-normal probability diagram. There are only small differences between the particle size distributions of the two samples. The geometric mean sizes are 0.130 mm for catalyst 1 and 0.125 mm for catalyst 2.

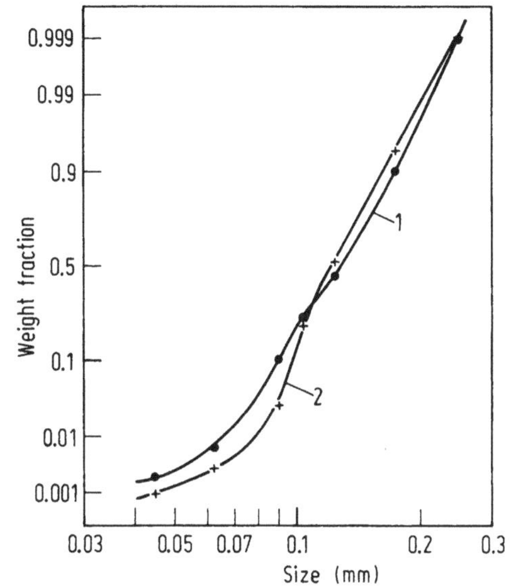

Figure 18. Log-normal probability plots of the weight distributions of particle size for catalysts 1 and 2, as defined in Figures 12 and 13

This textural comparison is completed by a scanning electron microscopy analysis. Some micrographs are presented in Figure 19. They show a large difference in the particle shapes of the two samples: the catalyst 1 particles are nearly spherical while the catalyst 2 particles appear like sharp-cornered flint-stones of various shapes. This observation makes somewhat questionnable the direct comparison of particle size distributions when the particles are so dissimilarly shaped. But, for the moment, there is no simple answer to this question. Finally, it must be pointed out that the high magnification pictures in Figure 19 do not show any macroporosity within the particles, which is in perfect agreement with the existence of a plateau in the pore volume distributions as illustrated in Figure 16.

C. Conclusion

These few examples given to illustrate the various methods described in this chapter, clearly demonstrate the possibility of coming to a complete and coherent description of the textural characteristics of a catalyst, provided the measurements are carefully carried out and the results critically analyzed and compared.

Figure 19. Scanning electron micrographs obtained on the two catalysts as defined in Figures 12 and 13. Catalyst 1: micrographs A, B and C. Catalyst 2: micrographs D, E and F. Magnifications appear on the micrographs

References

1. Brunauer, S., Emmett, P. H., Teller, E.: J. Amer. Chem. Soc. **60**, 309 (1938)
2. IUPAC Physical Chemistry Division, Commission on Colloid and Surface Chemistry: Definitions, Terminology and Symbols in Colloid and Surface Chemistry, Part II: Heterogeneous Catalysis, Rules approved 1975
3. Bering, B. P., Dubinin, M. M., Serpinsky, V. V.: J. Colloid Interface Sci. **21**, 378 (1966)
4. Satterfield, C. N.: Mass Transfer in Heterogeneous Catalysis. Cambridge: MIT Press 1970
5. L'Homme, G. A.: Ind. Chim. Belge **35**, 179 and 291 (1970)
6. Germain, A., Crine, M., Marchot, P.: In. Chemical Engineering of Gas-Liquid-Solid Catalyst Reactions. L'Homme, G. A. (ed.). Liège: Cebedoc 1979, pp. 187—217
7. Baker, F. S., Sing, K. S. W.: J. Colloid. Interface Sci. **55**, 605 (1976)
8. Lemcoff, N. O., Sing, K. S. W.: J. Colloid Interface Sci. **61**, 227 (1977)
9. Gregg, S. J., Sing, K. S. W.: The Adsorption of Gases on Porous Solids. In: Surface and Colloid Science. Matijevic, E. (ed.). New York: Interscience 1976, Vol. 9, pp. 231 to 359
10. Gregg, S. J., Sing, K. S. W.: Adsorption, Surface Area and Porosity. London: Academic Press 1967
11. Van Dongen, R. H., Kaspersma, J. H., de Boer, J. H.: Surface Sci. **28**, 237 (1971)
12. Hagymassy, J. Jr., Odler, I., Yudenfreund, M., Skalny, J., Brunauer, S.: J. Colloid Interface Sci. **38** 20 (1972)
13. Odler, I., Hagymassy, J. Jr., Yudenfreund, M., Hanna, K. M., Brunauer, S.: J. Colloid Interface Sci. **38** 265 (1972)
14. Mikhail, R. Sh., Guindy, M. N., Hanafi, S.: J. Colloid Interface Sci. **70** 282 (1979)
15. Nicolaon, G. A.: Bull. Soc. Chim. Fr. **1969**, 91
16. Nicolaon, G. A., Teichner, S. J.: J. Chim. Phys. **65**, 870 (1968)
17. Brunauer, S., Deming, L. S., Deming, W. S., Teller, E.: J. Amer. Chem. Soc. **62**, 1723 (1940)
18. Brunauer, S.: The Adsorption of Gases and Vapours. Princeton: University Press 1943
19. Germain, J. E.: Catalyse hétérogène. Paris: Monographie Dunod 1959
20. Young, D. M., Crowell, A. D.: Physical Adsorption of Gases. London: Butterworths 1962
21. de Boer, J. H.: The shapes of capillaries. In: The Structure and Properties of Porous Materials. Everett, D. H. and Stone, F. S. (eds.). London: Butterworths 1958, pp. 68–94
22. Lippens, B. C.: Structure and Texture of Aluminas. Thesis Delft 1961
23. Thomas, J. M., Thomas, W. J.: Introduction to the Principles of Heterogeneous Catalysis. London: Academic Press 1967
24. Lecloux, A.: Exploitation des isothermes d'adsorption et de désorption d'azote pour l'étude de la texture des solides poreux. In: Comptes rendus de la semaine d'étude de la Catalyse. D'Or, L., Derouane, E. G. (eds.) Liège: Congrès et Colloques de l'Université 1971 pp. 169–209
25. McClellan, A. L., Harnsberger, H. F.: J. Colloid Interface Sci. **23**, 577 (1967)
26. Sing, K. S. W.: Adsorption at the Gas/Solid Interface. In: Colloid Science. Everett, D. H., (ed.). London: The Chemical Society (1973) Volume 1 pp. 1–48
27. Brunauer, S.: J. Colloid Interface, Sci. **41** 612 (1972)
28. Brunauer, S.: J. Colloid Interface, Sci. **39** 435 (1972)
29. Dubinin, M. M.: In: Proc. Int. Symp. on Pore Structure and Properties of Materials. Modry, S. (ed.). Prague 1973, Vol. 4, p. C 27
30. Bering, B. P., Dubinin, M. M., Serpinsky, V. V.: J. Colloid Interface, Sci. **38** 185 (1972)
31. Pomeshechikov, V. D., Pozdeev, V. V.: Kinet. Katal. **12**, 794 (1971)
32. Rand, B., Marsh, H.: J. Colloid Interface Sci. **40**. 121 (1972)
33. Brunauer, S.: Int. Symp. on Pore Structure and Properties of Materials. Prague 1973. Discussion.
34. Dubinin, M. M.: Chem. Rev. **60**, 235 (1960)
35. Dubinin, M. M.: J. Colloid Interface Sci. **23**, 487 (1967)
36. Polanyi, M., Goldman, F.: Z. Phys. Chem. **A 132**, 313 (1928)

37. Müller, W.: Z. Phys. Chem. Neue Folge **83**, 118 (1973)
38. Reucroft, P. J., Simpson, W. H., Jonas, L. A.: J. Phys. Chem. **75**, 3526 (1971)
39. Huber, V., Stoeckli, F., Houriet, J.-P.: J. Colloid Interface Sci. **67**, 195 (1978)
40. Rand, B.: J. Colloid Interface Sci. **56**, 337 (1976)
41. Shull, C. G.: J. Amer. Chem. Soc. **70**, 1410 (1948)
42. Cranston, R. W., Inkley, F. A.: Advan. Catal. **9**, 143 (1957)
43. Pierce, C.: J. Phys. Chem. **57**, 149 (1953)
44. Lippens, B. C., de Boer, J. H.: J. Catal. **4**, 319 (1965)
45. Sing, K. S. W.: Chem. Ind. (London) **1968**, 1520
46. Sing, K. S. W.: In: Surface Area Determination. Everett, D. H., Ottewill, R. H. (eds.). London: Butterworths 1970 p. 25
47. Carruthers, J. D., Cutting, P. A., Day, R. E., Harris, M. R., Mitchell, S. A., Sing, K. S. W.: Chem. Ind. (London) **1968**, 1772
48. Payne, D. A., Sing, K. S. W.: Chem. Ind. (London) **1969**, 918
49. Payne, D. A., Sing, K. S. W., Turk, D. H.: J. Colloid Interface Sci. **43**, 287 (1973)
50. Lecloux, A.: Mém. Soc. Roy. Sci. Liège (6) **1** (4), 169 (1971)
51. Lecloux, A.: In: Surface Area Determination. Everett, D. H., Otterwill, R. H. (eds). London: Butterworths 1970 p. 21
52. Lecloux, A., Pirard, J. P.: J. Colloid Interface Sci. **70**, 265 (1979)
53. de Boer, J. H., Linsen, B. G., Osinga, Th. J.: J. Catal. **4**, 643 (1965)
54. de Boer, J. H., Linsen, B. G., van der Plas, Th., Zondervan, G. J.: J. Catal. **4**, 649 (1965)
55. Nicolaon, G. A., Teichner, S. J.: J. Colloid Interface Sci. **38**, 172 (1972)
56. Parfitt, G. D., Sing, K. S. W., Urwin, D.: J. Colloid Interface Sci. **53**, 187 (1975)
57. Lamon, T. G.: J. Colloid Interface Sci. **56**, 116 (1976)
58. Mikhail, R. Sh., Brunauer, S., Bodor, E. E.: J. Colloid Interface Sci. **26**, 45 and 54 (1968)
59. de Boer, J. H.: In: Surface Area Determination. Everett, D. H., Ottewill, R. H. (eds.). London: Butterworths 1970 p. 7
60. Koberstein, E., Voll, M.: Z. Phys. Chem. Neue Folge: **71**, 275 (1970)
61. Brunauer, S.: Chem. Eng. Progress, Symposium Series N° 96, Vol. 65, 1 (1969)
62. Shalny, J., Bodor, E. E., Brunauer, S.: J. Colloid Interface Sci. **37**, 476 (1971)
63. Hagymassy, J. Jr., Brunauer, S.: J. Colloid Interface Sci. **33**, 317 (1970)
64. Brunauer, S., Skalny, J., Odler, I.: In: Proc. Int. Symp. on Pore Structure and Properties of Materials. Modry, S. (ed.). Prague 1973, Vol. 4, p. C-3
65. Bhambhani, M. R., Cutting, P. A., Sing, K. S. W., Turk, D. H.: J. Colloid Interface Sci. **38**, 109 (1972)
66. Brown, M. G., Cadenhead, D. A.: J. Colloid Interface Sci. **70**, 139 (1979)
67. Sing, K. S. W.: Chem. Ind. (London) **1967**, 829
68. Bye, G. C., Sing, K. S. W.: Chem. Ind. (London) **1967**, 1139
69. Mieville, R. L.: J. Colloid Interface Sci. **41**, 371 (1972)
70. Johnson, M. F.: J. Catal. **52**, 425 (1978)
71. Lee, J. A., Newnham, C. E.: J. Colloid Interface Sci. **56**, 391 (1976)
72. Gregg, S. J., Tayyab, M. M.: J. Chem. Soc. Faraday Trans. I, **1978**, 348
73. Deryagin, B. V.: Acta Physicochim. URSS **12**, 139 (1940)
74. Broekhoff, J. C. P., de Boer, J. H.: J. Catal. **9**, 8 and 15 (1967), **10**, 153, 368, 377 and 391 (1968)
75. Broekhoff, J. C. P., Linsen, B. G.: Studies on Pore Systems in Adsorbents and Catalysts. In: Physical and Chemical Aspects of Adsorbents and Catalysts. Linsen, B. G. (ed.) London and New York: Academic Press 1970 pp. 1–62
76. Broekhoff, J. C. P.: Mesopore Determination from Nitrogen Sorption Isotherms: Fundamentals, Scope, Limitations. In: Preparation of Catalysts II. Delmon, B., Grange, P., Jacobs, P., and Poncelet, G. (eds.). Amsterdam: Elsevier 1979, pp. 663–684
77. Dollimore, D., Heal, G. R.: J. Colloid Interface Sci. **33**, 508 (1970)
78. Kiselev, A. V.: In: The Structure and Properties of Porous Materials. Everett, D. H., Stone, F. S. (eds.). London: Butterworths 1957, p. 130
79. Brunauer, S., Mikhail, R. Sh., Bodor, E. E.: J. Colloid Interface Sci. **25**, 353 (1967)
80. Lecloux, A.: In: Proc. Int. Symp. on Pore Structure and Properties of Materials. Modry, S. (ed.). Prague 1973 Vol. 4 p. C 43

81. Barrett, E. P., Joyner, L. G., Halenda, P. H.: J. Amer. Chem. Soc. **73**, 373 (1951)
82. Everett, D. H., Smith, F. S.: Trans. Faraday Soc. **50**, 187 (1954)
83. Everett, D. H., Haynes, J. M.: Capillarity and Porous Materials: Equilibrium Properties. In: Colloid Science. Everett, D. H. (ed.). London: The Chemical Society (1973) Volume 1 pp. 123–172
84. Broekhoff, J. C. P., Brown, L. F., van Beek, W. P.: In: Proc. Int. Symp. on Pore Structure and Properties of Materials. Modry, S. (ed.). Prague 1973, Vol. 4, p. C 85
85. Broekhoff, J. C. P., van Beek, W. P.: J. Chem. Soc. Faraday Trans. I, **1979**, 36 and 42
86. Ksenzhek, O.: Russ. J. Phys. Chem. **37**, 691 (1963)
87. Scholten, J. J. F., Beers, A. M., Kiel, A. M.: J. Catal. **36**, 23 (1975)
88. De Wit, L. A., Scholten, J. J. F.: J. Catal. **36**, 30 (1975)
89. Pirard, J. P., Lecloux, A.: to be published
90. Washburn, E. W.: Proc. Nat. Acad. Sci. **7**, 115 (1921)
91. Kruyer, S.: Trans. Faraday Soc. **54**, 1758 (1958)
92. Frevel, L. K., Kressley, L. J.: Anal. Chem. **35**, 1492 (1963)
93. Mayer, R. P., Stowe, R. A.: J. Colloid Interface Sci. **20**, 893 (1965)
94. Ione, K. G., Karnaukhov, A. P.: Kinet. Katal. **12**, 179 (1971)
95. Rootare, H. M.: Aminco Lab. News **24** (3), 4A–4H (1968)
96. Orr, C.: Powder Technol. 3, 117 (1969/70)
97. De Wit, L. A., Scholten, J. J. F.: J. Catal. **36**, 36 (1975)
98. Baiker, A., Richarz, W.: Chem.-Ing.-Tech. **49**, 399 (1977)
99. Kossen, N. W. F., Heertjes, P. M.: Chem. Eng. Sci. **20**, 593 (1965)
100. Volfkovich, Yu. M., Shkolnikov, E. I.: Russian J. Phys. Chem. **52**, 113 (1978)
101. Liabastre, A. A., Orr, C.: J. Colloid Interface Sci. **64**, 1 (1978)
102. Roberts, N. K.: J. Chem. Soc. **1964**, 1907
103. Rootare, H. M., Prenzlow, C. F.: J. Phys. Chem. **71**, 2733 (1967)
104. Guyer, A. Jr., Böhlen, B., Guyer, A.: Helvetica Chim. Acta **42** (6), 2103 (1959)
105. Winslow, N. M., Shapiro, J. J.: ASTM Bulletin, Feb. 1959, p. 39
106. Baker, D. J.: J. of Physics E.: Sci. Instrum. **4**, 388 (1971)
107. Sarakhov, A. I.: Russian J. Phys. Chem. **37**, 242 (1963)
108. Winslow, D. N.: J. Colloid Interface Sci. **67**, 42 (1978)
109. Ione, K. G., Karnaukhov, A. P., Kvon, E. E.: Kinet. Katal. **12**, 457 (1971)
110. Smith, L. B., Keyes, F. G.: Proc. Amer. Acad. Arts. Sci. **69**, 310 (1934)
111. Wotzak, G. P.: AIChE Journal **20**, 197 (1974)
112. Lee, J. A., Maskell, W. C.: Powder Technol. **7**, 259 (1973)
113. Svata, M.: Powder Technol. **5**, 345 (1971/72)
114. Allen, T.: In: Particle Size Measurement. Williams, J. C. (ed.). London: Chapman and Hall 1975, pp. 1–30
115. Fooks, J. C.: Brit. Chem. Eng. **15**, 799 (1970)
116. Hulley, B. J.: Chem. Eng. (London) **1970**, CE 410
117. Wadel, H.: Physics **5**, 281 (1934)
118. Heywood, H.: J. Pharm. Pharmacol., Suppl. **15**, 56 T (1963)
119. Heywood, H.: Harold Heywood Memorial Lectures, Loughborough Univ., England, 1973
120. Allen, T.: In: Particle Size Measurement. Williams, J. C. (ed.). London: Chapman and Hall 1975, pp. 93–112
121. I.S.O. 565: Test Sieves, Woven Metal Wire Cloth and Perforate plate Nominal sizes of apertures (1972)
122. I.S.O. 2591: Test Sieving (1973)
123. Parfitt, G.: Dispersion of Powders in Liquids. London: Applied Science Publishers 1973
124. Allen, T.: In: Particle Size Measurement. Williams, J. C. (ed.). London: Chapman and Hall 1975, pp. 178–190
125. Andreasen, A. H. M., Lundberg, J. J. V.: Ber. Dent. Keram. Ges., **11**, 312 (1930)
126. Berg, S.: In: Symposium on Particle Size Measurement, Sp. Publ. No. 234, A.S.T.M. 1958 p. 143
127. Leschonski, K.: Staub., **22**, 475 (1962)

128. Allen, T.: In: Particle Size Measurement. Williams, J. C. (ed.). London: Chapman and Hall 1975, pp. 203–206
129. Conlin, S. G., et al.: J. Sci. Instrum. **44**, 606 (1967)
130. Oliver, J. P., Hickin, G. K., Orr, C.: US Patent 3, 449, 567 (1969)
131. Gaudin, A. M., Schumann, R., Schlechter, A. W.: J. Phys. Chem., **46**, 903 (1942)
132. Bostock, W.: J. Sci. Instrum., **29**, 209 (1952)
133. Edwald, P.: Ind. Eng. Chem., analyt. ed., **14**, 66 (1942)
134. Bachmann, D., Gerstenberg, H.: Chem. Eng. Tech. **8**, 598 (1957)
135. Leschonski, K.: Staub., **22**, 475 (1962)
136. Cohen, L.: Instrum. Practice, **13**, 1036 (1959)
137. Friedrich, W.: Staub, **19**, 281 (1959)
138. Suito, E., Arakawa, M.: Bull. Inst. Chem. Res., Kyoto Univ., **23**, 7 (1950)
139. Marshall, C. E.: Proc. Roy. Soc., Ser. A, **126**, 427 (1930)
140. Kaye, B. H.: British Patent No 895,222 (1962)
141. Naumann, D., Seydel, K. J.: Plaste Kaut., **15**, 136 (1969)
142. Statham, B. R.: Proc. Soc. Analyt. Chem. **9**, 40 (1972)
143. Atherton, E., Cooper, A. C.: British Patent 983,760 (1962)
144. Carr, W.: J. Oil Colour Chem. Ass. **53**, 1 and 81 (1970), **54**, 155 (1971)
145. Norton, F. H., Spiel, S. J.: J. Amer. Ceram. Soc. **21**, 89 (1938)
146. Jacobson, A. E., Sullivan, W. F.: Ind. Eng. Chem. analyt. ed., **18**, 360 (1946)
147. Slater, C., Cohen, L.: J. Sci. Instrum., **39**, 614 (1962)
148. Hildreth, J. D., Patterson, D.: J. Soc. Dyers Colourists, **80**, 474 (1964)
149. Allen, T., Svarovsky, L.: Proc. Soc. Analyt. Chem. **9**, 38 (1972)
150. Endter, F., Gebauer, H.: Optik, **13**, 87 (1956)
151. Becker, P.: J. Colloid Interface Sci. **19**, 468 (1964)
152. Chatfield, E. J.: J. Sci. Instrum. **44**, 615 (1967)
153. Taylor, W. K.: Brit. J. Appl. Phys. **1954** suppl. 3, S 173
154. Morgan, B. B.: Research (London) **10**, 271 (1957)
155. Roberts, F., Young, J. Z.: Nature **169**, 962 (1952)

Chapter 5

Solid Acid and Base Catalysts

K. Tanabe

Department of Chemistry, Faculty of Science
Hokkaido University,
Sapporo, Japan

Contents

1. Introduction

Solid acids have been extensively used as catalysts or catalyst carriers in petroleum chemistry and organic syntheses for many years. Use of solid acid catalysts provides the following several advantages compared with use of liquid acid catalysts:
1) High catalytic activity and selectivity are frequently observed;
2) Solid acid catalysts do not corrode reaction vessels or reactors;
3) Repeated use of solid acid catalysts is possible;
4) Separation of a solid acid catalyst from a reaction mixture is easy;
5) There is no problem for the disposal of used solid acid catalysts, though the disposal of liquid acid catalysts requires much money for treatment to make it environmentally safe.

The importance of solid acids as catalysts has caused an extensive and quantitative study of the acidic property on catalyst surfaces, the structure of acid sites, and their catalytic action, for the last thirty years. On the basis of the accumulated knowledge of solid acids, it is becoming possible to design and develop highly active and selective solid acid catalysts for particular reactions.

Recently, many kinds of solid bases have been found and used as catalysts for various reactions. The nature of basic sites has been elucidated in some cases. Advances in the study of solid base catalysts are leading to the success in interpretation of catalytic action in terms of acid-base bifunctional catalysis. In fact, some solid acids are known to exhibit more or less basic properties, while some solid bases have acidic properties to some extent. Thus, solid acids and bases are classified conventionally by their main (or predominant) surface property.

In this chapter, the types of solid acids and bases, the acidic and basic properties, and the structure of acidic and basic sites are dealt with. The chemical principles by which acidic and basic properties (Brönsted and Lewis type, strength, and amount) are estimated, and the mechanism regarding the generation of acidity and basicity are also described. How acidic and basic property are controlled chemically is discussed in connection with the preparation method and the pretreatment condition of solid acids and bases. Catalytic activity and selectivity of solid acids and bases are limited to the extent necessary for general account of the above subjects.

2. Definition and Classification of Solid Acids and Bases

In general terms, a solid acid may be understood as a solid on which the color of a basic indicator changes, or as a solid on which a base is chemically adsorbed. More strictly, following both the Brönsted and Lewis definitions, a solid acid shows a tendency to donate a proton or to accept an electron pair, whereas a solid base tends to accept a proton or to donate an electron pair. These definitions are adequate for an understanding of the acid-base

Table 1. Solid acids

1. Natural clay minerals: kaolinite, bentonite, attapulgite, montmorillonite, clarit, fuller's earth, zeolites
2. Mounted acids: H_2SO_4, H_3PO_4, $CH_2(COOH)_2$ mounted on silica, quartz sand, alumina or diatomaceous earth
3. Cation exchange resins
4. Charcoal heat-treated at 573 K
5. Metal oxides and sulfides: ZnO, CdO, Al_2O_3, CeO_2, ThO_2, TiO_2, ZrO_2, SnO_2, PbO, As_2O_3, Bi_2O_3, Sb_2O_5, V_2O_5, Cr_2O_3, MoO_3, WO_3, CdS, ZnS
6. Metal salts: $MgSO_4$, $CaSO_4$, $SrSO_4$, $BaSO_4$, $CuSO_4$, $ZnSO_4$, $CdSO_4$, $Al_2(SO_4)_3$, $FeSO_4$, $Fe_2(SO_4)_3$, $CoSO_4$, $NiSO_4$, $Cr_2(SO_4)_3$, $KHSO_4$, K_2SO_4, $(NH_4)_2SO_4$, $Zn(NO_3)_2$, $Ca(NO_3)_2$, $Bi(NO_3)_3$, $Fe(NO_3)_3$, $CaCO_3$, BPO_4, $AlPO_4$, $CrPO_4$, $FePO_4$, $Cu_3(PO_4)_2$, $Zn_3(PO_4)_2$, Mg_3(PO $Mg_3(PO_4)_2$, $Ti_3(PO_4)_4$, $Zr_3(PO_4)_4$, $Ni_3(PO_4)_2$, AgCl, CuCl, $CaCl_2$, $AlCl_3$, $TiCl_3$, $SnCl_2$, CaF_2, BaF_2, $AgClO_4$, $Mg(ClO_4)_2$
7. Mixed oxides: SiO_2—Al_2O_3, SiO_2—TiO_2, SiO_2—SnO_2, SiO_2—ZrO_2, SiO_2—BeO, SiO_2—MgO, SiO_2—CaO, SiO_2—SrO, SiO_2—ZnO, SiO_2—Ga_2O_3, SiO_2—Y_2O_3, SiO_2—La_2O_3, SiO_2—MoO_3, SiO_2—WO_3, SiO_2—V_2O_5, SiO_2—ThO_2, Al_2O_3—MgO, Al_2O_3—ZnO, Al_2O_3—CdO, Al_2O_3—B_2O_3, Al_2O_3—ThO_2, Al_2O_3—TiO_2, Al_2O_3—ZrO_2, Al_2O_3—V_2O_5, Al_2O_3—MoO_3, Al_2O_3—WO_3, Al_2O_3—Cr_2O_3, Al_2O_3—Mn_2O_3, Al_2O_3—Fe_2O_3, Al_2O_3—Co_3O_4, Al_2O_3—NiO, TiO_2—CuO, TiO_2—MgO, TiO_2—ZnO, TiO_2—CdO, TiO_2—ZrO_2, TiO_2—SnO_2, TiO_2—Bi_2O_3, TiO_2—Sb_2O_5, TiO_2—V_2O_5, TiO_2—Cr_2O_3, TiO_2—MoO_3, TiO_2—WO_3, TiO_2—Mn_2O_3, TiO_2—Fe_2O_3, TiO_2—Co_3O_4, TiO_2—NiO, ZrO_2—CdO, ZnO—MgO, ZnO—Fe_2O_3, MoO_3—CoO—Al_2O_3, MoO_3—NiO—Al_2O_3, TiO_2—SiO_2—MgO, MoO_3—Al_2O_3—MgO

Table 2. Solid bases

1. Mounted bases: NaOH, KOH mounted on silica or alumina; alkali metal and alkaline earth metal dispersed on silica, alumina, carbon, K_2CO_3 or in oil; NR_3, NH_3, KNH_2 on alumina; Li_2CO_3 on silica
2. Anion exchange resins
3. Charcoal heat-treated at 1173 K or activated with N_2O, NH_3 or $ZnCl_2$—NH_4Cl—CO_2
4. Metal oxides: BeO, MgO, CaO, SrO, BaO, ZnO, Al_2O_3, Y_2O_3, La_2O_3, CeO_2, ThO_2, TiO_2, ZrO_2, SnO_2
5. Metal salts: Na_2CO_3, K_2CO_3, $KHCO_3$, $KNaCO_3$, $CaCO_3$, $SrCO_3$, $BaCO_3$, $(NH_4)_2CO_3$, $Na_2WO_4 \cdot 2\,H_2O$, KCN
6. Mixed oxides: SiO_2—MgO, SiO_2—CaO, SiO_2—SrO, SiO_2—BaO, SiO_2—ZnO, SiO_2—Al_2O_3, SiO_2—ThO_2, SiO_2—TiO_2, SiO_2—ZrO_2, SiO_2—MoO_3, SiO_2—WO_3, Al_2O_3—MgO, Al_2O_3—ThO_2, Al_2O_3—TiO_2, Al_2O_3—ZrO_2, Al_2O_3—MoO_3, Al_2O_3—WO_3, ZrO_2—ZnO, ZrO_2—TiO_2, TiO_2—MgO

phenomena shown by various solids, and are convenient for a clear description of solid acid and base catalysis.

In accordance with the above definitions, a summarized list of solid acids and bases is given in Tables 1 and 2. The first group of solid acids in Table 1 includes naturally occurring clay minerals. The main constituents are silica and alumina. The well-known solid acid, synthetic silica-alumina is listed in the VII'th Group, which also includes the many oxide mixtures which have recently been found to display acidic properties and catalytic activity. In the V'th and VI'th Groups are included many inorganic chemicals such as metal oxides, sulfides, sulfates, nitrates, phosphates and halides. Many in these groups have recently been found to show characteristic selectivities as catalysts.

Of the solid bases listed in Table 2, special mention should be made of the alkaline earth metal oxides in the IV'th Group, and mixed metal oxides in the VI'th Group, whose basic properties and catalytic action have been recently found to be striking and interesting.

3. Principle of Measurement of Acidic and Basic Property

A complete description of acidic and basic properties on solid surfaces requires the determination of the acid and base strength, and of the amount and nature (Brönsted or Lewis type) of the acidic and basic sites.

A. Acidic Property

1. Strength and Amount of Solid Acid

When measuring the strength of a solid acid or base, it should be recognized that activity coefficients for species on the solid are unknown. Therefore, acidity and basicity functions for the solid are not properly defined thermodynamically. Nevertheless, the acidity and basicity functions are clearly valuable in a relative sense, while the absolute values are also useful provided the above limitations are recognized and numerical accuracy is not overstated.

The acid strength of a solid is defined as the ability of the surface to convert an adsorbed neutral base into its conjugate acid. If the reaction proceeds by means of proton transfer from the surface to the adsorbate, the acid strength is expressed by the Hammett acidity function H_0 [1],

$$H_0 = pK_a + \log [B]/[BH^+] \,, \qquad (1)$$

where [B] and $[BH^+]$ are respectively the concentrations of the neutral base (basic indicator) and its conjugate acid and pK_a is pK_{BH^+}. If the reaction takes place by means of electron pair transfer from the adsorbate to the surface, H_0 is expressed by

$$H_0 = pK_a + \log [B]/[AB] \,, \qquad (2)$$

where [AB] is the concentration of the neutral base which reacted with the Lewis acid or electron pair acceptor, A.

The amount of acid on a solid is usually expressed as the number or mmol of acid sites per unit weight or per unit surface area of the solid, and is obtained by measuring the amount of a base which reacts with the solid acid. It is also sometimes loosely called "acidity".

For the determination of strength and amount of a solid acid, there are two main methods; amine titration method using indicators and gaseous base adsorption method.

a) Amine titration method using indicators

The color of suitable indicators adsorbed on a surface will give a measure of its acid strength: if the color is that of the acid form of the indicator, then the value of the H_0 function of the surface is equal to or lower than the pK_a of the conjugate acid of the indicator. Lower values of H_0 correspond to greater acid strength. Thus, for indicators undergoing color changes in this way, the lower the pK_a, the greater is the acid strength of the solid. For example, a solid which gives a yellow coloration with benzalacetophenone ($pK_a = -5.6$), but is colorless with anthraquinone ($pK_a = -8.2$), has an acid strength H_0 which lies between -5.6 and -8.2. A solid having $H_0 \leqq -8.2$ will change all indicators in Table 3 from the basic to the acidic colors, whereas one which changed none of them would have an acid strength of $H_0 > +6.8$.

The determination is easily made by placing the sample in powder form into a test tube, adding non-polar solvent containing indicator, and shaking briefly [2]. The solvents used are benzene, isooctane, decalin and cyclohexane. H_R indicators (a series of arylalcohols which react with acids to form carbonium ions) are also available for acid strength determinations in certain specialized applications [3, 4]. The acid strength of a super solid acid which is very sensitive to moisture can be determined by observing the color change of an indicator whose vapor is adsorbed on a solid sample through a breakable seal in a vacuum system at room temperature [5]. The indicators used for the determination are liquid at room temperature as indicated in Table 3.

This method readily enables a determination of relative acid strengths. However, the usual color change is difficult to observe when the sample is itself colored or dark, but the difficulty can be minimized by mixing a white substance of known acidity with the sample [2, 6] or by employing the spectrophotometric method [2, 6].

The amount of a solid acid can be measured by amine titration immediately after a determination of acid strength by the above method [2]. The method consists of titrating a solid acid suspended in benzene with n-butylamine, using p-dimethylaminoazobenzene as an indicator. The yellow basic form of the indicator changes to its red acidic form when adsorbed on the solid acid. Thus, the titers of n-butylamine required to restore the yellow color give a measure of the number of acid sites on the surface. Since the indicator which has been adsorbed on the acid sites must be replaced by the titrating base, the basic strength of the latter should be higher than the former.

The amount of acid indicated in this way by using p-dimethylaminoazobenzene ($pK_a = +3.3$) is actually the amount of those sites having acid

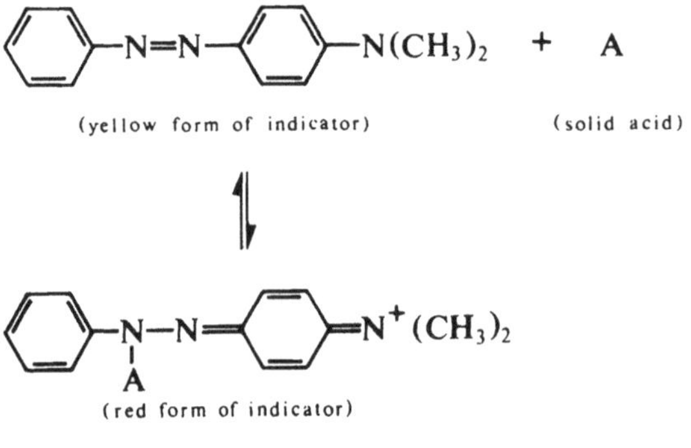

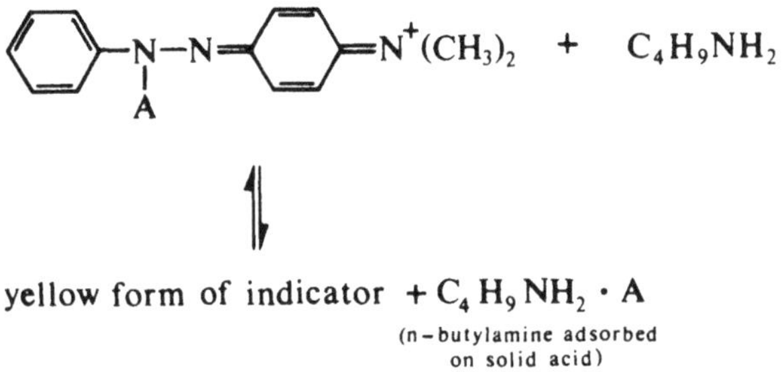

strengths with $H_0 \leqq +3.3$. The use of various indicators with different pK_a values (see Table 3) enables a determination of the amount of acid *at various acid strengths* by amine titration. The experimental details such as the effects of titration time, volume of added indicator, powder size, and moisture on measured acid amount are given in a book [2]. As an example, the acid strength and amount of MoO_3—SiO_2 having compositions of 9/1, 1/1, and 1/9 as well as those of SiO_2 and MoO_3, when calcined at 773 K in air, are shown in Figure 1. Although SiO_2 and MoO_3 showed negligible amounts of acid sites stronger than $H_0 \leqq 3.3$, SiO_2—MoO_3 with composition 9/1 and 1/1 showed large amounts of the acid sites [7]. A good correlation between the acid amount at $H_0 \leqq +3.3$ and the rate of the depolymerization of paraldehyde is seen in Figure 1. Calorimetric titration of a solid acid with amine is also available for the estimation of the acid amount of a colored or dark sample [2, 6].

The amine titration method gives the sum of the amounts of both Brönsted and Lewis acid, since both electron pair acceptors and proton donors on the surface will react with either the electron pair (—N̤=) of the indicator or that of the amine (≡N:) to form a coordination bond.

b) Gaseous base adsorption method

When gaseous bases are adsorbed on acid sites, a base adsorbed on a strong acid site is more stable than one adsorbed on a weak acid site, and is more difficult to desorb. As elevated temperatures stimulate evacuation of the

Table 3. Basic indicators used for the measurement of acid strength

Indicators	Color		pK_a[a]	$[H_2SO_4]^b$/%
	Base-form	Acid-form		
Neutral red	yellow	red	+6.8	8×10^{-8}
Methyl red	yellow	red	+4.8	—
Phenylazonaphthylamine	yellow	red	+4.0	5×10^{-5}
p-Dimethylaminoazobenzene	yellow	red	+3.3	3×10^{-4}
2-Amino-5-azotoluene	yellow	red	+2.0	5×10^{-3}
Benzeneazodiphenylamine	yellow	purple	+1.5	2×10^{-2}
Crystal violet	blue	yellow	+0.8	0.1
p-Nitrobenzeneazo-(p'-nitro-dephenylamine)	orange	purple	+0.43	—
Dicinnamalacetone	yellow	red	−3.0	48
Benzalacetophenone	colorless	yellow	−5.6	71
Anthraquinone	colorless	yellow	−8.2	90
2,4,6-Trinitroaniline	colorless	yellow	−10.10	98
p-Nitrotoluene	colorless	yellow	−11.35	[c]
m-Nitrotoluene	colorless	yellow	−11.99	[c]
p-Nitrofluorobenzene	colorless	yellow	−12.44	[c]
p-Nitrochlorobenzene	colorless	yellow	−12.70	[c]
m-Nitrochlorobenzene	colorless	yellow	−13.16	[c]
2,4-Dinitrotoluene	colorless	yellow	−13.75	[c]
2,4-Dinitrofluorobenzene	colorless	yellow	−14.52	[c]
1,3,5-Trinitrotoluene	colorless	yellow	−16.04	[c]

[a] pK_a of the conjugate acid, BH^+, of indicator, B, $(= pK_{BH^+})$

[b] wt. percent of H_2SO_4 in sulfuric acid solution which has the acid strength corresponding to the respective pK_a.

[c] The indicator is liquid at room temperature and the acid strength corresponding to the indicator is higher than the acid strength of 100 percent H_2SO_4

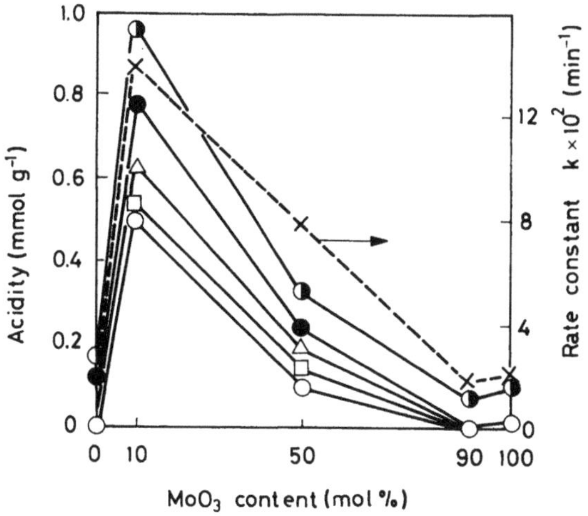

Figure 1. Correlation of the acidity of SiO_2-MoO_3 with the activity for depolymerization of paraldehyde: (○) $H_0 \leqq -3$, (□) $H_0 \leqq 1.5$, (△) $H_0 \leqq 3.3$, (●) $H_0 \leqq 4.8$, (◑) $H_0 \leqq 6.8$, (×) catalytic activity

adsorbed bases from acid sites, those at weaker sites will be evacuated preferentially. Thus, the proportion of adsorbed base evacuated at various temperatures can give a measure of acid strength.

The amount of a gaseous base which a solid acid can adsorb chemically from the gaseous phase is a measure of the amount of acid on its surface. After a solid sample is put in a quartz spring balance and evacuated, the vapor of an organic base may be introduced for adsorption. When prolonged subsequent evacuation produces no further decrease in sample weight, then the base which is retained upon the sample is understood to be chemically adsorbed. [8]

The amounts of ammonia, n-butylamine and pyridine remained on the surfaces of TiO_2—SiO_2 (1/1) and SiO_2—Al_2O_3 (Al_2O_3 20 wt. percent) after evacuating at various temperatures are shown in Figure 2 [9]. The fact that the amounts of basic molecules retained on TiO_2—SiO_2 were larger than those on SiO_2—Al_2O_3 at all temperatures suggests that the TiO_2—SiO_2 possesses larger acid amount and higher acid strength than SiO_2—Al_2O_3.

The heat of adsorption of various bases is also clearly a measure of the acid strength on the solid surface [2]. Differential thermal analysis and thermogravimetry of desorption of basic molecules is available for the estimation of the acid amount together with the acid strength of a solid [2].

Ammonia, n-butylamine, and pyridine have been extensively used as gaseous bases for the determination of strength and amount of a solid acid [2]. However, ammonia and n-butylamine in which hydrogen atoms are attached on nitrogen atom have a tendency to dissociate (e.g., $NH_3 \rightarrow NH_2^- + H^+$) and to adsorb on both acidic and basic sites depending on the kinds of solids and the adsorption condition. In this regard, triethylamine which is much

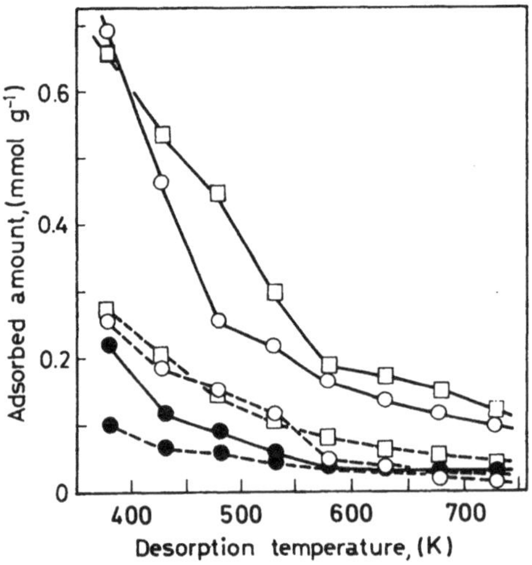

Figure 2. Amounts of ammonia, n-butylamine and pyridine remained on catalyst surface after evacuating at various temperatures (—) TiO_2—SiO_2, (- - -) SiO_2—Al_2O_3, (□) ammonia, (○) n-butylamine, (●) pyridine

more difficult to dissociate is recommended to be used as an adsorbate [10]. We have also to be careful in the use of pyridine, because pyridine has been recently found to adsorb on strong basic sites to form an anion radical of pyridine [11].

Methods utilizing adsorption and desorption of gaseous bases have the advantage that the acid amount for a solid at high temperatures (several hundred degrees centigrade), or under its actual working conditions as a catalyst, can be determined. The methods also apply even to colored samples. They suffer from the disadvantage that it is difficult to distinguish between chemical and physical adsorption and to differentiate between the amounts of acid at various acid strengths.

c) Estimation of acidity from catalytic activity

Catalytic activity has been used as a measure of acidity and acid strength [2]. Recently, the activity for the dehydration of isopropyl alcohol or the isomerization of butene in the presence of an excess of air is reported to be a good measure of acidity of some oxidation catalysts whose surface areas are so small that a gas adsorption method is difficult for the determination of acidity [12–15]. In the case of the isomerization of 1-butene, it is necessary to poison basic sites with the products formed by oxidation of a mixture of air and butene, which is passed through catalyst at 470–570 K before the activity measurement, since the isomerization is catalyzed not only by acidic sites, but also by basic sites [16]. Fairly good correlations are found between the acidity measured by adsorption of ammonia or pyridine and the activity for dehydration of isopropyl alcohol and isomerization of 1-butene [14].

Kinetics of the coisomerization of cis-2-butene-d_0/d_8 gives an insight as to whether the active sites on SiO_2—MoO_3 [17], MoO_3 [17], and WO_3 [18] are acidic or basic.

2. Brönsted and Lewis Acid Sites

The methods for determining the strength and amount of acid described in the foregoing sections (3.A.1.a, b, and c) do not distinguish between Brönsted acid sites and Lewis acid sites. The acid amount which is measured is the sum of the amounts of Brönsted acid and Lewis acid at a certain acid strength. In order to elucidate the catalytic actions of solid acids, it is often necessary to distinguish between Brönsted acids and Lewis acids.

Infrared spectroscopic studies of ammonia and pyridine adsorbed on solid surfaces make it possible to distinguish between Brönsted and Lewis acid and to assess the amounts of Brönsted and Lewis acid independently. Basila and Kantner showed that the modes in which ammonia is adsorbed on SiO_2—Al_2O_3 are as physically adsorbed NH_3, as coordinately bonded NH_3, and as NH_4^+, each of which can be detected by means of their absorption bands [19]. Their investigations of the relative intensities of the corresponding bands showed a ratio of Lewis to Brönsted acid sites of 4/1. The spectrum of pyridine coordinately bonded to the surface is very different from that of the pyridinium ion as shown in Table 4 [20]. This fact permits differentiation

Table 4. Infrared bands of pyridine on solid acids in the 1,400–1,700 cm^{-1} region[a]

Hydrogen bonded pyridine	Coordinately bonded pyridine	Pyridinium ion
1,400–1,447 (vs)	1,447–1,460 (vs)	
1,485–1,490 (w)	1,488–1,503 (v)	1,485–1,500 (vs)
		1,540 (s)
1,580–1,600 (s)	~1,580 (v)	
	1,600–1,633 (s)	~1,620 (s)
		~1,640 (s)

[a] Band intensities: vs — very strong; s — strong; w — weak; v — variable

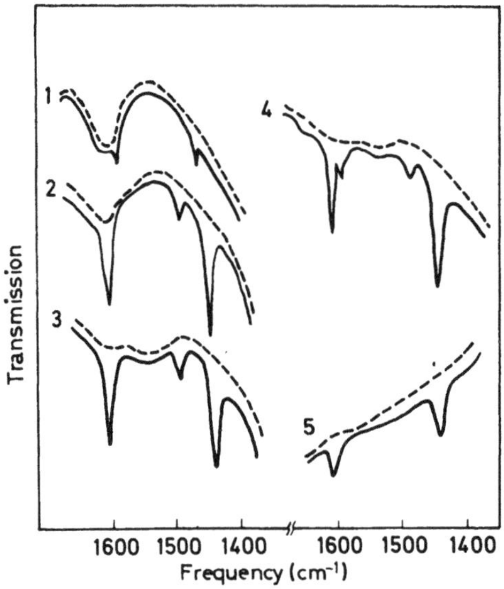

Figure 3. Infrared spectra of pyridine adsorbed on $ZnO—SiO_2$, 1: SiO_2, 2: $ZnO—SiO_2$ (1/9), 3: $ZnO—SiO_2$ (3/7), 4: $ZnO—SiO_2$ (9/1), 5: ZnO, Brokene lines: backgrounds

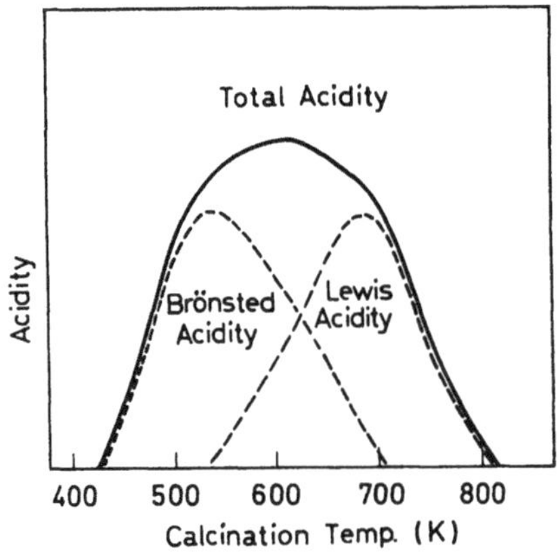

Figure 4. Acidity of $NiSO_4 \cdot x H_2O$ calcined at various temperatures

between acid types on the surface of a solid acid. Figure 3 shows the infrared spectra of pyridine adsorbed on SiO_2—ZnO of various compositions which had been calcined at 773 K for 3 hr in air [21]. The bands at 1,450, 1,490, and 1,610 cm^{-1} which are observed on all the mixed oxides are characteristic bands of pyridine coordinately bonded to Lewis acid sites. On all samples, no bands were detected at 1,540 cm^{-1} which is due to pyridinium ion formed by the adsorption on Brönsted acid sites. Therefore, the kind of acid sites on the mixed oxides is Lewis type. The Brönsted and Lewis acidity of nickel sulfate heat-treated at various temperatures, as derived from the infrared spectra of adsorbed pyridine is shown in Figure 4 [22]. A maximum of Brönsted acidity appears when the sulfate is heat-treated at 523 K, while that of Lewis acidity appears at higher temperature (570–670 K). The sum of the two acidity curves gives the total acid amount which can be measured by amine titration method (cf. 3.A.1.a).

Infrared spectroscopic method using pyridine as an adsorbate is extensively used and is considered to be a most reliable method, though there are many other methods to distinguish between Brönsted and Lewis acid [2].

B. Basic Property

The basic strength of a solid surface is defined as the ability of the surface to convert an adsorbed electrically neutral acid to its conjugate base, i.e. the ability of the surface to donate an electron pair to an adsorbed acid. The amount of base (basic sites) on a solid is usually expressed as the number (or mmol) of basic sites per unit weight or per unit surface area of the solid. It is also sometimes more loosely called "basicity." There are two main methods for the measurement of strength and amount of basic sites; benzoic acid titration method using indicators and gaseous acid adsorption method.

1. Benzoic Acid Titration Method Using Indicators

When an electrically neutral acid indicator is adsorbed on a solid base from a non-polar solution, the color of the acid indicator is changed to that of its conjugate base, provided that the solid has the necessary basic strength to impart electron pairs to the acid. Thus, it is generally possible to determine the basic strength by observing the color changes of acid indicators over a range of $pK_a = pK_{BH}$ values.

For the reaction of an acid indicator BH with a solid base $\bar{B}$,

$$BH + \bar{B} \rightleftharpoons B^- + \bar{B}H^+ \tag{3}$$

the basic strength H_- of $\bar{B}$ is given by an equation similar to equation 1,

$$H_- = pK_a\,(=pK_{BH}) + \log [B^-]/[BH] \tag{4}$$

where [BH] is the concentration of the acidic form of the indicator and $[B^-]$ the concentration of the basic form.

The first perceptible change in the color of an acid indicator occurs when about 10 percent of the adsorbed layer of indicator is in the basic form, i.e. when the ratio $[B^-]/[BH]$ reaches 0.1/0.9 (= 0.11). Further increase in the intensity of the color is only perceptible to the naked eye when about 90 percent of the indicator is in the basic form, i.e. $[B^-]/[BH] = 0.9/0.1$ (= 9). Thus the initial color change and the subsequent change in intensity are observed at values of H_- equal to $pK_{BH} - 1$ and $pK_{BH} + 1$ respectively. If we assume that the intermediate color appears when the basic form reaches 50 percent, i.e. when $[B^-]/[BH] = 1$, we have $H_- = pK_{BH}$.

According to this assumption, the approximate value of the basic strength on the surface is given by the pK_{BH} value of the adsorbed indicator at which the intermediate color appears [23]. Indicators which lend themselves to this method are listed in Table 5 [24, 25]. Non-polar solvents such as benzene and isooctane are used for the indicators.

The amount of basic sites can be measured by titrating a suspension in benzene of a solid on which an indicator has been adsorbed in its conjugate basic form, with benzoic acid dissolved in benzene. The benzoic acid titers are a measure of the amount of basic sites (in mmol g^{-1} or mmol m^{-2}) having a basic strength corresponding to the pK_{BH} value of the indicator used [2].

The strength and amount of basic sites on MgO—TiO_2 of different compositions which were measured by benzoic acid titration method using indicators, are shown in Figure 5 [26]. MgO, MgO—TiO_2 (9/1 and 1/1) showed a basic strength of $H_- = 17.2$, whereas MgO—TiO_2 (1/9) and TiO_2 did not show any basic property. The basicities at $H_- \geqq 12.2$ and 15.0 of MgO—TiO_2 (9/1) are higher than those of MgO. The basicity maxima become sharper when the basicity per unit surface area is plotted. Figure 5, which also shows the acidic property measured by the method described in 3.A.1.a., reveals that MgO and MgO—TiO_2 (9/1) have only the basic property and TiO_2 (1/1) possesses both acidic and basic properties.

2. *Gaseous Acid Adsorption Method*

The principle of this method is the same as that of gaseous base adsorption method (3.A.1.b) and all of the latter method can be applied. As adsorbates,

Table 5. Indicators used for the measurement of basic properties

Indicators	Color		pK_a [a]
	Acid-form	Base-form	
Bromothymol blue	yellow	green	7.2
Phenolphthalein	colorless	red	9.3
2,4,6-Trinitroaniline	yellow	redish-orange	12.2
2,4-Dinitroaniline	yellow	violet	15.0
4-Chloro-2-nitroaniline	yellow	orange	17.2
4-Nitroaniline	yellow	orange	18.4
4-Chloroaniline	colorless	pink [b]	26.5 [c]

[a] pK_a of indicator, BH, (= pK_{BH})
[b] The color disappears with the addition of benzoic acid
[c] This value was estimated from the data of Stewart, R. and Dolman, D.: Can, J. Chem. **45**, 925 (1967)

acidic molecules such as carbon dioxide, nitric oxide and phenol vapor have been used. The adsorption of phenol [27] is not necessarily good for the measurement of basic property, because phenol is easily dissociated to adsorb on both acidic and basic sites [28] and hence acidic property affects the adsorption of phenol. Nitric oxide is used for the measurement of unusually strong basic sites [29]. The amount of carbon dioxide irreversibly adsorbed is a good measure of the amount of basic sites on solid surfaces. The amounts of carbon dioxide irreversibly adsorbed at 373 K on ZrO_2 pretreated at various temperatures are shown in Figure 6 [30]. In the case of CaO, carbon dioxide is reported to adsorb on the basic site as a unidentate complex when the pressure of carbon dioxide is relatively high, but on both acidic and basic sites as a bidentate complex when the pressure is low (cf.

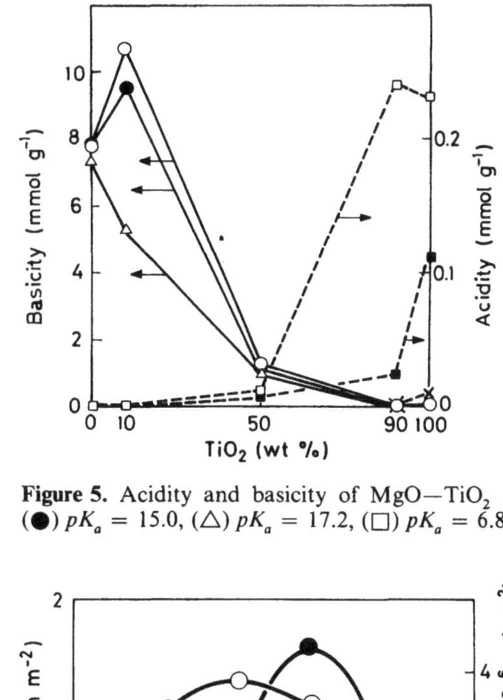

Figure 5. Acidity and basicity of MgO—TiO_2 of different compositions: (○) $pK_a = 12.2$, (●) $pK_a = 15.0$, (△) $pK_a = 17.2$, (□) $pK_a = 6.8$, (■) $pK_a = 4.8$, (×) $pK_a = 1.5$

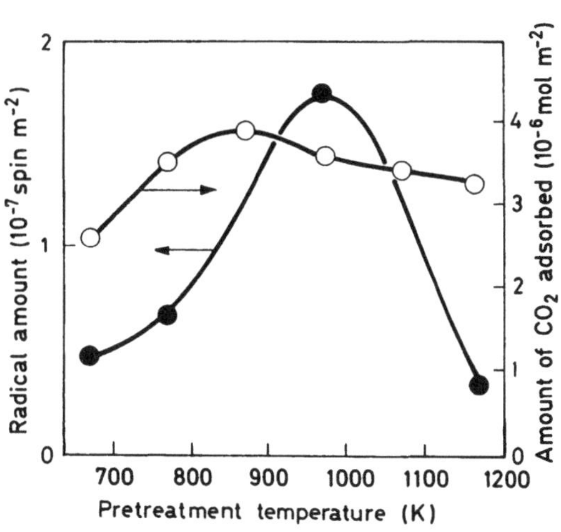

Figure 6. Amounts of diphenylnitroxide radicals (●) and CO_2 irreversibly adsorbed at 373 K (○) on ZrO_2 pretreated at various temperatures

Figure 7) [31]. However, only a unidentate complex of carbon dioxide is formed over ZrO_2 regardless of the pressure of carbon dioxide. Ai has recently found a good correlation between the basicity of $Co_3O_4—K_2O$ measured by carbon dioxide adsorption and the oxidation activity for n-hexane, phenol, and methanol [32].

unidentate complex bidentate complex **Figure 7.** Adsorbed states of CO_2 on CaO

Diphenylamine can be used to measure the amount of strong basic sites. When diphenylamine is adsorbed on ZrO_2, diphenylnitroxide radicals are produced on ZrO_2 from diphenylamine and oxygen [30]. The signal amplitude of the ESR increases when diphenylamine is added to the ZrO_2 that had adsorbed triphenylamine followed by admission of O_2. This suggests that the sites for the formation of diphenylnitroxide radicals are different from the sites for the formation of triphenylamine cation radicals, because the ionization potential of triphenylamine is lower than that of diphenylamine. Therefore, it is considered that the diphenylnitroxide radicals are formed on CaO by an action of basic sites as follows.

$$(C_6H_5)_2N{-}H \xrightarrow{\text{basic site}} (C_6H_5)_2N^- + H^+ \xrightarrow{O_2} (C_6H_5)_2N{-}O^{\cdot} + {}^{\cdot}OH$$

The concentration of diphenylnitroxide radicals is taken as the concentration of basic sites, since the radicals could not desorb from the sites unless a polar molecule is added. On the other hand, triphenylamine cation radicals is considered to form on oxidizing sites (one electron acceptor sites). Since the pK_a value of diphenylamine is 23, only strong basic sites could be measured by this method. The basicity on ZrO_2 pretreated at various temperatures measured by this method is different from the basicity calculated from the amount of CO_2 irreversibly adsorbed as shown in Figure 6.

3. Estimation of Basicity From Catalytic Activity

As mentioned in 3.A.1.c, the catalytic activity for dehydration of isopropyl alcohol to propylene (r_p) is proportional to the acidity of a catalyst.

$$r_p = k \cdot \text{acidity} \tag{5}$$

On the other hand, the activity for dehydrogenation of isopropyl alcohol to acetone (r_a) is assumed to be proportional to the acidity and basicity of a catalyst, since the dehydrogenation is considered to proceed by a concerted mechanism, for example [13, 15]:

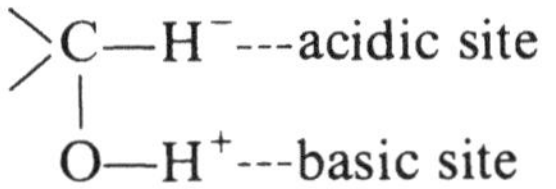

$$r_a = k' \cdot \text{acidity} \cdot \text{basicity} \tag{6}$$

From equations 5 and 6, the following equation is derived,

$$\text{basicity} = k'' \cdot r_a/r_p, \quad (7)$$

where k, k', and k'' are constants.

Thus, r_a/r_p can be used as a measure of the basicity of a catalyst. In fact, a good correlation is found between r_a/r_p and the amount of carbon dioxide irreversibly adsorbed [14, 15].

This method can be well applied for the basicity measurement of some oxidation catalysts such as V_2O_5—K_2SO_4—H_2SO_4 whose surface area is so small (about 0.7 $m^2 g^{-1}$) that the accurate measurement of the amount of carbon dioxide irreversibly adsorbed is not easy [13].

4. Method by XPS

The O_{1s} binding energy of metal oxides, which can be measured by x-ray photoelectron spectroscopy (XPS), is also a measure of basic strength of metal oxides, since the electron pair donating ability of oxides is assumed to be expressed by the O_{1s} binding energy. The order of basic strength determined by this method is as follows [33]: La_2O_3 (529.0 eV) > Sm_2O_3 (529.2) > CeO_2 (529.4) = Dy_2O_3 (529.4) > Y_2O_3 (529.5) > Fe_2O_3 (530.3) > Al_2O_3 (531.8) > GeO_2 (532.4) > P_2O_5 (532.4) > SiO_2 (533.1). The metal oxides whose binding energy is less than 529.5 eV are reported to be catalytically active for the selective formation of l-olefin from secondary alcohol [33].

C. Representative Parameter, $H_{o,max}$, of Acid-Base Property

As described in 3.A.1.a and 3.B.1, acid strength (H_0) is expressed by pK_a values of the conjugate acids of basic indicators, while base strength (H_-) by pK_a values of acidic indicators. Since the indicators used for the basicity measurement are different from those used for acidity measurement (cf. Tables 3 and 5), it was impossible to determine the acid-base strength distribution on a common scale. Recently, a new method which determines the basicity at various base strengths of solid samples by using a series of Hammett indicators as shown in Table 3 has been presented [34]. By this method, both acidic and basic property can be determined *on a common H_0 scale*, where the strength of basic sites is expressed by the H_0 of the conjugate acidic sites. It was found that the strongest H_0 values of the acidic sites was approximately equal to the strongest H_0 value of the basic sites [35]. The equal strongest H_0 was termed "$H_{0,max}$" which is a practical parameter to represent an acid-base property on solid surfaces.

Before discussing the significance and usefulness of $H_{0,max}$, we shall study the principle of the method of expressing basic property by an H_0 scale.

1. Basic Property Expressed by H_0 Scale

The acidity and acid strength of a solid can be determined by amine titration method using a series of Hammett basic indicators, B, listed in Table 3, as mentioned in 3.A.1.a. When a solid has no acid sites of $H_0 \leqq pK_{BH^+}$, the color of the basic indicator does not change. In this case, if a standard solution of Brönsted acid in benzene is added gradually, the color of the basic indicator on the surface will change to the color of its conjugate acid. The color change is taken as the end-point of the titration. At the end-point, the acid strength H_0 of the resultant solid, which was formed by the addition of Brönsted acid to the original solid, is equal to the pK_{BH^+} of the indicator used. As basic sites are neutralized by Brönsted acid at the end-point, the titers of Brönsted acid required for the neutralization should give a measure of the number of basic sites (basicity) on the surface. During the titration, stronger basic sites are neutralized earlier and weaker ones later and weaker basic sites require stronger acids for the neutralization. Therefore, it can be assumed that the weakest basic sites have been finally neutralized by an acid having an acid strength of $H_0 = pK_{BH^+}$.

The proton donating ability of the solid at the end-point of titration is considered to be either due to the conjugate acids which were formed by the proton transfer from Brönsted acid solution to the original solid or due to the Brönsted acid which was physically adsorbed on the surface during the titration. The proton donating ability of both the conjugate acid and the Brönsted acid used for titration is assumed to be equal. Since the weakest basic sites form the strongest conjugate acids, the acid strength, H_0, of the conjugate acid of the weakest basic sites should be equal to or greater than the pK_{BH^+} of the indicator used.

Thus, "basic strength H_0" of basic sites is defined as the acid strength, H_0, of the conjugate acids of the basic sites. We shall express the function H_0 used previously by "acid strength H_0" in the case where it is necessary to distinguish between this and "basic strength H_0". As the basicity at "basic strength H_0" $= pK_{BH^+}$ is easily determined by using a series of basic indicators as described above, the distribution of basic strength of a solid as well as that of acid strength can be expressed by a common scale of acid-base strength. The use of the function H_0 for basic strength is neither surprising nor curious, because the basic strengths of the organic compounds in homogeneous solution are usually expressed by pK_a's of the conjugate acids. It should be noticed that the measurement of the basicity when the basic strength H_0 is equal to or greater than a pK_{BH^+} value is possible only when there are no acid sites whose acid strength is equal to or less than the same pK_{BH^+} value.

Figure 8 shows the results of acid-base strength distributions on a common H_0 scale of some solids [34], where the acidity at various acid strengths was measured by the method described in 3.A.1.a., while the basicity at various basic strengths by titrating the solid suspended in benzene with a 0.1 N solution of trichloroacetic acid in benzene using the same indicators as those used for acidity measurement [34]. The acidity at an H_0 value shows the number of acid sites whose acid strength is equal to or less than the H_0 value and the basicity at an H_0 value shows the number of basic sites whose basic strength is equal to or greater than the H_0 value.

Titanium oxide exhibited high basicity at basic strength $H_0 \geqq 1.5$, but low acidity at acid strength $H_0 \leqq 6.8$, while $MgSO_4$ showed high acidity

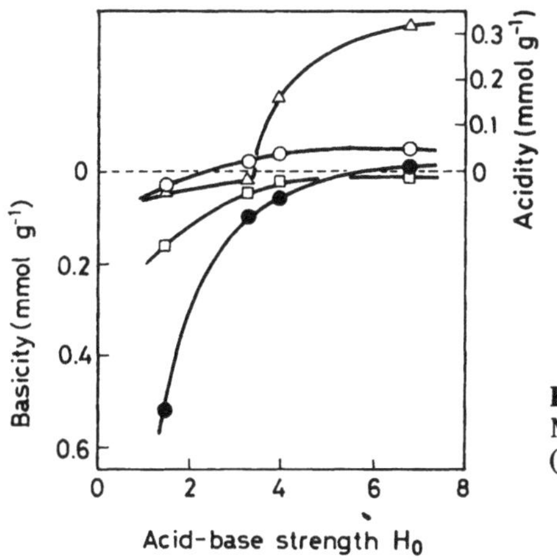

Figure 8. Acid-base strength distribution of MoO_3 (○), TiO_2 (●), V_2O_5 (□), and $MgSO_4$ (△)

at acid strength $H_0 \leqq 6.8$ but low basicity at basic strength $H_0 \geqq 1.5$. Both the acidic and basic sites of equal strength do not coexist on the same solid surface. Therefore, the measurement is to determine a significant acid-base strength distribution of a given solid in the full range of the H_0 scale.

2. Significance and Usefulness of $H_{0,\,max}$

As seen in Figure 8, the acid-base strength distribution curves intersect at a point on the abscissa where acidity = basicity = 0. Hence, the strongest H_0 value of the acid sites is equal to the strongest H_0 value of the basic sites. $H_{0,\,max}$ is defined as the H_0 value at a point of intersection, which expresses the equal strongest H_0 value of both acidic and basic sites. Each $H_{0,\,max}$ value, which was determined from a point of intersection of each acid-base strength distribution curve and the abscissa, is given in Table 6. A unique $H_{0,\,max}$ is found for every solid. The $H_{0,\,max}$ value changes on calcination. For example, the $H_{0,\,max}$ values of $MgSO_4 \cdot 7H_2O$ calcined at 573, 673, 793, and 943 K are −3—0, 3.4, 3.3, and 3.5, respectively. Since $MgSO_4 \cdot 7H_2O$ without calcination shows $H_{0,\,max}$ of 6.0, the solid calcined at 573 K has the minimum $H_{0,\,max}$ [34]. On the other hand, $H_{0,\,max}$ of TiO_2 does not much change on calcination and the variances were less than 1.0 unit of H_0 scale [34].

$H_{0,\,max}$ can be regarded as a practical parameter to represent an acid-base property on solids, which is sensitive to the surface structure. A solid with a large positive $H_{0,\,max}$ has strong basic sites and weak acidic sites. Thus, basic sites mainly play an important role. On the other hand, a solid with a large negative $H_{0,\,max}$ has strong acidic sites and weak basic sites. In this case, acid sites become often important.

The role of $H_{0,\,max}$ as a structurally sensitive parameter should be distinguished from the intrinsic parameters such as an effective charge (γ) on a combined oxygen ($O^{\gamma-}$), which represents the overall acid-base property of solids [36]. In Figure 9, $H_{0,\,max}$ values are compared with γ calculated from Sanderson's electronegativities [37]. The correlation of $H_{0,\,max}$ with γ is

Table 6. Acidities, basicities, $H_{0,max}$, and γ of various solids

Solids	Basicity, mmol g^{-1}				Acidity, mmol g^{-1}					$H_{0,max}$	γ
	$H_0 \geqq 1.5$	$H_0 \geqq 3.3$	$H_0 \geqq 4.0$	$H_0 \geqq 6.8$	$H_0 \leqq -3$	$H_0 \leqq 1.5$	$H_0 \leqq 3.3$	$H_0 \leqq 4.0$	$H_0 \leqq 6.8$		
activated Al_2O_3	0.30	0.10	0.06	0.01	0	0	0	0	0	8.0	0.31
γ-Al_2O_3	0.43	0.22	0.16	0.01	0	0	0	0	0	7.2	0.31
ZrO_2	0.08	0.03	0.02	0.01	0	0	0	0	0	9.5	0.44
TiO_2	0.52	0.10	0.06	0	0	0	0	0	0.01	5.5	0.39
B_2O_3	0.27	0.04	0.02	0	0	0	0	0	0.01	8.0	0.24
ZnO		0.07	0.05	0	0	0	0	0	0.005	6.4	0.29
BaO		0.09	0.06	0.05	0	0	0	0	0	15	0.67
MoO_3	0.03	0	0	0		0	0.02	0.04	0.05	2.1	0.28
$MgSO_4$ [a]	0.03	0.02	0		0	0	0	0.16	0.32	3.4	0.20
$MgWO_4$	0.23	0.03	0		0	0	0	0	0.05	4.0	0.34
Ta_2O_5	0				0	0.02	0.05	0.06	0.07	2.0	0.39
WO_3	0				0	0.01	0.06	0.04	0.14	1.3	0.31
Bi_2O_3	0.05	0.07	0.03	0	0	0	0	0	0.003	6.6	0.20
V_2O_5	0.16	0.05	0.02	0.01	0	0	0	0	0	8.5	0.31
SrO	0.14	0.04	0.02	0.01	0	0	0	0	0	9	0.62
$ZnSO_4 \cdot H_2O$	0				0	0.07	0.07	0.20	0.30	1.5	0.14
$CuSO_4$					0	0.11	0.22	0.22	0.28	0	0.16
$CaSO_4 \cdot 0.5\ H_2O$	0.03	0.01	0.01	0	0	0	0	0	0.003	6.0	0.27
$MnSO_4$ [b]	0				0	0.14	0.28	0.30	0.30	0.2	0.19
$Al_2(SO_4)_3$	0				0	0.13	0.20	0.43	0.67	−1.0	0.14
$AlPO_4$	0.61	0.08	0.07	0	0	0	0	0	0.01	6.0	0.21
$Zn_3(PO_4)_2 \cdot 4\ H_2O$	0.64	0.04	0.04	0	0	0	0	0	0.02	5.2	0.20
$CaWO_4$	0.07	0.01	0.02	0	0	0	0	0	0.01	5.0	0.41
$Na_2WO_4 \cdot 2\ H_2O$	0.50	0.03	0.06	0.03	0	0	0	0	0	12	0.58
$CaCO_3$	0.14	0.03	0.02	0	0	0	0	0	0.01	6.0	0.33
$Ba(OH)_2$	0.13	0.03	0.02	0.01	0	0	0	0	0	9.0	0.46
$Mg(OH)_2$	0.09	0.03	0.02	0	0	0	0	0	0	6.8	0.32
$NiSO_4$ [c]	0.46	0			0	0	0.44	0.43		2.0	0.19

[a] $MgSO_4 \cdot 7\ H_2O$ was calcined at 673 K, 3 h. [b] $MnSO_4$ was calcined at 523 K, 4 h. [c] $NiSO_4 \cdot 7\ H_2O$ was calcined at 573 K, 4 h.

rough, since $H_{0,max}$ on the solid of a definite γ deviates sensitively according to surface conditions. However, the deviation seems not to serious, and is within a restricted range.

No practical parameters other than zero point of charge (ZPC) has been introduced to evaluate the acid-base properties of various solids [36]. It should be noticed that $H_{0,max}$ is a more useful parameter than ZPC which does not directly express the effect of crystalline water or traces of moisture,

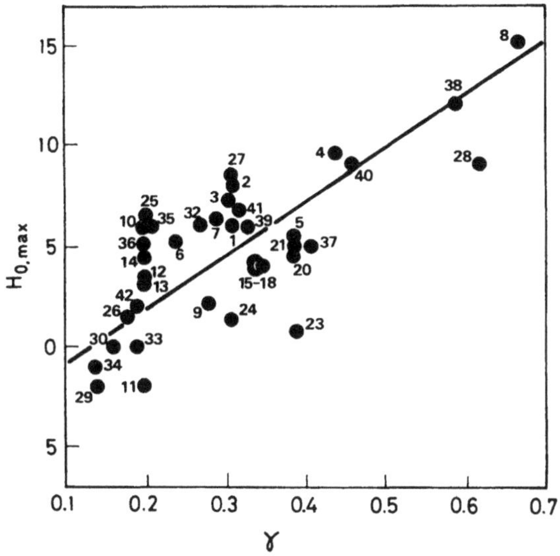

Figure 9. Comparison of $H_{0,max}$ values of solids to effective negative charge (γ). 1: Al_2O_3, 2: active Al_2O_3, 3: γ-Al_2O_3, 4: ZrO_2, 5: TiO_2, 6: B_2O_3, 7: ZnO, 8: BaO, 9: MoO_3, 20: TiO_2 calcined at 573 K, 21: TiO_2 calcined at 773 K, 22: ThO_2, 23: Ta_2O_5, 24: WO_3, 25: Bi_2O_3, 26: Sb_2O_3, 27: V_2O_5, 28: SrO. Other numbers: sulfates, phosphates, carbonates, hydroxides, etc. [35]

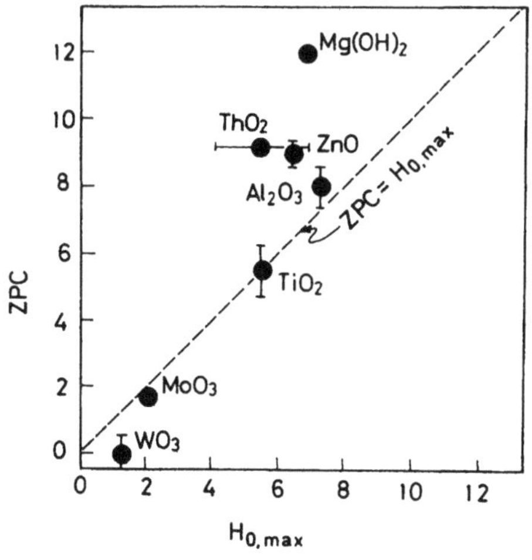

Figure 10. Comparison of $H_{0,max}$ to zero point of charge (ZPC)

because $H_{0,max}$ is determined in a non-polar solvent, while ZPC is determined generally in aqueous solution. $H_{0,max}$ expresses the actual surface property which is sensitive to the structure and water or moisture content, in addition to the intrinsic nature of the compounds.

In Figure 10, $H_{0,max}$ is compared to ZPC. The correlation is not particularly good. This is considered to be due to the difference in the nature of the solvents used for each measurement, since non-polar solvent is used for the measurement of $H_{0,max}$, whereas water is used for the measurement of ZPC. If the water content of solids is the same in each system, $H_{0,max}$ should be equal to ZPC.

4. Mechanism of the Generation of Acidity and Basicity

A. Metal Oxides

The metal oxides whose acidic and/or basic properties have been determined are
MgO [38–40], CaO [38–40], SrO [39, 41], BaO [39, 42], ZnO [21, 43], B_2O_3 [2], Al_2O_3 [2], Y_2O_3 [44], La_2O_3 [44], CeO_2 [44], ThO_2 [45], SiO_2 [2], SnO_2 [46], TiO_2 [43, 47], ZrO_2 [30], As_2O_3 [48], Sb_2O_5 [48], Bi_2O_3 [48], V_2O_5 [2], Cr_2O_3 [2], MoO_3 [7], and WO_3 [18].
Structural studies of acidic and basic sites have been made only for Al_2O_3, MgO, and CaO.

1. Alumina

Acidic and basic sites on the surface of alumina can be pictured according to the scheme shown below [49].

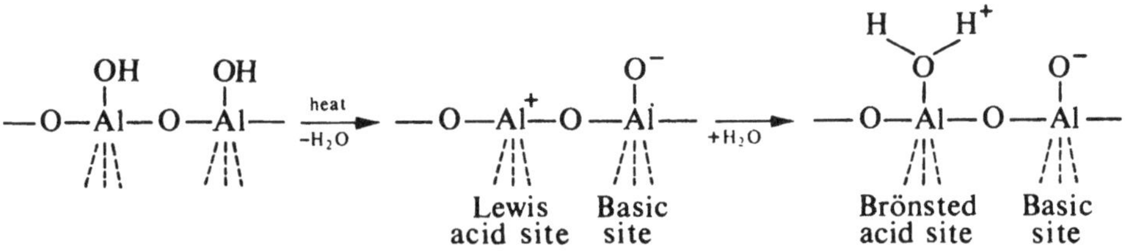

The Lewis acid site is visualized as an incompletely coordinated aluminum atom formed by dehydration, and the weak Brönsted site as a Lewis site which has adsorbed moisture, while the basic site is considered to be a negatively charged oxygen atom. Calcined alumina does not exhibit any basic property by the indicator method mentioned in 3.B.1, but does when it is exposed to moisture [50]. The effect of moisture may be interpreted as follows. The electronegativity of the Lewis site is weakend by adsorption of water, since an electron pair from the oxygen atom of the water molecule is donated to the Lewis site. Thus, the negative charge of oxygen at a basic site becomes higher when water is adsorbed on a Lewis site due to a weaker inductive effect of the aluminum atom.

For the hydroxyl ions as acidic and basic sites on the surface of γ-alumina heat-treated at 1073 K, a detailed scheme was presented by Peri [51]. As shown in Figure 11, five types of isolated hydroxyl groups labelled A to E are identifiable. Each of them has a different nearest neighbor configuration, and the groups cover 10 percent of the surface shown (5 out of a possible 49 sites). The five absorption maxima which are observed in the infrared spectra (at 3800, 3780, 3744, 3733 and 3700 cm^{-1}) are associated with sites A, D, B, E and C, respectively. Each site has a different local charge density: type A with 4 O^{2-} ions as neighbors is the most negative (a basic site), and type C lacking neighbours is the most positive (an acidic site) [51].

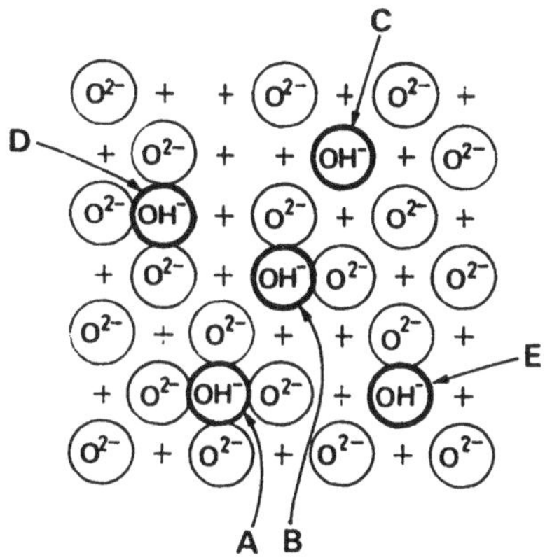

Figure 11. Suggested scheme for acidic and basic hydroxyl groups on γ-Al_2O_3. Letters A to E identify the different types of isolated hydroxyl ions, and "+" denotes an Al^{3+} ion on the layer below the surface

The state of hydroxyl group over γ-Al_2O_3 was also studied by NMR. Three dielectric relaxations were found and inferred to be related to loosely adsorbed water, a surface hydroxyl group which can be easily dehydrated, and some other kind of hydroxyl group with a weak interaction with its neighbors, respectively [52].

There is a suggestion that the acid sites of alumina which was dehydrated at high temperature may be formed by lattice distortions [53, 54]. Recently, it has been reported that acidic and basic strength of alumina is controlled by the distance between acidic site (Al^{3+}) and basic site (O^{2-}) which changes depending on the preparation method [55].

It is well known that the addition of halogen to Al_2O_3 or SiO_2 enhances the acid strength of the solid, whereas the addition of potassium or sodium enhances the basic strength [2]. Recently, the effect has been studies by a semi-empirical SCF-MO method using the parametrization of CNDO type. In the case of halogenated silica gel, where $(OH)_3Si{-}O{-}Si(OH)_2X$ (X = OH, F, Cl, and Br) was used as a model, the bonding interaction between the hydrogen of the OH group and halogen was concluded to be essential to the enhancement of acid strength by halogenation of silica gel [56]. In the case of sodium added to alumina, the electron density of oxygen becomes higher in a Al—O—Na group than in a Al—O—H group [57].

2. Calcium Oxide and Magnesium Oxide

Calcium oxide shows a maximum basicity when $Ca(OH)_2$ is calcined at 773 K and a maximum amount of reducing sites (one electron donor sites) when calcined at 973 K. The activity for conversion of benzaldehyde into benzylbenzoate which is known to be catalyzed by bases correlates well with the basicity, while polymerization of styrene which is considered to proceed via an anion radical intermediate correlates with the amount of reducing sites (sites to form anion radicals) [58]. It was recently found that CaO and MgO evacuated at 1073 K and 1373 K respectively, exhibited activities for hydrogenation of olefins [59–61]. From the infrared study on adsorbed acidic molecules such as alcohol, aldehyde, chloroform, and toluene, a large part (90 percent) of the basic sites on CaO was concluded to be the surface O^{2-} which is strongly basic and a small part of the basic sites to be isolated OH group which is weakly basic [58]. On the other hand, the reducing sites were concluded to be point defects or edge dislocations by an ESR study [58].

Evidence for the above conclusions about the basic sites and the reducing sites will be described briefly. The infrared absorption band at 3765–3770 cm^{-1} of CaO which appears by calcination at around 673 K (see Figure 12) shifts

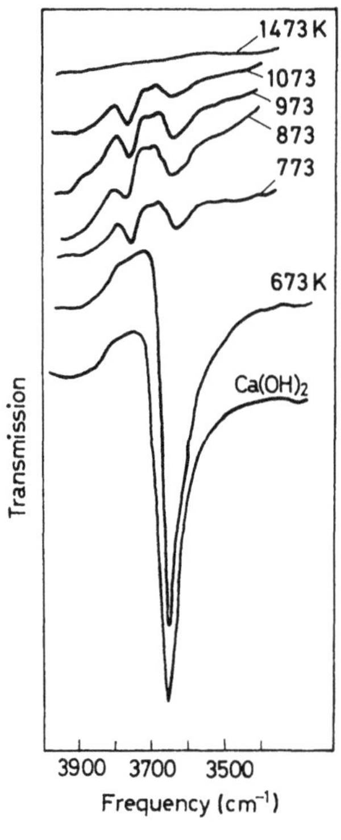

Figure 12. Infrared spectra of $Ca(OH)_2$ calcined in air at various temperatures for 3 hr

to a lower frequency and becomes broad by the adsorption of benzaldehyde (Figure 13) and isopropyl alcohol. These facts indicate that the band is due to isolated OH on the surface. Since the band appears at a higher frequency than 3750 cm^{-1} at which the absorption of neutral OH group of silanol is usually observed, the nature of the isolated OH is considered to be basic. This basic OH group is only a part of the basic sites on dehydrated CaO, because the amount of basic sites of the sample calcined at 773 K is so large that benzoic acid covers the surface almost completely, despite the fact that a large amount of water is dehydrated. Hence, a large part of basic sites is considered to be surface O^{2-}.

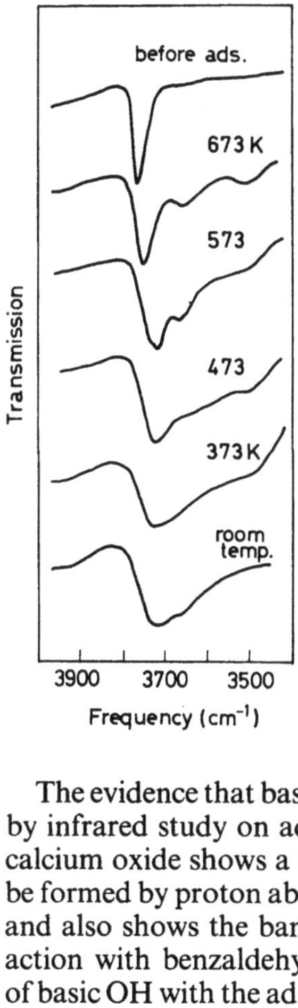

Figure 13. Infrared spectra of benzaldehyde adsorbed on CaO calcined at 773 K in vacuo and evacuated at various temperatures

The evidence that basic sites consist of surface O^{2-} and basic OH is provided by infrared study on adsorbed acidic molecules. As shown in Figure 13, the calcium oxide shows a new band of OH at 3650 cm^{-1} which is considered to be formed by proton abstraction by surface O^{2-} from adsorbed benzaldehyde, and also shows the band of weakly basic OH which is in fairly strong interaction with benzaldehyde. In the case of isopropyl alcohol, the interaction of basic OH with the adsorbate is not so strong, but the OH band at 3640 cm^{-1} which was formed by proton transfer from isopropyl alcohol to surface O^{2-} was mainly observed. On the other hand, the basic OH band does not change when chloroform is adsorbed, and only the OH band formed by proton abstraction of surface O^{2-} was observed. The nature of basic sites for the above two acidic substances might be pictured as in Figure 14.

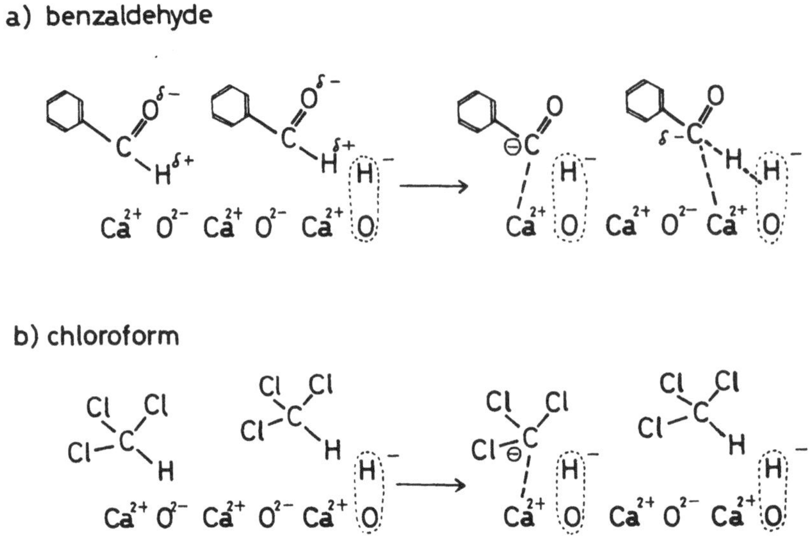

Figure 14. Model of adsorption state of benzaldehyde and chloroform on CaO. OH^- in broken oval denotes surface hydroxide ion.

In the case of the adsorption of very weakly acidic toluene ($pK_a = 35$), the observed small broadening of the band at 3765 cm^{-1} is due to physical adsorption, indicating that the surface O^{2-} cannot accept a proton from such a weak acid. The fact that the newly formed OH band at 3640–3650 cm^{-1} observed in cases of benzaldehyde, isopropyl alcohol, and chloroform was not observed in this case does confirm that the OH was formed by proton transfer from adsorbates to surface O^{2-}, strongly basic site.

The types of basic sites on the surface of partially dehydrated magnesium hydroxide was proposed by Krylov et al. [62] on the basis of the study on adsorption and isotope exchange of carbon dioxide. These are strongly basic sites derived from the O^{2-} ions adjacent to the surface OH groups, and surface OH groups constituting weakly basic sites.

The maximum amount of the anion radical of nitrobenzene which was formed on CaO calcined in a vacuum at 773 K occupies only about 0.1 percent of the surface. If we assume that the radicals are homogeneously dispersed on the surface, the distance between the radicals should be several hundred Å. Since, however, the distance is calculated to be about 13 Å according to

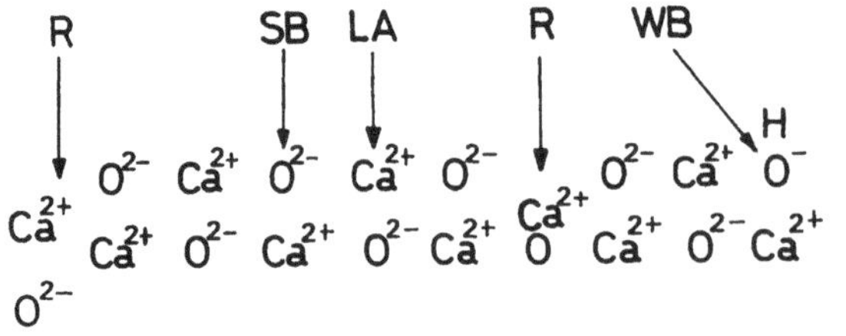

Figure 15. Surface model of CaO. R: reducing site, SB: strong basic site, LA: Lewis acid site, WB: weak basic site

Van Vleck's equation, the reducing site is considered to be localized on the surface. The existence of the localized reducing site and the fact that the anion radical was not formed on commercial calcium oxide calcined at 773 K, but formed when calcium hydroxide, which was prepared by adding water to the oxide, was heated at 773 K seem to indicate that the reducing site is associated with the surface state after dehydration. A possible reducing site might be a terminal of dislocation or a point defect appearing on the surface by dehydration of the hydroxide. The surface model of CaO is shown in Figure 15.

B. Mixed Metal Oxides

A new hypothesis regarding the acidity generation of binary oxides has been proposed by Tanabe et al., which predicts what kinds of binary oxides will show acidic properties (Brönsted or Lewis acid) and gives an insight on the structure of the acid sites [63]. According to the hypothesis, the acidity generation is caused by an excess of a negative or positive charge in a model structure of a binary oxide. The model structure is pictured according to the following two postulates: i) The coordination number of a positive element of a metal oxide, C_1, and that of a second metal oxide, C_2, are maintained even when mixed; ii) The coordination number of a negative element (oxygen) of a major component oxide is retained for all the oxygens in a binary oxide.

For example, the structure of TiO_2—SiO_2, where TiO_2 is the major component oxide, and that of SiO_2—TiO_2, where SiO_2 is the major component oxide, are shown in Figures 16a and 16b. In Figures 16a and 16b, the co-

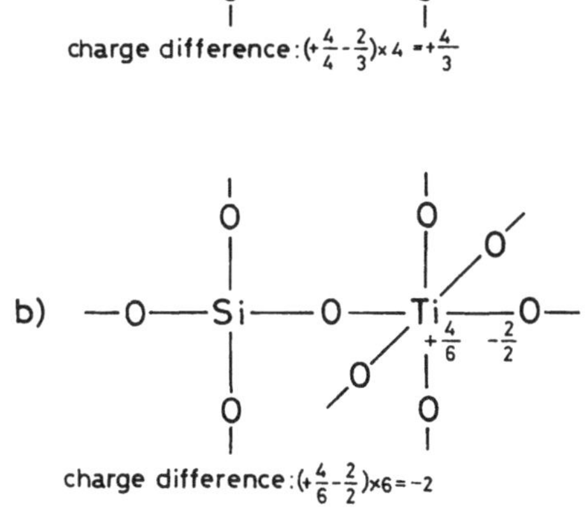

Figure 16. Model structures of TiO_2—SiO_2 pictured according to postulates i) and ii) **a** when TiO_2 is major oxide; **b** when SiO_2 is major oxide

ordination numbers of the positive elements in the component single oxides remain 4 for Si and 6 for Ti when they are mixed, whereas those of the negative elements should be 3 and 2 respectively, according to postulates i) and ii) above. In the case of Figure 16a, the four positive charges of the silicon atom are distributed to four bonds, i.e., a positive charge is distributed to each bond, while the two negative charges of the oxygen atom are distributed to three bonds, i.e., $-2/3$ of a valence unit is distributed to each bond. The difference in charge for one bond is $+1 - 2/3 = +1/3$, and for all the bonds the valence unit of $+1/3 \times 4 = +4/3$ is excess. In this case, the Lewis acidity is assumed to appear upon the presence of an excess of positive charge. In Figure 16b, four positive charges of the titanium atom are distributed to six bonds, i.e. $+4/6$ of a valence unit to each bond, while two negative charges of the oxygen atom are distributed to two bonds, i.e. a negative charge to each bond. The charge difference for each bond is $+4/6 - 1 = -1/3$, and for all the bonds the valence unit of $-1/3 \times 6 = -2$ is excess. In this case, Brönsted acidity is assumed to appear, because two protons are considered to associate with six oxygens to keep electric neutrality. In any case, TiO_2—SiO_2 is expected to show an acidic property because of the excess of a positive or negative charge. In fact, it exhibited a very high acidity [9].

Let us examine another example. In ZnO—ZrO_2, there is no excess charge in any part of its composition according to our model structure written by postulates i) and ii), as illustrated in Figure 17. Therefore, the binary oxide is not expected to show any acidic property. This prediction agrees with an

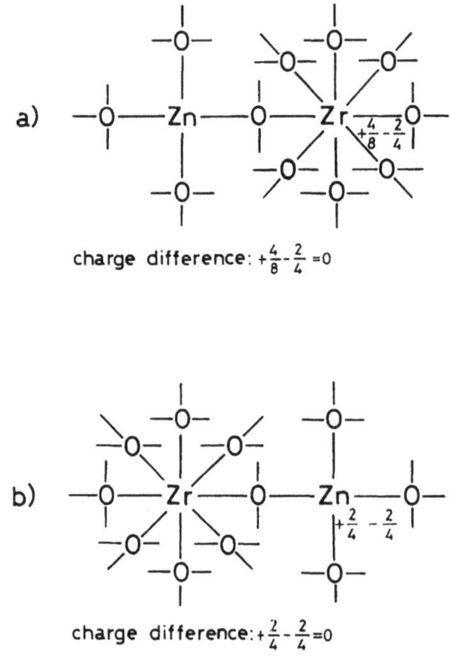

Figure 17. Model structures of ZnO—ZrO_2 pictured according to postulates i) and ii). **a** when ZnO is major oxide; **b** when ZrO_2 is major oxide

experimental result that ZnO—ZrO_2 does not show acidity larger than the sum of the acidities of the component oxides [64].

The validity of the hypothesis was examined for 31 kinds of binary oxides. The case where the hypothesis predicts that a binary oxide should generate acidity is marked by an open circle in the fifth column of Table 7. On the other hand, the case where a binary oxide should not generate acidity is shown by a cross in the same column. The experimental results cited from the literature are shown in the next column, where open circles mark when

Table 7. Validity of hypotheses for acidity prediction
V: valence of positive element, *C*: coordination number of positive element

Mixed-Oxides		$\alpha = V/C$		Acidity increase			Validity of hypotheses	
1	2	α_1	α_2	Hypotheses		Experimental results		
				Thomas	Tanabe et al.		Thomas	Ours
TiO_2	CuO	4/6	2/4	×	○	○	×	○
TiO_2	MgO		2/6	○	○	○	○	○
TiO_2	ZnO		2/4	×	○	○	×	○
TiO_2	CdO		2/6	○	○	○	○	○
TiO_2	Al_2O_3		3/6	○	○	○	○	○
TiO_2	SiO_2		4/4	×	○	○	×	○
TiO_2	ZrO_2		4/8	×	○	○	×	○
TiO_2	PbO		2/8	×	○	○	×	○
TiO_2	Bi_2O_3		3/6	○	○	○	○	○
TiO_2	Fe_2O_3		3/6	○	○	○	○	○
ZnO	MgO	2/4	2/6	×	○	○	×	○
ZnO	Al_2O_3		3/6	×	×	○	×	×
ZnO	SiO_2		4/4	○	○	○	○	○
ZnO	ZrO_2		4/8	○	×	×	×	○
ZnO	PbO		2/8	○	○	×	×	×
ZnO	Sb_2O_3		3/6	×	×	×	○	○
ZnO	Bi_2O_3		3/6	×	×	×	○	○
Al_2O_3	MgO	3/6	2/6	○	○	○	○	○
Al_2O_3	B_2O_3		3/3	○	○	○	○	○
Al_2O_3	ZrO_2		4/8	×	×	○	×	×
Al_2O_3	Sb_2O_3		3/6	×	×	×	○	○
Al_2O_3	Bi_2O_3		3/6	×	×	×	○	○
SiO_2	BeO	4/4	2/4	○	○	○	○	○
SiO_2	MgO		2/6	×	○	○	×	○
SiO_2	CaO		2/6	×	○	○	×	○
SiO_2	SrO		2/6	×	○	?	?	?
SiO_2	BaO		2/6	×	○	?	?	?
SiO_2	Ga_2O_3		3/6	×	○	○	×	○
SiO_2	Al_2O_3		3/4, 3/6	○	○	○	○	○
SiO_2	Y_2O_3		3/6	×	○	○	×	○
SiO_2	La_2O_3		3/6	×	○	○	×	○
SiO_2	ZrO_2		4/8	○	○	○	○	○
ZrO_2	CdO	4/8	2/6	×	○	○	×	○

Thomas' hypothesis 15/31 = 48% correct.
Tanabe et al.'s hypothesis 28/31 = 90% correct.

the acid amounts at a certain acid strength per unit surface area of any binary oxides is larger than the sum of the acid amounts at the same acid strength divided by the sum of the surface areas of the component single oxides, while crosses mark when no acidity is generated. The results indicate that new acid sites which differ from those of single oxides are created on the surface of 25 species of binary oxides. Cases where the result predicted by the hypothesis agrees with the experimental result is marked by open circles in the last column of Table 7, and the cases of disagreement, by a cross. As may be seen in the table, an agreement of the prediction of Tanabe et al.'s hypothesis with the experimental results was found for 28 of the 31 kinds of binary oxides. Thus, the validity of their hypothesis is 90 percent. However, the validity of an old but well known Thomas' hypothesis [65] is only 48 percent.

Though Thomas' hypothesis cannot be applied to Lewis acids [65], Tanabe et al.'s hypothesis also predicts the type of acid sites (Brönsted or Lewis), as has been mentioned above. According to the hypothesis, TiO_2—ZnO should show Brönsted acidity when TiO_2 is a major component oxide and Lewis acidity when ZnO is a major component oxide. An infrared study of pyridine adsorbed on the binary oxide revealed that TiO_2 including 5 percent ZnO exhibited Brönsted acidity alone, while ZnO including 5 percent TiO_2 exhibited Lewis acidity alone [43].

As may be seen in the model structures (Figures 16 and 17) pictured according to postulates i) and ii), the hypothesis is applicable to chemically-mixed binary oxides, but not to mechanically-mixed oxides. Since the binary oxides given in Table 7 were prepared by calcining mixtures of co-precipitated hydroxides at a high temperature (770 K), they are not mechanically-mixed oxides. The x-ray diffraction diagrams of the binary oxides showed no or only weak diffraction lines, and almost all of them were to be amorphous. Thus, the structures are different from those of the component single oxides.

The Tanabe et al.'s hypothesis predicts which combinations of oxides in the periodic table will generate acidity and at what compositions the Brönsted or Lewis acidity will appear, but it does not predict the acid strength. The prediction of acid strength will be described in 4.B.

Recently, Seiyama has presented a different model for the acidity generation of binary metal oxides [66]. He assumes that acidity appears at the boundary where two oxides contact. In the case of ZnO—ZrO_2 (Figure 18), the oxygen by which ZnO is combined with ZrO_2 has a negative charge, since $+2/4$ charge of Zn and $+4/8$ charge of Zr are distributed to the boundary oxygen which has two negative charges and, hence, the charge difference, Δ, around the boundary oxygen becomes -1. Therefore, Brönsted acid site should appear according to the same argument as mentioned in Tanabe et al.'s hypothesis. In fact, the acidity generation of binary oxides such as ZnO—ZrO_2 strongly depends on the preparation method. The acidity generation of ZnO—ZrO_2 was observed in Seiyama's experiment [66], but not in Tanabe et al.'s experiments [63]. Seiyama's model may be applied to binary oxides in which chemical mixing is not adequate, while Tanabe et al.'s model can be applied to amorphous binary oxides.

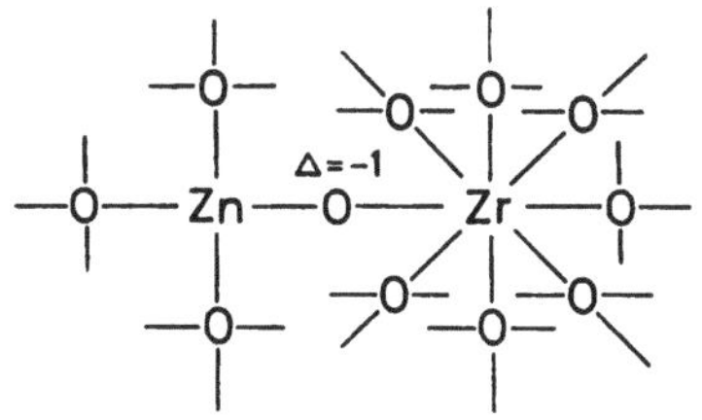

charge difference: $(+\frac{2}{4}+\frac{4}{8})-2=-1\ (\Delta)$

Figure 18. Model of acidity generation at the boundary between two oxides [66]

The structures of acidic and basic sites on decationized zeolites (NH_4-X and NH_4-Y) have been studied by many workers. The model proposed by Hall et al. [67] is shown in the diagram below. On this scheme, an absorption

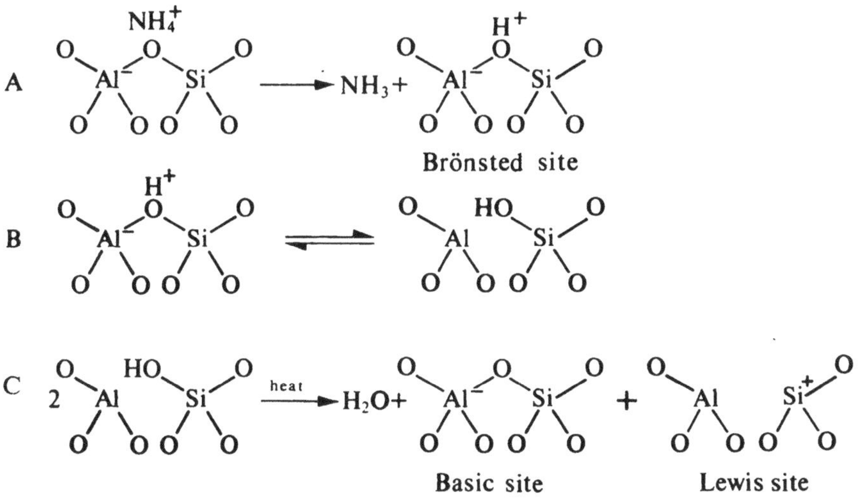

band which was found at 3660 cm^{-1} is attributed to the OH stretching vibration of Si—OH. Addition of ammonia decreases the intensity of this band, and simultaneously that characteristic of NH_4^+ appears. Upon evacuation at 773 K, the OH band further decreases in intensity and strongly electrophilic sites (Lewis acid sites) can be identified. In diagram B the aluminum atom of three-fold coordination is also considered to be a rather weak Lewis acid site. Turkevich et al. [68] observed the ESR spectra of trinitrobenzene adsorbed on decationized zeolite NH_4-Y, and found that electron donor sites (Lewis sites in the broad sense) increase with increasing dehydration temperatures. The diagram below illustrates the mechanism proposed by Ward for the formation of acid sites on zeolite Y, exchanged with a divalent alkaline earth metal [69].

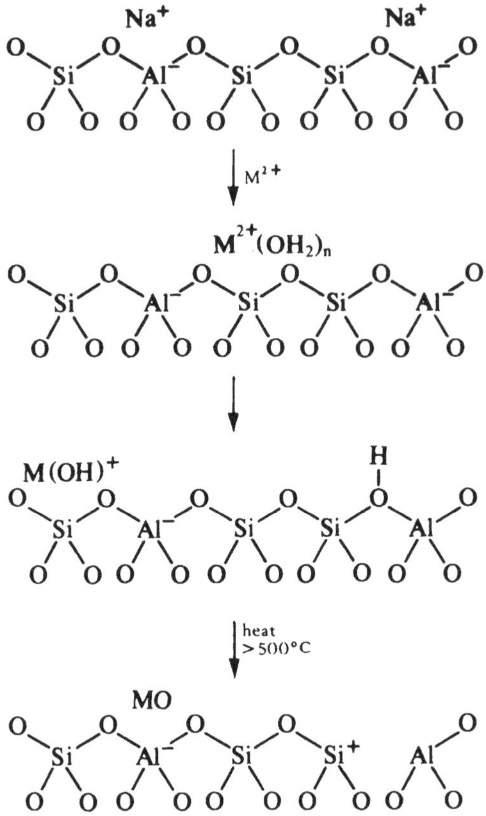

C. Metal Salts

Among metal salts, the structure of acidic and basic sites of nickel sulfate has been extensively studied. On the basis of studies with infrared, electron spin resonance, x-ray, nuclear magnetic resonance and Mössbauer effect techniques, Tanabe et al. proposed that the acid site is formed by an empty orbital of the nickel ion which appears in an incompletely dehydrated metastable transition structure (see (II) in Figure 19) [70–72]. This configuration is intermediate between the monohydrate (I) and anhydrous (III) forms. In this transitional form, nickel is pentavalent, and has a vacant sp^3d^2 orbital. This vacant orbital, and the resultant affinity for an electron pair, accounts for the Lewis acid properties of nickel sulfate and its catalytic activity. The suggested structure is both strained and unstable, although the crystal network and the retained water do have some stabilizing influence. The Brönsted acidity arises from two sources. One is the water which is coordinated directly with a nickel ion in the above transitional form. The nickel tends to attract the oxygen atom, thus freeing a hydrogen ion. The other source is the surface water, acidified by the inductive effect of the neighboring cationic Lewis acid sites. The temperature dependence of the Brönsted and Lewis acidity,

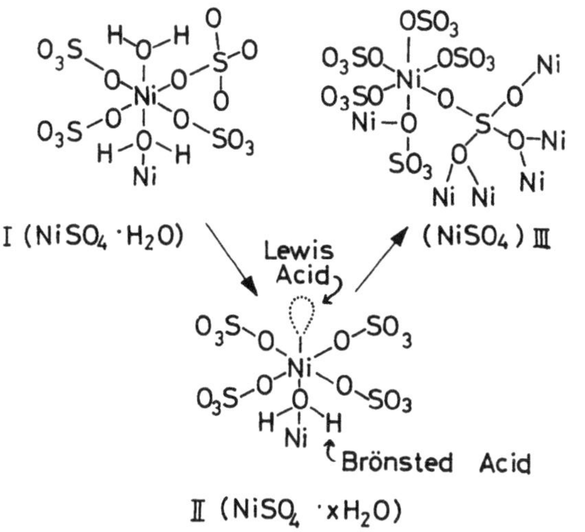

Figure 19. Acid sites on $NiSO_4 \cdot xH_2O$

as derived from the infrared spectra of adsorbed pyridine, is shown in Figure 4. Brönsted acid first appears when the vacant orbital of the nickel ion is formed by dehydration, and the amount increases progressively with increase in dehydration temperature. Since, however, the amount of water of hydration decreases as the temperature of heat treatment is increased, a temperature is eventually reached at which the Brönsted acidity begins to decline. Lewis acidity also increases with increasing heat treatment temperature, but only begins to decline when the metastable structure with the vacant orbital collapses and changes to the stable anhydrous structure at higher temperatures. The sum of the two acidity curves gives the total acid amount which can be measured by the amine titration method (3.A.1.a.). $MgSO_4$, $MnSO_4$, $FeSO_4$, $CoSO_4$, $CuSO_4$ and $ZnSO_4$, which have monohydrate structures similar to that of nickel sulfate, are also considered to have acid centers which are comparable in structure with those of nickel sulfate [73].

Compression of sulfate hydrates increases their surface acidities in the cases of $CdSO_4 \cdot 8\ H_2O$, $Ce_2(SO_4)_3 \cdot 8\ H_2O$, $Fe_2(SO_4)_3 \cdot x\ H_2O$, $Al_2(SO_4)_3 \cdot 18\ H_2O$, and $KHSO_4$,
while $BaSO_4$, $PbSO_4$, K_2SO_4, and $MgSO_4 \cdot 7\ H_2O$
exhibit no acidity at all [74], even at pressures 3000 kg cm^{-2}. Aluminum sulfate increases in acidity upon compression, but does not produce any acid sites at an acid strength of $H_0 = -3$ by application of pressures as high as 4,200 kg cm^{-2}. Thus compression does not generate as strong acid sites as does heat treatment. The weak acid sites on aluminum sulfate hydrate formed on compression are mainly Brönsted acid sites due to removal of anion water. The octahedron surrounding the Al^{3+} ion is deformed so that the distances between the Al^{3+} ion and the coordinated water become unequal, giving rise to two types of coordinated water, H_2O (α) and H_2O (β). O (α) is closer to Al^{3+} than before compression, thus increasing its covalent character, so that H_2O (α) tends to develop a protonic positive charge [74].

Strontium carbonate and $BaCO_3$ were found to increase their basicities on compression [75]. The basicity increase is considered due to the change in symmetry of the carbonate ion.

5. Chemical Control of Acidic and Basic Property

The acidic and basic properties as well as the activity of a catalyst are varied by different preparative methods and pretreatment conditions. The effects of preparation method and pretreatment on the acid-base properties will be described first in this section. The principal factors which control chemically the acid-base properties will be discussed.

A. Effect of Preparation Method and Pretreatment

A large effect of the preparation method on the acid-base properties is mainly caused by the difference in the kinds of starting materials used in catalyst preparation, and the difference in precipitation procedures of the starting materials. Acid-base properties of a catalyst surface are also influenced by the pretreatment condition.

1. Difference in Starting Materials

The surface property and catalytic activity of MgO (I) prepared by evacuating magnesium hydroxide, $Mg(OH)_2$, is different from those of MgO (II) prepared by evacuating magnesium carbonate hydroxide, $4\,MgCO_3 \cdot Mg(OH)_2 \cdot 5\,H_2O$. In the alkylation of phenol with methanol over MgO (I), phenol is selectively alkylated at the ortho position to produce *o*-cresol and 2,6-xylenol. The orthoselectivity of MgO (I) is higher than 98 percent, while that of MgO (II) is 73 percent [76]. For the isomerization of cis-2-butene, the maximum activity

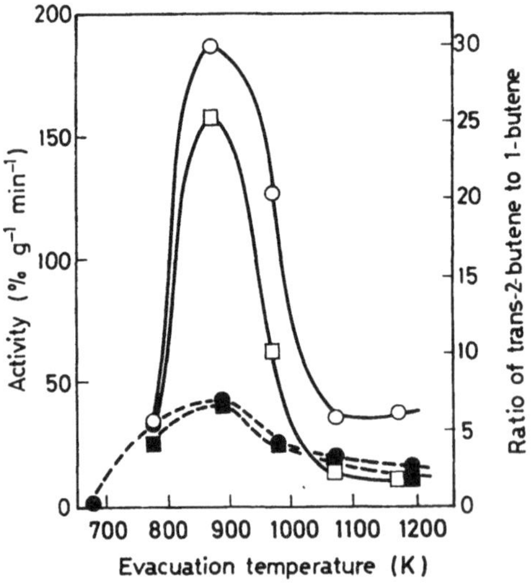

Figure 20. Activity and selectivity for isomerization of cis-2-butene. Activity (○) and selectivity (□) of MgO (I). Activity (●) and selectivity (■) of MgO (II)

and selectivity (ratio of trans-2-butene to 1-butene) of MgO (I) is about four times higher than those of Mg (II) as shown in Figure 20 [76]. The difference in the catalytic action is considered to be due to the difference in regularity of the surface atom arrangement. The surface structure is less regular or more heterogeneous on MgO (II) than on Mg (I), which is suggested from the fact that the IR bands of CO_2 species are broader for MgO (II) than for MgO (I). The strength of the electrostatic field which can be obtained from the *g* values in ESR spectra of the adsorbed oxygen species, O_2^-, is also found to be different [77].

A more striking effect was observed in the selectivity of CaO catalysts for the reaction of 3-carene. Double bond isomerization takes place over CaO prepared from $CaCO_3$, the selectivity for the formation of 2-carene being 96 percent, while dehydrogenation occurs over CaO prepared from $Ca(OH)_2$ [78].

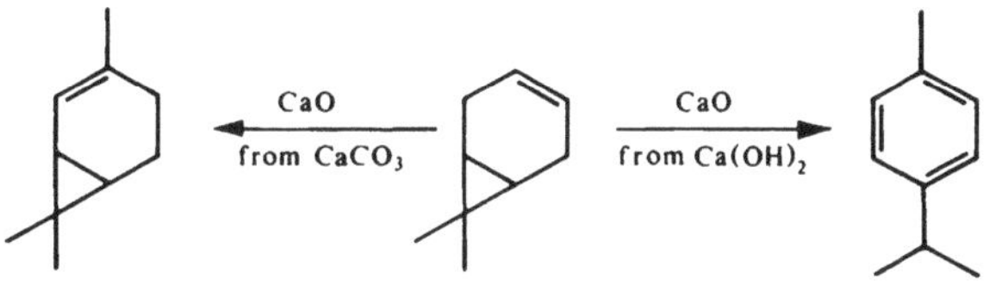

The selectivity of ThO_2 catalysts for the formation of 1-octene from 2-octanol is strongly dependent on the kinds of thorium salts, the preparation method, and the pretreatment as shown in Table 8 [79]. ThO_2 prepared by hydrolysis of the nitrate with ammonia water followed by calcination at 873 K in a hydrogen atomosphere showed high selectivity (99 percent) for 1-octene formation, while the selectivity of ThO_2 prepared by thermal decomposition was 50 percent. The catalyst prepared from the chloride gave a low selectivity (25 percent). Hydrogen treatment makes ThO_2 more active for the dehydration of 2-octanol, whereas oxygen treatment makes it

Table 8. Effect of preparation method on selectivity of ThO_2 catalysts for formation of octenes from 2-octanol

Th salts	Pretreatment temp./K	Octenes/mol %		
		1-octene	trans-2-octene	cis-2-octene
Nitrate[a]	H_2, 873	99	trace	trace
Nitrate[b]	O_2, 873	34	36	30
Nitrate[b]	H_2, 873	50	26	24
Oxalate	H_2, 873	72	12	16
Oxalate	O_2, 873	43	27	30
Carbonate	H_2, 873	82	10	8
Carbonate	H_2, 823	86	7	7
Chloride	H_2, 873	25	30	45

[a] Precipitation method with NH_4OH
[b] Thermal decomposition method

more active for the dehydrogenation. The preferential dehydration with ThO_2 pretreated with hydrogen is due to the acidity increase on the catalyst surface, which is evidenced by poisoning experiments with pyridine [80].

Most metal oxides and complex oxides are usually prepared from their chlorides or sulfates by a precipitation method. In those preparation, about 0.5 wt. percent of anions remain in the precipitates, even if the precipitates are washed thoroughly with distilled water until no anions are detected in the washings with silver nitrate or barium chloride. In particular, residual sulfate anion is not removed by calcination at 773 K for 3 hr. Thus, the effect of anions on acidic and catalytic properties was examined by adding ammonium chloride or sulfate to pure TiO_2, TiO_2—ZnO and ZnO. The addition of NH_4Cl caused a decrease in the acidity per unit surface area and in the acid strength of TiO_2, but an increase in those of ZnO and TiO_2—ZnO. The addition of $(NH_4)_2SO_4$ increased the acidity of TiO_2—ZnO, though it did not increase the acidity of TiO_2. The specific surface areas of all oxides increased on the addition of the salts except that of TiO_2 on the addition of NH_4Cl. In Table 9 are shown the effects of the addition of NH_4Cl and $(NH_4)_2SO_4$ on the catalytic activity and selectivity for the isomerization of 1-butene [81]. The addition of $(NH_4)_2SO_4$ increased the activities of all oxides, especially the activity of TiO_2. On the other hand, the addition of NH_4Cl decreased the activity of TiO_2 and TiO_2—ZnO.

Table 9. Effect of the addition of NH_4Cl or $(NH_4)_2SO_4$ on activity and selectivity for isomerization of 1-butene

Catalyst	Additives	Evacuation temperature/K	Activity % min^{-1} g^{-1}	Ratio of cis/trans	Surface area m^2 g
TiO_2	no	723	2.6	5	42
	no	423	0.9	3	—
TiO_2—ZnO(III)	no	723	4.1	5	37
	no	423	1.1	4	—
ZnO	no	723	0	—	0.4
	no	423	0	—	—
TiO_2	NH_4Cl	723	1.9	5	43
	NH_4Cl	423	1.4	3	—
TiO_2—ZnO(III)	NH_4Cl	723	0	—	76
	NH_4Cl	423	0.5	2	—
ZnO	NH_4Cl	723	0	—	1.1
	NH_4Cl	423	0	—	—
TiO_2	$(NH_4)_2SO_4$	723	240	1	103
	$(NH_4)_2SO_4$	423	520	1	—
TiO_2—ZnO(III)	$(NH_4)_2SO_4$	723	41	2	65
	$(NH_4)_2SO_4$	423	82	2	—
ZnO	$(NH_4)_2SO_4$	723	0.2	3	0.5
	$(NH_4)_2SO_4$	423	0	—	—

Reaction temperature; 423 K. Initial pressure of 1-butene; 13 kPa. Ratio of cis/trans; the value extrapolated to 0% conversion.

2. Difference in Preparation Procedure

Let us compare the acidic property of TiO_2—ZnO prepared by a usual heterogeneous co-precipitation method with that prepared by a homogeneous co-precipitation method. In a usual co-precipitation method where ammonia water is used as a precipitating reagent, the concentration of ammonium hydroxide differs locally during the formation of precipitates, that is, precipitates are formed in a heterogeneous concentration of hydroxide ion. However, if excess of urea is used instead of ammonia, precipitates are formed from a homogeneous solution of hydroxide ion, since urea decomposes to supply ammonium hydroxide at elevated temperatures according to the following equilibria [81],

$$(NH_2)_2CO + H_2O \rightleftharpoons 2\,NH_3 + CO_2$$

$$NH_3 + H_2O \rightleftharpoons NH_4OH$$

As shown in Figure 21, TiO_2—ZnO prepared by the homogeneous method exhibits much larger acidity in the range of $-3 < H_0 \leqq +6.8$ and lower acid strength than TiO_2—ZnO prepared by the heterogeneous method [43]. A similar tendency is also observed for TiO_2—SiO_2 [82]. The catalytic activity of TiO_2—SiO_2 prepared by the homogeneous method for isomerization was about two times higher than that of TiO_2—SiO_2 prepared by the heterogeneous method. The difference in the activity is approximately equal to the difference in the acidity in the range of $-3 < H_0 \leqq +3.3$, indicating that the acid sites having this acid strength are active sites.

It might be said from the above results that the heterogeneous precipitation method using aqueous ammonia produces strong acid sites and the homogeneous precipitation method using urea forms a large amount of relatively weak acid sites. Another distinctive feature of the homogeneous precipitation

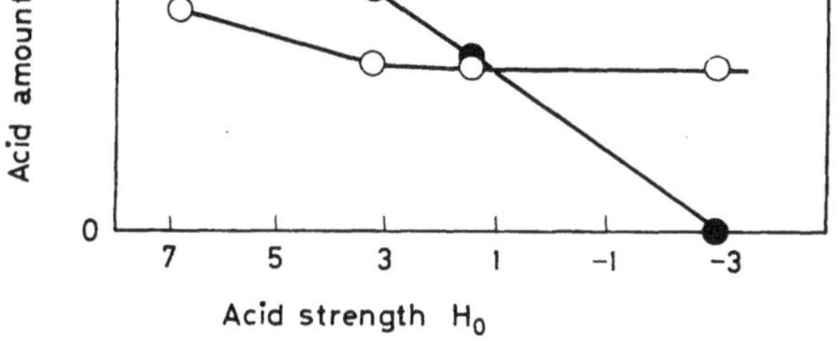

Figure 21. Acidity and acid strength of TiO_2—ZnO prepared by homogeneous (●) and heterogeneous (○) coprecipitation methods

Table 10. Preparation methods of aluminas and surface areas and catalytic activities of aluminas evacuated at 873 K

Al_2O_3	Preparation		Area	Surface activity[a]
	Starting material	Method	$m^2\ g^{-1}$	
I	Isopropoxide	hydrolysis	236	9.9
II	$Al(NO_3)_3$	homogeneous precipitation (urea)	110	4.2
III	$Al(NO_3)_3$	heterogeneous precipitation (ammonia)	156	12.8
IV	$Al(NO_3)_3$	direct thermal decomposition	135	0
V	$Al_2(SO_4)_3$	homogeneous precipitation (urea)	11	0
VI	$Al_2(SO_4)_3$	heterogeneous precipitation (ammonia)	216	3.9
VII	$AlCl_3$	homogeneous precipitation (urea)	147	27.4
VIII	$AlCl_3$	heterogeneous precipitation (ammonia)	248	10.9
IX	$AlCl_3$	direct thermal decomposition	103	1.1
X	$AlBr_3$	homogeneous precipitation (urea)	173	3.8
XI	$NaAlO_2$	heterogeneous precipitation (CO_2)	194	0
XII	high purity α-Al_2O_3 (Sumitomo Chemical Co., Ltd., AKP 20)		5	0
XIII	γ-Al_2O_3 (Nishio Industries CO., Ltd., AE 11)		—	6.5

[a] activity at room temperature for CH_4—D_2 exchanges; arbitrary units per unit mass of catalyst.

method is that this method allows the preparation of mixed oxides having reproducible acidic properties.

An interesting effect of the preparation method and preparation procedure on the surface area of Al_2O_3, and the activity for the exchange reaction between methane and deuterium is given in Table 10. In this case, the distance between acidic sites and basic sites seems to control the catalytic activity [55].

3. *Effect of Pretreatment*

As mentioned in 4.A.2, basic sites, reducing sites and the sites active for hydrogenation of olefins appear on the surface of CaO when $Ca(OH)_2$ is heated at different temperatures. Figure 22 shows how the generation of three kinds of the sites depends on temperature of heat treatment. In the cases of SiO_2—TiO_2 [83], and TiO_2—SnO_2 [84], strongly acidic property appears when calcined in air at 773 K, whereas strong reducing property appears when calcined in a vacuum at the same temperature.

The activity and selectivity of TiO_2 and ThO_2 [45] for the isomerization of 1-butene are found to increase on pretreatment with hydrogen, but decrease on pretreatment with oxygen, while the activity of TiO_2—SiO_2 increases on oxygen pretreatment, but decreases on hydrogen pretreatment (Table 11) [83]. These phenomena are interpreted to be caused by the appearance and disappearance of acidic sites and reducing sites [83]. It is also known that hydrogen pretreatment increases the acidities of TiO_2—WO_3, TiO_2—NiO, Al_2O_3—NiO, Al_2O_3—CdO, and Al_2O_3—Cr_2O_3, but decreases the acidities of TiO_2—MoO_3, TiO_2—Fe_2O_3, TiO_2—Cr_2O_3, TiO_2—V_2O_5, TiO_2—Co_3O_4, TiO_2—CuO, and Al_2O_3—Fe_2O_3 [85].

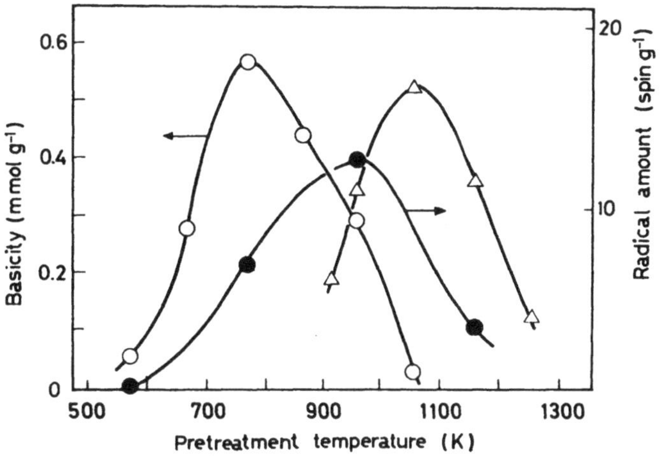

Figure 22. Change of surface property on CaO with change of pretreatment temperature. (○) basicity, (●) amount of reducing sites, (△) activity for hydrogenation of propylene

Table 11. Changes in the activity and the ratio of cis/trans-2-butene in the isomerization of 1-butene with various treatments and poisonings after evacuation at 123 K

	TiO_2		TiO_2-SiO_2 (9/1)	
	Activity (% min^{-1} g^{-1})	Ratio of cis/trans-2-butene	Activity (% min^{-1} g^{-1})	Ratio of cis/trans-2-butene
Untreated	13.3	6.1	13.2	2.2
Poisoned with NH_3	3.0	9.0	2.8	2.2
Poisoned with CO_2	14.2	6.8	14.9	2.3
Treated with O_2	7.0	2.9	27.7	1.9
Treated with H_2	15.8	11.0	11.2	3.2

B. General Rules

Although acid-base properties on solid surface change depending on the preparation method and pretreatment as mentioned above, it is controlled principally by the electronegativity of the metal ions of the solid. Thus, the correlation of acid-base properties with electronegativity has been observed for a series of metal oxides, complex oxides or solid metal salts, when the same kinds of solids are prepared by the same method and the same pretreatment. Let us first consider the correlation of the acidic property of a metal ion with its electronegativity. In general, metal ions show acidic properties and the acid strength of a metal ion in aqueous solution is expressed by the dissociation constant (pK_a) of its hydrated form as shown below.

$$M(H_2O)_m^+ + H_2O \overset{K_a}{\rightleftharpoons} [M(H_2O)_{m-1}(OH)]^{(\alpha-1)+} + H_3O^+$$

Since the pK_a of $M(H_2O)_m^+$ is related to the electrogenativity of M, the acid strength of a metal ion is correlated with the electronegativity.

The electronegativity, χ_i, of a metal ion is proposed by Tanaka et al. [86] and Misono et al. [87] as shown below.

$$\chi_i = (1 + 2Z)\,\chi_o\,,$$

$$\chi_i = \chi_o + \Big(\sum_i I_i\Big)^{\frac{1}{2}},$$

where χ_o is χ_i of metal, Z is a charge of metal ion, and I_i is ionization potential of M^{i-1} to M^i.

In fact, a fairly good correlation between pK_a values of hydrated metal ions and electronegativity, χ_i, of metal ions is found as shown in Figure 23 [86]. In the case of metal oxides, it can be said in general that the oxides of the metals of the upper part in the periodic table possess higher Lewis acid strength and lower base strength compared with those of the lower part in the same group, and the oxides of the metals of the left side in the periodic table have higher base strength and lower Lewis acid strength compared with those of the right hand side in the same period. The electronegativity of metal ion mainly governs the acid-base strength of the metal oxides, provided that the crystal structure is the same. Thus, the order of base strength of alkaline earth metal oxides is determined experimentally as BaO > SrO > CaO > MgO [39]. Hence, the order of the Lewis acid strength is the reverse. The acid strength decreases in the order of $ZrO_2 > Y_2O_3 > SrO$. Y_2O_3, La_2O_3 and CeO_2 which are less basic and more acidic than alkaline

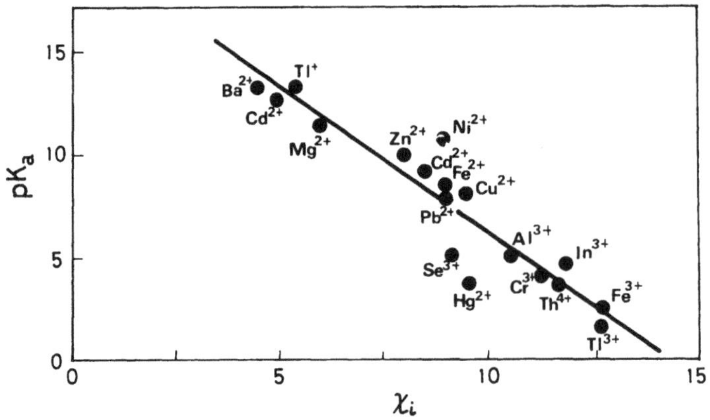

Figure 23. Dissociation constants (pK_a) of hydrated metal ions vs. electronegativity (χ_i) of metal ions

earth oxides are less active for isomerization of 1-butene, though they show the nature of base catalysis [44]. ZrO_2 and ThO_2 which are more acidic and less basic than Y_2O_3, La_2O_3 and CeO_2 exhibit an acid-base bifunctional catalysis [30, 45]. Thus, both oxides are highly selective for the formation of 1-butene from 2-butanol [88] and highly active for H-D exchange reaction between methyl group of isopropyl alcohol and surface OH groups, though strongly acidic catalysts (Al_2O_3, $SiO_2—Al_2O_3$) and strongly basic catalysts (MgO, CaO) were all inactive [89]. It is interesting to note that ZrO_2 possesses not only acidic properties, but also oxidizing-reducing properties [30]. TiO_2 has acidic properties [43] and reducing properties [43]. SnO_2 has a high reducing property [84]. On the other hand, MoO_3 [7] and WO_3 [18], in the VI Group of periodic table, have fairly strong acidic properties.

For binary oxides, the highest acid strengths are found to increase with the increase of the algebraically averaged electronegativities and decrease with the increase of partial negative charge of oxygen as shown in Figures 24 [64] and 25 [90], respectively. The correlations in those figures as well as that in Figure 9, which are useful for predicting the acid strengths of unknown oxides and binary oxides, indicate that electronegativity or partial negative charge of oxygen controls the acid strength.

Figure 26 shows that the catalytic action of binary oxides including SnO_2 for the oxidation of propylene is controlled by the electronegativity of the

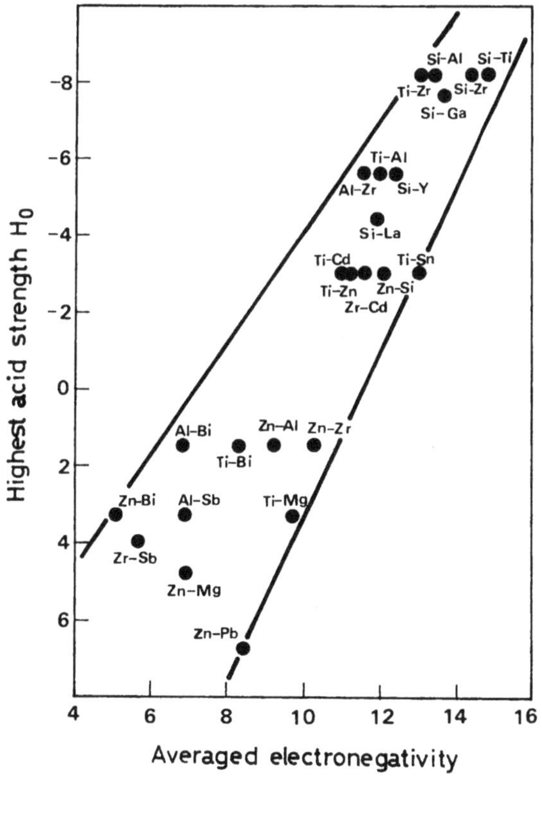

Figure 24. Highest acid strength vs. averaged electronegativity of metal ions of binary oxides (molar ratio = 1)

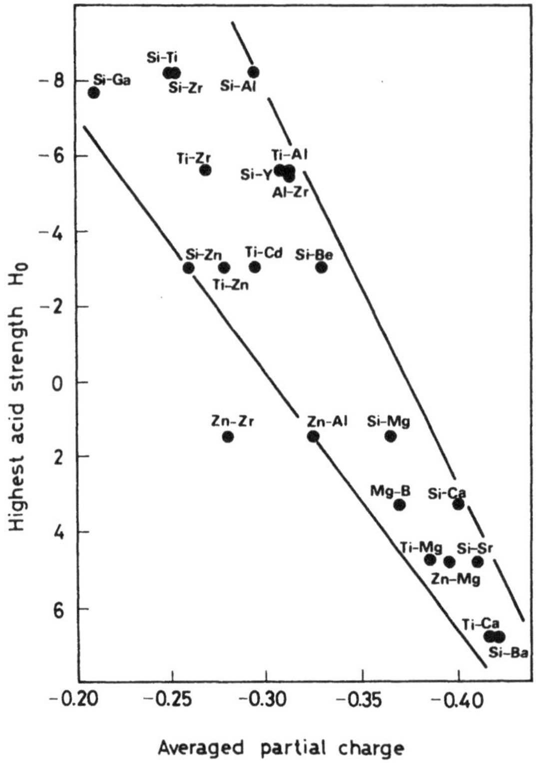

Figure 25. Highest acid strength vs. averaged partial negative charge of oxygens of binary oxides (molar ratio = 1)

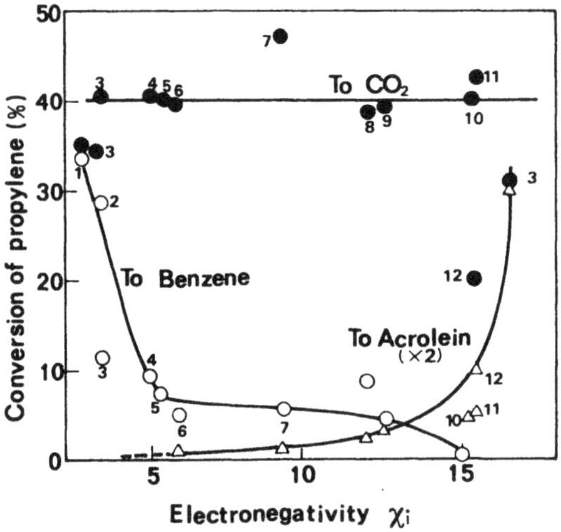

Figure 26. Catalytic activity of various binary oxides including SnO_2 for oxidation of propylene vs. electronegativity of metal oxides which are mixed with SnO_2.
1: SnO_2—K_2O, 2: SnO_2—Na_2O, 3: SnO_2—Li_2O, 4: SnO_2—BaO, 5: SnO_2—CaO, 6: SnO_2—MgO, 7: SnO_2—Cr_2O_3, 8: SnO_2, 9: SnO_2—Sb_2O_4, 10: SnO_2—WO_3, 11: SnO_2—As_2O_5, 12: SnO_2—P_2O_5, 13: SnO_2—MoO_3

oxides which are mixed with SnO_2 [91]. The higher the electronegativity (acid strength) of metal ions, the larger the amount of acrolein formed and the lower the electronegativity, the larger the amount of benzene formed.

The basicity increase caused by mixing two metal oxides has been reported for Al_2O_3—MgO [92, 93], TiO_2—MgO [26], SiO_2—ZnO [21], TiO_2—ZrO_2 [94], and Al_2O_3—ZnO [95]. However, no general rules about the basicity increase have yet been found.

The acid-base properties of hydroxyl groups is also controlled by the electronegativity of surface metal ions [66]. A hydroxyl group is dissociated to exhibit an acidic or basic property as shown below.

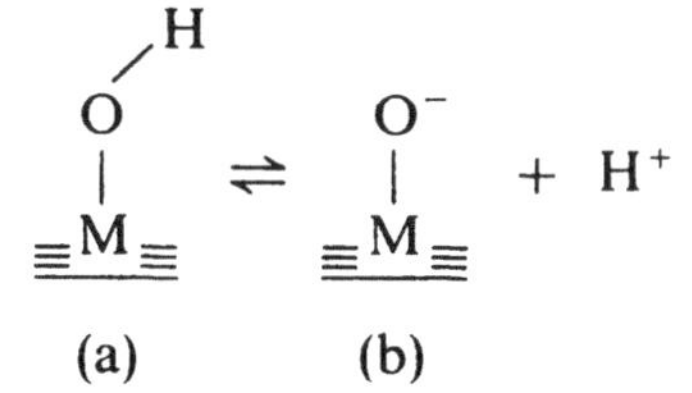

(8)

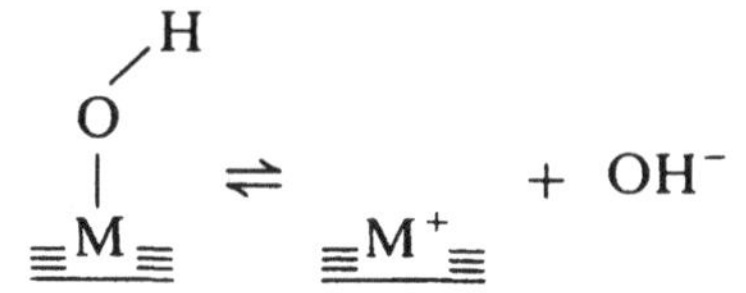

(9)

where M is a metal ion on the surface. When the electronegativity of M is high, the O—H bond is weakened and dissociated to liberate a proton, as shown by scheme (8), since M strongly attracts the electron pair of the oxygen. On the contrary, when the electronegativity is low, the M—O bond is dissociated to liberate a hydroxide ion as illustrated by scheme (9). The dissociation of M—O—H is also influenced by the acid-base strength of a reactant molecule.

Acknowledgment

The author is indebted to Dr. H. Hattori, Dr. T. Yamaguchi, and Mrs. Kaneko for their help for preparation of the manuscript.

References

1. Hammett, L. P., Deyrup, A. J.: J. Amer. Chem. Soc. **54**, 2721 (1932)
2. Tanabe, K.: Solid Acids and Bases. Tokyo: Kodansha and New York-London: Academic Press 1970
3. Hirschler, A. E.: J. Catal. **2**, 428 (1963)
4. Ohnishi, R., Ishikura, S., Tanabe, K.: Nippon Kagaku Kaishi **1973**, 183
5. Tanabe, K., Hattori, H.: Chem. Lett. **1976**, 625
6. Forni, L.: Advan. Catal. **8**, 65 (1974)
7. Maruyama, K., Hattori, H., Tanabe, K.: Bull. Chem. Soc. Jap. **50**, 86 (1977)
8. Richardson, R. L., Benson, W.: J. Phys. Chem. **61**, 405 (1957)
9. Itoh, M., Hattori, H., Tanabe, K.: J. Catal. **35**, 225 (1974)
10. Shimazu, K., Hattori, H., Tanabe, K.: J. Catal. **48**, 302 (1977)

11. Iizuka, T., Tanabe, K.: Bull. Chem. Soc. Jap. **48**, 2527 (1975)
12. Ai, M., Suzuki, S.: J. Catal. **30**, 362 (1973)
13. Ai, M.: Bull. Jap. Petrol. Inst. **18**, 50 (1976)
14. Ai, M.: Bull. Chem. Soc. Jap. **49**, 1328 (1976)
15. Ai, M.: ibid. **50**, 355, 2579 (1977)
16. Ai, M.: Yuki Gosei Kyokai-shi **35**, 201 (1977)
17. Hattori, H., Maruyama, K., Tanabe, K.: Bull. Chem. Soc. Jap. **50**, 2181 (1977)
18. Hattori, H., Asada, N., Tanabe, K.: ibid. **51**, 1704 (1978)
19. Basila, M. R., Kantner, T. R.: J. Phys. Chem. **71**, 467 (1967)
20. Parry, E. P.: J. Catal. **2**, 371 (1963)
21. Sumiyoshi, T., Tanabe, K., Hattori, H.: Bull. Jap. Petrol. Inst. **17**, 65 (1975)
22. Hattori, H., Miyashita, S., Tanabe, K.: Bull. Chem. Soc. Jap. **44**, 893 (1971)
23. Malinowski, S., Szczepanska, S.: J. Catal. **2**, 310 (1963)
24. Tanabe, K., Yamaguchi, T.: J. Res. Inst. Catal. Hokkaido Univ. **11**, 179 (1964)
25. Take, J., Kikuchi, N., Yoneda, Y.: J. Catal. **21**, 164 (1971)
26. Tanabe, K. et al.: J. Catal. **53**, 1 (1978)
27. Krylov, O. V., Fokina, E. A.: Problemy Kinet. Katal., Acad. Nauk, USSR, **8**, 248 (1955)
28. Tanabe, K., Nishizaki, T.: Proc. Sixth Internat. Congr. Catalysis (London, 1976), **2**, 863 (1977)
29. Iizuka, T. et al.: Chem. Lett. **1976**, 803
30. Nakano, Y. et al.: J. Catal. **57**, 1 (1979)
31. Fukuda, Y., Tanabe, K.: Bull. Chem. Soc. Jap. **46**, 1616 (1973)
32. Ai, M.: J. Catal. **54**, 223 (1978)
33. Vinek, H. et al.: J. Chem. Soc. Faraday Trans. I **1977**, 734
34. Yamanaka, T., Tanabe, K.: J. Phys. Chem. **79**, 2409 (1975)
35. Yamanaka, T., Tanabe, K.: ibid. **80**, 1723 (1976)
36. Tanaka, K., Ozaki, A.: J. Catal. **8**, 1 (1967)
37. Sanderson, R. T.: Inorganic Chemistry (Translation into Japanese), New York: Reinhold 1967, Tokyo: Hirokawa 1971
38. Hattori, H., Yoshii, N., Tanabe, K.: Proc. Fifth Internat. Congr. Catalysis (Palm Beach, 1972), **10** 233 (1973)
39. Tanabe, K., Fukuda, Y.: React. Kinet. Catal. Lett. **1**, 21 (1974)
40. Tanabe, K., Saito, K.: J. Catal. **35**, 247 (1974)
41. Mohri, M., Tanabe, K., Hattori, H.: J. Catal. **32**, 144 (1974)
42. Hattori, H., Maruyama, K., Tanabe, K.: J. Catal. **44**, 50 (1976)
43. Tanabe, K. et al.: Bull. Chem. Soc. Jap. **45**, 47 (1972)
44. Fukuda, Y., Hattori, H., Tanabe, K.: ibid. **51**, 3150 (1978)
45. Imizu, Y. et al.: ibid. **50**, 1040 (1977)
46. Itoh, M., Hattori, H., Tanabe, K.: J. Catal. **43**, 192 (1976)
47. Tanabe, K. et al.: J. Catal. **53**, 1 (1978)
48. Tanabe, K. et al.: J. Res. Inst. Catal. Hokkaido Univ. **19**, 185 (1972)
49. Hindin, S. G., Weller, S. W.: J. Phys. Chem. **60**, 1501 (1956)
50. Yamadaya, M. et al.: Shokubai **7**, 313 (1965)
51. Peri, J. B.: J. Phys. Chem. **69**, 211 (1965)
52. Kondo, S. et al.: Bull. Chem. Soc. Jap. **46**, 1362 (1973)
53. Echigoya, E.: Dissert. Tokyo Inst. Technology, 1957
54. Cornelius, E. B. et al.: J. Phys. Chem. **59**, 809 (1955)
55. Tanabe, K., Utiyama, M., Hattori, H.: Symp. Sci. Catalysis and its Application in Ind., Sindri/India, 1979
56. Yoshida, S., Tai, S., Tarama, K.: J. Catal. **45**, 242 (1976)
57. Malinowski, S.: private communication
58. Iizuka, T. et al.: J. Catal. **22**, 130 (1971)
59. Hattori, H., Tanaka, Y., Tanabe, K.: Chem. Lett. **1975**, 659
60. Tanaka, Y., Hattori, H., Tanabe, K.: ibid. **1976**, 1973
61. Hattori, H., Tanaka, Y., Tanabe, K.: J. Amer. Chem. Soc. **98**, 4652 (1976)
62. Krylov, O. V. et al.: Kinet. Katal. **6**, 128 (1965)
63. Tanabe, K. et al.: Bull. Chem. Soc. Jap. **47**, 1064 (1974)

64. Shibata, K. et al.: ibid. **46**, 2985 (1973)
65. Thomas, C. L.: Ind. Eng. Chem. **41**, 2564 (1949)
66. Seiyama, T.: Metal Oxides and their Catalytic Actions. Tokyo: Kodansha Scientific 1978
67. Uytterhoven, J. B., Christner, L. G., Ha, W. K.: J. Phys. Chem. **69**, 3463 (1965)
68. Turkevich, J.: Japan-USA. Seminar Catalytic Sci., Tokyo and Kyoto, Prepr. No. A-3 (1968)
69. Ward, J. W.: J. Catal. **9**, 225 (1967); **10**, 34 (1968); **11**, 238, 251 (1968)
70. Takeshita, T. et al.: J. Phys. Chem. **69**, 4077 (1965)
71. Tanabe, K., Takeshita, T.: Adv. Catal. **17**, 315 (1967)
72. Takeshita, T., Ohnishi, R., Tanabe, K.: Catal. Rev. **8**, 29 (1973)
73. Coing-Boyat, J., Bassi, G.: C.R. Acad. Sci., Paris **256**, 1482 (1963)
74. Kawakami, T., Konno, A., Ogino, Y.: Bull. Chem. Soc. Jap. **44**, 1772 (1971)
75. Baba, S., Ogino, Y.: ibid. **46**, 127 (1973)
76. Hattori, H. et al.: Bull. Chem. Soc. Jap. **49**, 969 (1976)
77. Iizuka, T., Saito, M., Tanabe, K.: Shokubai **18**, 114 (1976)
78. Tanabe, K., Shimazu, K., Hattori, H.: Chem. Lett. **1975**, 507
79. Davis, B. H.: J. Catal. **25**, 81 (1972)
80. Hensel, J., Pines, H.: J. Catal. **24**, 197 (1972)
81. Willard, H. H., Tang, N. K.: J. Amer. Chem. Soc. **59**, 1190 (1937)
82. Tanabe, K. et al.: Preparation of Catalysts. (Delmon, B., Jacobs, P. A., Poncelet, G., ed.), Amsterdam: Elsevier 1976, pp. 65–74
83. Hattori, H., Itoh, M., Tanabe, K.: J. Catal. **38**, 172 (1975)
84. Itoh, M., Hattori, H., Tanabe, K.: J. Catal. **43**, 192 (1976)
85. Shibata, K., Kiyoura, T., Tanabe, K.: J. Res. Inst. Catal., Hokkaido Univ. **18**, 189 (1970)
86. Tanaka, K., Ozaki, A.: J. Catal. **8**, 2 (1967)
87. Misono, M. et al.: J. Inorg. Nucl. Chem. **29**, 2685 (1967)
88. Yamaguchi, T., Sasaki, H., Tanabe, K.: Chem. Lett. **1973**, 1017
89. Yamaguchi, T. et al.: Chem. Lett. **1976**, 677
90. Tanabe, K.: General rules regarding acidity of metal oxides. In: Metal oxides and complex oxides (Tanabe, K., Seiyama, T., Hueki, K., ed.), Tokyo: Kodansha Scientific 1978, pp. 407–414
91. Seiyama, T. et al.: J. Catal. **24**, 76 (1972)
92. Miyata, S. et al.: Nippon Kagaku Zasshi **92**, 514 (1971)
93. Yamagata, N. et al.: J. Catal. **47**, 358 (1977)
94. Arata, K., Akutagawa, S., Tanabe, K.: Bull. Chem. Soc. Jap. **49**, 390 (1976)
95. Tanabe, K. et al.: J. Catal. **57**, 35 (1979)

Subject Index

Bei Fragen zur Produktsicherheit wenden Sie sich bitte an:
If you have any questions regarding product safety,
please contact:

Walter de Gruyter GmbH
Genthiner Straße 13
10785 Berlin
productsafety@degruyterbrill.com